Theoretical Foundations of Asset Pricing

This text provides an advanced introduction to the modeling of competitive financial markets, encompassing arbitrage and equilibrium pricing of financial contracts, as well as optimal lifetime consumption and portfolio choice. Notable features include its coverage of recursive utility in discrete and continuous time and several results not previously available in book form. Each chapter concludes with a set of exercises, with solutions available to verified instructors.

Ideal as a graduate-level course text, this book can also serve as a valuable reference for researchers and finance industry practitioners. Readers with a finance focus can use the text to build analytical foundations for a significant component of the economics of financial markets, while readers with a mathematics focus will find a well-motivated introduction to basic tools of stochastic analysis and convex analysis.

COSTIS SKIADAS is the Harold L. Stuart Professor of Finance at Northwestern University, Illinois, where he has served as chairman of the Finance department. He has made research contributions on foundational aspects of the topics covered in this text. He previously authored *Asset Pricing Theory* (2009).

Theoretical Foundations of Asset Pricing

COSTIS SKIADAS
Northwestern University, Illinois

CAMBRIDGE
UNIVERSITY PRESS

Shaftesbury Road, Cambridge CB2 8EA, United Kingdom

One Liberty Plaza, 20th Floor, New York, NY 10006, USA

477 Williamstown Road, Port Melbourne, VIC 3207, Australia

314–321, 3rd Floor, Plot 3, Splendor Forum, Jasola District Centre,
New Delhi – 110025, India

103 Penang Road, #05–06/07, Visioncrest Commercial, Singapore 238467

Cambridge University Press is part of Cambridge University Press & Assessment, a
department of the University of Cambridge.

We share the University's mission to contribute to society through the pursuit of
education, learning and research at the highest international levels of excellence.

www.cambridge.org
Information on this title: www.cambridge.org/9781009439039

DOI: 10.1017/9781009439077

When citing this work, please include a reference to the DOI 10.1017/9781009439077

First published 2025

A catalogue record for this publication is available from the British Library.

A Cataloging-in-Publication data record for this book is available from the Library of
Congress.

ISBN 978-1-009-43903-9 Hardback

Additional resources for this publication at www.cambridge.org/skiadas

To my wife, Robin

Contents

Preface

This text is the distillation of material I teach to doctoral students at Northwestern University as part of an advanced introduction to the theoretical foundations of what is traditionally known as "asset pricing." The focus is on the modeling of competitive financial markets, encompassing arbitrage and equilibrium pricing of financial contracts, as well as optimal consumption and portfolio choice. I have relied on lectures for broader context and on this text for a self-contained statement of a theory with classical roots in Walrasian competitive analysis, extended by Arrow and Debreu to include time and uncertainty, and further developed in Finance to emphasize the role of arbitrage arguments and the tools of modern stochastic analysis. The presentation places equal emphasis on sound economics and well-motivated methodology. Readers with an economics focus can use this text to build analytical foundations for a significant component of the economics of financial markets, while readers with a mathematics focus can use the text as a well-motivated introduction to basic tools of stochastic analysis and convex analysis. Despite its introductory orientation, the book includes results I could only find in research papers, or are refined versions of original results of my last monograph.

The progression of topics can be thought of as increasing in scope and decreasing in realism. Arbitrage arguments are presented first, followed by characterizations of optimality and competitive equilibrium. Arbitrage arguments utilize the assumption that the market does not allow incremental cash flows that are desirable in the narrow sense of arbitrage. Optimality is then introduced as a refinement of the no-arbitrage assumption. The notion of a desirable cash flow is enlarged through preferences and the idea of an arbitrage is extended to allow for multiagent transactions through profitable market-making opportunities. The exclusion of such market-making opportunities refines the traditional concept of Pareto optimality and leads to a version of classical competitive welfare analysis that is better suited to financial markets, emphasizing endogenous forces for market creation. Restrictions on preferences are initially minimal, reflecting the fact that competitive equilibrium notions are robust to preference structure, and are gradually strengthened in order to express ideas of how market prices and optimal consumption–portfolio choice relate to preferences for smoothing across time and states of the world. The auxiliary Appendix A on additive utility forms and risk aversion presents some fundamental but not well-known arguments, drawing on material that, to my knowledge, has only been buried in research papers.

On the methodological side, a self-contained introduction to probabilistic methods starts with a rigorous treatment on a finite information tree and concludes with an introduction to the continuous-time theory, which omits several technical details but leverages the thorough understanding of the tools on a finite tree. In an approximate numerical sense, the continuous-time model is presented as a simplified special case of a high-frequency finite-information model. Tools like martingale representations, Girsanov change-of-measure arguments, the Ito calculus, and forward and backward stochastic differential equations are hopefully demystified in this way, providing an entry point to a literature which is typically obfuscated by the requirements of set-theoretic rigor. The optimality and equilibrium theory emphasizes a unified geometric viewpoint and convex analysis methods, making this course complementary to a macroeconomics course emphasizing dynamic programming methods. Appendix B provides a succinct rigorous statement of the background convex analysis, which can also be thought of as an introductory mini course on the functional analytic approach to optimization theory.

I have been using this text in a class of mostly second-year doctoral students taking a Finance course for the first time. I emphasize to my students that the text organizes foundational insights that are essential but far from sufficient for understanding actual financial markets. Class discussion can touch on topics such as the role of collateral, limits to arbitrage, and the inadequacy of consumption-based asset pricing. I have resisted the temptation to include such discussions in the text, which is intended only for core material in a state of completion meant to last. It should be up to the instructor and follow-up courses to offer broader perspective and avenues to research, depending on the interests of the audience. Although the material is formally self-contained, a background on basic linear algebra is essential and some prior exposure to probability theory and graduate-level introductory economics is helpful. I ask students to read Appendix B on convex analysis as preparation for my class, with emphasis on geometric understanding. I recommend two books to complement this one: Back (2017) on the more elementary side (but including topics not covered here), and Duffie (2001) for more on the continuous-time theory in a language and notation that will look familiar to readers of this book.

This text is consistent in approach with my older book (Skiadas (2009)), but differs in some significant ways (as well as in numerous smaller improvements that would take too long to list here). The treatment has been simplified around a central conceptual development, but also extended to include an introduction to continuous-time methodology, a section on preferences with habit formation or durability of consumption, as well as several other smaller additions, such as an explanation that certain multiple-prior utility forms are equivalent to more familiar recursive utility forms. The result is a more compact, panoramic and cohesive book. Some formal structures, such as those underlying the discussion of optimality, have been simplified and more-efficiently presented. Several results and proofs are improved in this text or are missing from the older book. The material is directly presented in a dynamic setting, as opposed to the traditional practice of first considering the static theory. The probabilistic foundations are pedagogically interweaved into the main material, as opposed to a disconnected appendix. Some of the older book's material that is not essential to the main narrative has been converted into exercises or omitted. The exercises are now better integrated with the main text and classroom tested, with detailed solutions available to instructors on www.cambridge.org/skiadas. Some of the peripheral theory on additive utility structures has been significantly refined and extended, and pulled into Appendix A. The convex analysis of Appendix B has also been significantly improved and extended.

Acknowledgments I was lucky to take my first steps in the area under the guidance of Darrell Duffie. I am deeply grateful for his encouragement, inspiration and friendship. Significant parts of this book's presentation are based on my subsequent research, which was always motivated by a desire to understand the fundamentals. I am grateful to my coauthors, especially Mark Schroder, who made this process so much more enjoyable. I thank Kunpeng Zhou and especially Xuning Ding for catching erroneous statements in earlier versions. Many others have helped weed out misprints, including Ariel Lanza, Huidi Lin, Nicolas Min, Junko Oguri, Raul Riva, Matheus Sampaio, Timothy Seida, Lior Shabtai, Rui Sousa, Yudan Ying, Brandon Zborowski, Mark Zhao and Weijia Zhao.

1

Market and Arbitrage Pricing

An arbitrage is a trade that results in a positive incremental cash flow, that is, some inflow at some time in some contingency, but no possible outflow. This chapter introduces a formal notion of a market and associated constructs, and lays the foundations for pricing arguments based on the assumption of the lack of arbitrage opportunities. Throughout the text, $\equiv$ stands for "equal by definition" and terms in boldface are being defined.

1.1 Uncertainty and Information

We begin with a formal representation of uncertainty and a common information stream that is available to market participants over a finite time horizon. Throughout this text, we use set-theoretic notation common in graduate-level mathematics.

A **time** is an element of the set $\{0, 1, \ldots, T\}$, for a positive integer T that is fixed throughout. One of a finite number of possible states of the world, or contingencies, is realized by the **terminal time** T. These **states** are represented by the elements of a finite set Ω, the **state space**. We are not yet concerned with the likelihood of any one state occurring, but every contingency represented by a state in Ω is possible and every relevant contingency is represented by a state in Ω. The subsets of Ω are called **events**. Time-t information is represented by a **partition** $\mathcal{F}_t^0$ of Ω, defined as a set of mutually exclusive nonempty events whose union is Ω. All that is known at time t is what element of $\mathcal{F}_t^0$ contains the state realized at time T. At time 0 it is only known that the state to be revealed at time T is an element of Ω and therefore $\mathcal{F}_0^0 = \{\Omega\}$. At time T the state is revealed and therefore $\mathcal{F}_T^0 = \{\{\omega\} \mid \omega \in \Omega\}$. We assume perfect recall: If at some time the state is known to belong to a partition element, the same remains true at all subsequent times. More formally, we assume that if $u > t$, the partition $\mathcal{F}_u^0$ is a **refinement** of the partition $\mathcal{F}_t^0$, meaning that every event in $\mathcal{F}_t^0$ is the union of events in $\mathcal{F}_u^0$.

EXAMPLE 1.1.1 Information is generated by observing the outcome of a coin toss at each time $t = 1, 2, \ldots, T$. Let us encode heads with 1 and tails with -1. A state is a finite sequence $\omega = (\omega_1, \ldots, \omega_T)$, where $\omega_t \in \{1, -1\}$, and the state space is the cartesian product $\Omega \equiv \{1, -1\}^T$. At time $t > 0$ the first t coin toss outcomes $\bar{\omega}_1, \ldots, \bar{\omega}_t \in \{1, -1\}$ have been observed and it is therefore known that the state is an element of the event $\{\omega \in \Omega \mid \omega_1 = \bar{\omega}_1, \ldots, \omega_t = \bar{\omega}_t\}$. The partition $\mathcal{F}_t^0$ is the set of all these events as $(\bar{\omega}_1, \ldots, \bar{\omega}_t)$ ranges over $\{1, -1\}^t$. $\diamond$

It is mathematically convenient to also define the sets

$$\mathcal{F}_t = \left\{ F \mid F \text{ is a union of elements of } \mathcal{F}_t^0 \right\}, \quad t = 0, \ldots, T.$$

An event F belongs to $\mathcal{F}_t$ if and only if at time t it is known whether F contains the state to be revealed at time T.

EXAMPLE 1.1.2 Let $\Omega \equiv \{1, 2, 3, 4\}$ and $\mathcal{F}_1^0 \equiv \{\{1, 2\}, \{3\}, \{4\}\}$. Then $\mathcal{F}_1 = \{\emptyset, \{1, 2\}, \{3\}, \{4\}, \{1, 2, 3\}, \{1, 2, 4\}, \{3, 4\}, \Omega\}$. Suppose state 1 is realized at time $T \equiv 2$. At time 1 it is known that the state is either 1 or 2. From that it can be inferred whether every one of the events in $\mathcal{F}_1$ contains 1 or not, and these are all the events about which such a claim can be made. $\diamond$

Every $\mathcal{F}_t$ is an **algebra** of events, meaning that it contains $\emptyset$ and Ω, and it is closed relative to the formation of Boolean set operations: For all $A, B \in \mathcal{F}_t$, the **union** $A \cup B \equiv \{\omega \mid \omega \in A \text{ or } \omega \in B\}$, **intersection** $A \cap B \equiv \{\omega \mid \omega \in A \text{ and } \omega \in B\}$, and **set difference** $A \setminus B \equiv \{\omega \mid \omega \in A \text{ and } \omega \notin B\}$ are all elements of $\mathcal{F}_t$. In particular, for all $F \in \mathcal{F}_t$, the **complement** $F^c \equiv \Omega \setminus F$ is an element of $\mathcal{F}_t$. This definition of an algebra is of course redundant. For example, since $A \cap B = (A^c \cup B^c)^c$ and $A \setminus B = A \cap B^c$, an **algebra** (of events) is any nonempty set of events that is closed with respect to the formation of unions and complements. Besides providing convenient notation, algebras are key in formulating this text's theory in ways that can be interpreted in infinite state-space extensions, where algebras are not generated by partitions. Here we will take full advantage of the partition representation, even though most results will be stated in ways that are amenable to generalization.

The intersection of an arbitrary collection of algebras is also an algebra. (The reader can construct a simple example showing that the union of two algebras is not necessarily an algebra.) The algebra $\sigma(\mathcal{S})$ **generated** by a set of events $\mathcal{S}$ is the intersection of all algebras that include $\mathcal{S}$. It is straightforward to verify that $\mathcal{F}_t = \sigma(\mathcal{F}_t^0)$ for all t. Conversely, $\mathcal{F}_t^0$ can be recovered from $\mathcal{F}_t$ as the set of nonempty elements of $\mathcal{F}_t$ that do not

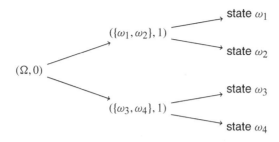

Figure 1.1 Information tree of Example 1.1.1 with $T = 2$. Each state ω_i can be identified with the terminal spot $(\{\omega_i\}, 2)$.

have a nonempty proper subset in $\mathcal{F}_t$. Note that $\mathcal{F}_u^0$ is a refinement of $\mathcal{F}_t^0$ if and only if $\mathcal{F}_t \subseteq \mathcal{F}_u$. This motivates the definition of a **filtration** as a time-indexed sequence $(\mathcal{F}_0, \mathcal{F}_1, \ldots, \mathcal{F}_T)$ of algebras of events, abbreviated to $\{\mathcal{F}_t\}$, such that $t < u$ implies $\mathcal{F}_t \subseteq \mathcal{F}_u$. We can therefore equivalently specify the information primitive of our model as a filtration $\{\mathcal{F}_t\}$ satisfying

$$\mathcal{F}_0 = \{\emptyset, \Omega\} \quad \text{and} \quad \mathcal{F}_T = 2^\Omega \text{ (the set of all subsets of } \Omega). \quad (1.1.1)$$

As illustrated in Figure 1.1, a filtration $\{\mathcal{F}_t\}$ can be thought of as an information tree, whose nodes correspond to what we call "spots." Formally, a **spot** of the filtration $\{\mathcal{F}_t\}$ is a pair (F, t) where $t \in \{0, \ldots, T\}$ and $F \in \mathcal{F}_t^0$. The root of the information tree corresponds to the **initial spot** $(\Omega, 0)$. A **terminal spot** takes the form $(\{\omega\}, T)$, where $\omega \in \Omega$, and can therefore be identified with the state ω as well as the unique path on the information tree from the initial spot to the given terminal spot. A nonterminal spot $(F, t-1)$, $t \in \{1, \ldots, T\}$, has **immediate successor spots** $(F_0, t), \ldots, (F_d, t)$, where $F_0, \ldots, F_d$ are the elements of $\mathcal{F}_t^0$ whose union is F.

We represent economic quantities, such as cash flows and prices, by stochastic processes, which are time-indexed sequences of random variables that are consistent with the information structure just introduced. We now formalize these notions and introduce related notation.

A **random variable** is a function of the form $x : \Omega \to \mathbb{R}$. A **stochastic process**, or simply **process**, can equivalently be defined as a time-indexed sequence $(x_0, x_1, \ldots, x_T)$ of random variables or as a function of the form $x : \Omega \times \{0, 1, \ldots, T\} \to \mathbb{R}$, where $x_t(\omega) = x(\omega, t)$ for $\omega \in \Omega$ and $t \in \{0, \ldots, T\}$. On occasion we write $x(t)$ instead of x_t. The function $x(\omega, \cdot) : \{0, \ldots, T\} \to \mathbb{R}$, for any fixed $\omega \in \Omega$, is a **path** of the process x. As is common in probability theory, we identify a scalar α and the process that is identically equal to α. For example, $x + \alpha$ denotes the process that takes the value $x(\omega, t) + \alpha$ at state ω and time t. Sums and products of processes are defined pointwise: $(x + yz)(\omega, t) \equiv x(\omega, t) + y(\omega, t) z(\omega, t)$. We write $x \le y$ to mean $x(\omega, t) \le y(\omega, t)$ for all (ω, t), and analogously for any other relation. A process is said to be **strictly positive** if it is valued in $(0, \infty)$. Analogous conventions apply to random variables (which can after all be viewed as processes with $T = 0$). Random variables are often used to define events in terms of predicates, as in $\{\omega \in \Omega \mid x(\omega) \le \alpha\}$. In such cases, we simplify the notation by eliminating the state variable, as in $\{x \le \alpha\}$. Another useful piece of notation is that of an **indicator function** 1_A of a set A, which takes the value 1 on A and 0 on the complement of A in the implied domain. If A is an event, then 1_A is the random variable

$$1_A(\omega) = \begin{cases} 1 & \text{if } \omega \in A, \\ 0 & \text{if } \omega \notin A. \end{cases}$$

If $A \subseteq \Omega \times \{0, \ldots, T\}$, the same expression, but with (ω, t) in place of ω, defines the process 1_A.

Throughout this text, we take as given an underlying filtration $\{\mathcal{F}_t\}$ on Ω satisfying (1.1.1). If a process is to represent an observed quantity, it cannot reveal more information than implied by the postulated filtration. For example, if $T = 1$ and $\Omega = \{0, 1\}$, the process $x_0(\omega) = x_1(\omega) = \omega$ is not consistent with the information structure, because observation of the realization of x_0 at time zero reveals the state ω. To formalize this type of informational constraint, we introduce the key notion of measurability with respect to an algebra. A random variable x is said to be **measurable** with respect to an algebra $\mathcal{A}$, or $\mathcal{A}$-**measurable**, if $\{x \le \alpha\} \in \mathcal{A}$ for all $\alpha \in \mathbb{R}$. If $\mathcal{A}$ is generated by the partition $\mathcal{A}^0 = \{A_1, \ldots, A_n\}$, then x is $\mathcal{A}$-measurable if and only if it can be expressed as $x = \sum_{i=1}^n \alpha_i 1_{A_i}$, where $\alpha_i \in \mathbb{R}$ is a constant value x takes on A_i. The set of $\mathcal{A}$-measurable random variables, which we denote by $L(\mathcal{A})$, is a linear subspace of $\mathbb{R}^\Omega$ that is also closed relative to nonlinear combinations of its elements in the following sense, where $f(x_1, \ldots, x_n)$ denotes the random variable that maps ω to $f(x_1(\omega), \ldots, x_n(\omega))$.

PROPOSITION 1.1.3 *If $x_1, \ldots, x_n \in L(\mathcal{A})$ then $f(x_1, \ldots, x_n) \in L(\mathcal{A})$ for all $f : \mathbb{R}^n \to \mathbb{R}$, and $\{(x_1, \ldots, x_n) \in S\} \in \mathcal{A}$ for all $S \subseteq \mathbb{R}^n$.*

PROOF If each x_i is constant over every element of $\mathcal{A}^0$, then so is $f(x_1, \ldots, x_n)$. Letting $f(x_1, \ldots, x_n) = 2 \times 1_{\{(x_1, \ldots, x_n) \in S^c\}}$ shows that $\{(x_1, \ldots, x_n) \in S\} = \{f(x_1, \ldots, x_n) \le 1\} \in \mathcal{A}$. $\square$

We formalize the requirement that a process respects the given information structure with the notion of adaptedness. The process x is said to be **adapted** (to the underlying filtration) if $x_t \in L(\mathcal{F}_t)$ for every time t. Since $\mathcal{F}_0 = \{\emptyset, \Omega\}$ and $\mathcal{F}_T = 2^\Omega$, the initial value of an adapted process is constant, while its terminal value can be any random variable. The space of all adapted processes, which we denote by $\mathcal{L}$, can be identified with the Euclidean space $\mathbb{R}^{1+N}$, where $1 + N$ is the total number of spots. To see how, consider any adapted process x. For any spot (F, t), the random variable x_t is constant over F, taking a value that we denote by $x(F, t)$ (that is, $x(\omega, t) = x(F, t)$ for all $\omega \in F$). If $\mathcal{F}_t^0 = \{F_1, \ldots, F_n\}$, then $x_t = \sum_{i=1}^n x(F_i, t) \, 1_{F_i}$. One can therefore regard x as an assignment of a real number to every spot of the information tree. The set of strictly positive adapted processes is denoted by $\mathcal{L}_{++}$ and is identified with $\mathbb{R}_{++}^{1+N}$.

Related to the notion of an adapted process is that of a **stopping time** defined as a function of the form $\tau \colon \Omega \to \{0, 1, \ldots, T\} \cup \{\infty\}$, provided that $\{\tau \leq t\} \in \mathcal{F}_t$ for every time t. The last restriction is equivalent to the adaptedness of the **indicator process** $1_{\{\tau \leq t\}}$, which takes the value zero prior to the (random) time τ, and the value one from time τ on (which on the event $\{\tau = \infty\}$ is never). By eliminating all spots where $1_{\{\tau \leq t\}}$ takes the value one, the stopping time τ can be visualized as a pruning of the information tree. A stopping time, or corresponding indicator process, announces the (first) arrival of an event which is consistent with the information stream encoded in the underlying filtration. For example, if x is an adapted process, then the first time that $x_t \geq 1$ defines a stopping time (with the value ∞ being assigned on the event that x remains valued below one). On the other hand, the first time that x reaches its path maximum is not generally a stopping time. For any process x and stopping time τ, the random variable x_τ or $x(\tau)$ is defined by letting $x_\tau(\omega) = x(\omega, \tau(\omega))$, with the convention $x(\omega, \infty) = 0$.

In applications, it is common to specify the filtration $\{\mathcal{F}_t\}$ as the information revealed, or generated, by given processes representing observable quantities. The algebra **generated** by a set S of random variables is the intersection of all algebras relative to which every $x \in S$ is measurable, and is denoted by $\sigma(S)$. If $S = \{x_1, \ldots, x_n\}$, $\sigma(x_1, \ldots, x_n) \equiv \sigma(S)$ is the same as the algebra generated by the partition of all nonempty events of the form $\{(x_1, \ldots, x_n) = \alpha\}$, where $\alpha \in \mathbb{R}^n$. We interpret $\sigma(x_1, \ldots, x_n)$ as the information that can be inferred by observing the realization of the random variables $x_1, \ldots, x_n$. Any other variable whose realization is revealed by this information must be determined as a function of the realization of $(x_1, \ldots, x_n)$.

PROPOSITION 1.1.4 *Given any random variables* $x_1, \ldots, x_n$, *a random variable* y *is* $\sigma(x_1, \ldots, x_n)$-*measurable if and only if there exists a function* $f : \mathbb{R}^n \to \mathbb{R}$ *such that* $y = f(x_1, \ldots, x_n)$.

PROOF That $f(x_1, \ldots, x_n)$ is $\sigma(x_1, \ldots, x_n)$-measurable follows from Proposition 1.1.3. Conversely, suppose y is $\sigma(x_1, \ldots, x_n)$-measurable and let $\{(x_1(\omega), \ldots, x_n(\omega)) \mid \omega \in \Omega\} = \{\alpha_1, \ldots, \alpha_m\} \subseteq \mathbb{R}^n$. The algebra $\sigma(x_1, \ldots, x_n)$ is generated by the partition $\{A_1, \ldots, A_m\}$, where $A_i \equiv \{(x_1, \ldots, x_n) = \alpha_i\}$. Let $y = \sum_{i=1}^{m} \beta_i 1_{A_i}$, where β_i is the constant value of y on A_i. Selecting any function $f : \mathbb{R}^n \to \mathbb{R}$ such that $\beta_i = f(\alpha_i)$ results in $y = f(x_1, \ldots, x_n)$. $\qquad\square$

A filtration is said to be **generated** by the processes $B^1, \ldots, B^d$, where each B_0^i is a constant, if for all t, $\mathcal{F}_t = \sigma(\{B_s^1, \ldots, B_s^d \mid s \le t\})$. Writing $B = (B^1, \ldots, B^d)'$, it follows from Proposition 1.1.4 that in this case a process x is adapted if and only if x_0 is constant and for every time $t > 0$, there exists a function $f(t, \cdot) : \mathbb{R}^{d \times t} \to \mathbb{R}$ such that $x(\omega, t) = f(t, B(\omega, 1), \ldots, B(\omega, t))$ for all $\omega \in \Omega$. In other words, x_t is a function of the path of B up to time t.

1.2 Market and Arbitrage

The fundamental role of financial markets is to reallocate resources across time and contingencies in a way that is beneficial to all market participants. Abstracting away from important implementation details, such as the need for appropriate collateral arrangements, we represent a financial market as a set of net incremental cash flows that are generated by trading financial contracts such as bonds, stocks, futures, options and swaps. For example, the purchase of a share of a stock at some spot generates the cash flow consisting of minus the stock price at the given spot followed by the dividend stream resulting from holding the share on the information subtree rooted at the given spot. Effectively, a market facilitates the exchange of funds across spots, be it saving, borrowing, hedging or speculation, or some combination thereof that cannot be cleanly categorized.

We implicitly assume that the terms of market exchanges are set competitively, meaning that there is a large number of market participants each of whom has negligible bargaining power, but who collectively influence market-clearing prices. A more formal definition of a market and related concepts follows. Recall that an underlying filtration $\{\mathcal{F}_t\}$ on Ω satisfying (1.1.1) is taken as given throughout.

DEFINITION 1.2.1 A **cash flow** is any adapted process. A **market** is any linear subspace X of the space $\mathcal{L}$ of all cash flows. A cash flow x is said to be **traded** in X if $x \in X$. A cash flow c is an **arbitrage** if $0 \neq c \geq 0$. A market X is **arbitrage-free** if it contains no arbitrage cash flow.

All cash flows are specified in some implicit unit of account that is fixed throughout. A market X represents the set of all net incremental cash flows that are available to market participants, typically by following some trading strategy over time, a notion that is formally defined later in this chapter. From the perspective of time zero, an agent can use the market X to modify a cash flow c to $c + x$ for any $x \in X$. Of course, the agent will add a traded cash flow x to c only if $c + x$ is preferred to c. An arbitrage cash flow x satisfies $x(F, t) \geq 0$ for every spot (F, t), and $x(F, t) > 0$ for some spot (F, t). An agent who prefers more to less prefers $c + x$ to c, for every c and every arbitrage cash flow x. If X contains an arbitrage, such an agent cannot find any c optimal given X. In a competitive equilibrium, to be defined formally in Chapter 3, agents follow optimal plans and therefore the market cannot contain arbitrage opportunities.

The assumption that a market is a linear subspace reflects the underlying assumption of perfect competition with no transaction costs or trading constraints. Traded cash flows can be reversed, scaled arbitrarily and combined. The discussion of market equilibrium in Chapter 3 provides an argument as to why market forces work to relax binding market constraints. Yet, frictions such as moral hazard, asymmetric information, search or information processing costs, concentrated market power and phenomena of market panics and runs can work to impede this process of increasingly complete markets, limiting the scope of the definition of traded cash flows as a linear subspace.[1] This being a first course, we will mostly confine ourselves to the case of linear markets, which is why linearity is part of a market's definition. Besides its pedagogical value, the perfect case is of practical value as an approximation in thinking about highly liquid and competitive markets in standardized financial contracts as long as the limitations of such an approximation are well recognized.

Fixing a reference market X in the background, we make a technical distinction between traded cash flows, which are the elements of X, and marketed cash flows, which are cash flows that can be obtained in the

---

[1] This text's conceptual framework and several of its arguments extend to allow for exogenous trading constraints, resulting in shapes of X that are not linear subspaces. The more interesting aspect of trading constraints, however, is their endogenous source in equilibrium.

market given sufficient time-zero cash. Recall that $1_{\Omega \times \{0\}}$ denotes the process that takes the value one at spot zero and the value zero everywhere else on the information tree.

DEFINITION 1.2.2 The cash flow c is **marketed** (in the market X at time zero) if $c - \alpha 1_{\Omega \times \{0\}} \in X$ for some $\alpha \in \mathbb{R}$. The market is **complete** if every cash flow is marketed.

A marketed cash flow c can be expressed as $c = \alpha 1_{\Omega \times \{0\}} + x$ for some traded cash flow x. The scalar α represents a time-zero price of c, which is unique if and only if $1_{\Omega \times \{0\}} \notin X$, a condition known as **the law of one price.** Clearly, an arbitrage-free market satisfies the law of one price but the converse is not generally true. For expositional simplicity, we will assume that the market is arbitrage-free, even where the law of one price would suffice.

DEFINITION 1.2.3 Assuming the market is arbitrage-free, the (time-zero) **present value** of a marketed cash flow c is the unique scalar α such that $c - \alpha 1_{\Omega \times \{0\}} \in X$.

Proceeding under the assumption that X is arbitrage-free, note that the set of marketed cash flows is the linear span of X and $1_{\Omega \times \{0\}}$, whose dimension is one more than that of X. The function that maps every marketed cash flow to its present value is linear and positive, where positivity means that the present value of every arbitrage cash flow is (strictly) positive. The set X is the kernel of this function, that is, the set of marketed cash flows of zero present value. If the market is complete, every cash flow is marketed and therefore has a uniquely defined present value. The dimension of X in this case is N, where $1 + N$ is the total number of spots on the information tree. In the following section we will show that if the market is not complete, then the present-value function on the set of marketed cash flows can be extended to all of $\mathcal{L}$ while retaining linearity and positivity, which leads to some useful mathematical representations of the present-value function. Such an extension is not unique, however.

We have defined the market from the perspective of time zero. A market $X_{F,t}$ can also be defined analogously from the perspective of any other spot (F, t) as a subset of

$$\mathcal{L}_{F,t} \equiv \left\{ x \in \mathcal{L} \mid x = x 1_{F \times \{t, \dots, T\}} \right\},$$

which is the set of cash flows that can only take nonzero values on the subtree rooted at spot (F, t). The following proposition introduces assumptions that allow us to construct $X_{F,t}$ in terms of the time-zero market X. Extending Definition 1.2.2, we say that a cash flow c is **marketed** at spot (F,t) in $X_{F,t}$ if $c1_{F\times\{t,...,T\}} - \alpha 1_{F\times\{t\}} \in X_{F,t}$ for some $\alpha \in \mathbb{R}$.

PROPOSITION 1.2.4 *Suppose that the (time-zero) market X is arbitrage-free, and for some spot (F,t), the set $X_{F,t}$ satisfies:*

(1) *(adaptedness)* $X_{F,t} \subseteq \mathcal{L}_{F,t}$.
(2) *(dynamic consistency)* $X_{F,t} \subseteq X$.
(3) *(liquidity) Every $x \in X$ is marketed at (F,t) in $X_{F,t}$.*

Then $X_{F,t} = X \cap \mathcal{L}_{F,t}$.

PROOF By adaptedness and dynamic consistency, $X_{F,t} \subseteq X \cap \mathcal{L}_{F,t}$. To show the reverse inclusion, suppose that $x \in X \cap \mathcal{L}_{F,t}$. By liquidity, there exist $y \in X_{F,t}$ and $\alpha \in \mathbb{R}$ such that $x = x1_{F\times\{t,...,T\}} = \alpha 1_{F\times\{t\}} + y$. By dynamic consistency, $y \in X$ and therefore $\alpha 1_{F\times\{t\}} = x - y \in X$. Since X is arbitrage-free, $\alpha = 0$ and $x = y$, and therefore $x \in X_{F,t}$. □

The assumptions of Proposition 1.2.4 have simple interpretations. The elements of $X_{F,t}$ represent cash flows available to a market participant at spot (F,t) through trading over the information subtree rooted at (F,t). Such cash flows are naturally viewed as elements of $\mathcal{L}_{F,t}$. The idea behind dynamic consistency is that at time zero a trader can have any cash flow x in $X_{F,t}$ by making a contingent plan to carry out the transactions that result in x if (F,t) materializes. The cash flow x is effectively also available to the agent at time zero, and must therefore be an element of the time-zero market X. Finally, liquidity, in the narrow technical sense used here, means that if at time zero the agent starts following a plan generating the cash flow $x \in X$, then at any future spot the agent can liquidate all positions and cancel all remaining cash flows. For example, suppose there is no uncertainty, $T = 2$ and at time zero the agent buys a bond for 99 (units of account) that pays 100 at time two, thus generating the cash flow $x = (-99, 0, 100)$. In a liquid market, the cash flow $(0, p, -100)$ is traded at time one for some price p, in other words, the same bond continues to be traded at some price.

Making adaptedness and dynamic consistency part of the definition, a **dynamic market** specifies a market $X_{F,t} \subseteq X \cap \mathcal{L}_{F,t}$ for every spot (F,t), with $X = X_{\Omega,0}$ being the time-zero market. The dynamic market is **liquid** if it satisfies the third condition of Proposition 1.2.4, and therefore $X_{F,t} = X \cap \mathcal{L}_{F,t}$ and X specifies the entire liquid dynamic market. Not every time-zero market is consistent with a liquid dynamic market, however, which motivates the following definition.

DEFINITION 1.2.5 The market X is **liquid** if for every spot (F,t), every $x \in X$ is marketed at spot (F,t) in the market $X \cap \mathcal{L}_{F,t}$.

This definition abuses terminology in the interest of simplicity, since liquidity is really a property of a dynamic market. It is entirely consistent to consider a dynamic market $\{X_{F,t}\}$, where $X_{\Omega,0} = X$ is a market satisfying the property of Definition 1.2.5, while for some nonzero spot (F,t), $X_{F,t}$ violates the liquidity property of Proposition 1.2.4, for example, $X_{F,t} = \{0\}$. Whenever we specify a *liquid market* X, however, we implicitly assume that at each spot (F,t), the market $X_{F,t} \equiv X \cap \mathcal{L}_{F,t}$ is available, and therefore a corresponding liquid dynamic market $\{X_{F,t}\}$ is specified by X.

Liquidity requires that an initially marketed cash flow continues to be marketed, but it does not require that every cash flow is marketed. On the other hand, the following simple proposition shows that a time-zero arbitrage-free complete market stays complete as uncertainty unfolds and in particular is liquid (or, more precisely, implies a liquid dynamic market).

PROPOSITION 1.2.6 *Suppose X is an arbitrage-free complete market. Then for every spot (F,t), all cash flows are marketed at (F,t) in $X \cap \mathcal{L}_{F,t}$.*

PROOF Consider any cash flow c. Using the assumption that X is complete, pick scalars α and β such that $c\mathbf{1}_{F\times\{t,...,T\}} - \alpha\mathbf{1}_{\Omega\times\{0\}} \in X$ and $\mathbf{1}_{F\times\{t\}} - \beta\mathbf{1}_{\Omega\times\{0\}} \in X$. Since X is arbitrage-free, $\beta > 0$. Linearity of X then implies that $c\mathbf{1}_{F\times\{t,...,T\}} - (\alpha/\beta)\mathbf{1}_{F\times\{t\}} \in X$. Therefore, c is marketed at (F,t) in $X \cap \mathcal{L}_{F,t}$. $\square$

1.3 Trading and Pricing of Financial Contracts

While market participants are ultimately interested in the incremental cash flows of the market X, they must undertake certain actions to generate these cash flows. In theory, every traded cash flow could be implemented as a buy-and-hold portfolio in contracts generating cash flows that form a linear basis of X. If there are $1 + N$ spots and every nonterminal spot has at least two immediate successors, then the dimension of a complete market is $N \geq 2^{T+1}$, rendering the assumption of competitively traded basic contracts unrealistic, even for moderate values of T. A key insight is that a small number of contracts can implement a high-dimensional market provided these same contracts can be traded at every spot of the filtration. Postponing a more formal explanation of this claim, in this section we define contracts and we discuss their relationship to a market and some straightforward implications of the no-arbitrage assumption for contract pricing.

We use the term "contract" in a narrow formal sense to mean a dividend stream together with a price process indicating at what price the dividend stream can be traded at every spot.

DEFINITION 1.3.1 A **contract** is a pair (δ, V) of adapted processes satisfying the convention

$$\delta_0 \equiv 0 \quad \text{and} \quad \delta_T \equiv V_T. \tag{1.3.1}$$

The process δ is the contract's **dividend process** and V is the contract's **value process** or **cum-dividend price process**. The contract's **ex-dividend price process** is $S \equiv V - \delta$.

We use the word "dividend" in a generalized sense to mean any cash flow that is the result of holding a long position in the contract up to time T and liquidating at time T. In applications, such payments may correspond to coupon payments and a face-value payment at maturity in the case of a bond, dividends prior to time T and the sale price cum dividend at time T in the case of a stock (whose actual dividend stream can extend beyond time T), net cash settlements in the case of a swap and so on. A contract entitles the owner (or long position) to a single dividend stream. An extension to include options, where the owner can select from a set of dividend streams, is discussed in Section 1.5.

The owner of a contract (δ, V) receives the dividend payment δ_t at time t. If the contract is bought at time t, either the buyer pays the seller V_t and receives the time-t dividend δ_t, or the buyer pays the seller the ex-dividend price S_t and the time-t dividend goes to the seller. The dividend convention (1.3.1) is made for notational simplicity and entails no loss of generality within the scope of the model presented here. One can think of the condition $\delta_T = V_T$ as reflecting the implicit assumption that the contract is liquidated at the terminal date T, resulting in the payment $V_T = S_T + \delta_T$. It makes no difference within the model how the value V_T is split between S_T and δ_T or how V_0 is split between S_0 and δ_0; for simplicity, we set $S_T = 0$ and $\delta_0 = 0$.

DEFINITION 1.3.2 A contract (δ, V) is **traded** at spot (F, t) in X if

$$x \equiv -V1_{F\times\{t\}} + \delta1_{F\times\{t,...,T\}} \in X. \tag{1.3.2}$$

The cash flow x is **generated by buying** the contract at spot (F, t), while $-x$ is **generated by selling** the contract at the same spot. The contract (δ, V) is **traded** (in X) if it is traded at every spot.

We proceed in the context of a given reference arbitrage-free market X. When we say that a contract is traded (at some spot), it is implied that the contract is traded in the reference market X. Note that a cash flow δ, where $\delta_0 = 0$, is marketed in X if and only if there is a contract (δ, V) that is traded at time zero, in which case V_0 is the present value of δ. This claim relies critically on the convention $\delta_T \equiv V_T$; V_0 is the present value of the dividend stream $\delta_1, \ldots, \delta_{T-1}$ plus the present value of the terminal payment V_T resulting from selling the contract at time T. Clearly, a model on the given filtration cannot make any predictions on how V_T relates to dividends paid after T.

REMARK 1.3.3 The trades generating an arbitrage as a consequence of the violation of a claimed arbitrage pricing relationship are instructive in that they suggest potential ways in which the pricing relationship can be violated in realistic applications due to frictions left out of the formal model. For example, suppose the contracts (δ, V) and (δ, V') are both traded, but $V_0 > V_0'$. The arbitrage $(V_0 - V_0')1_{\Omega\times\{0\}}$ results by selling the first contract and buying the second one. Suppose the former represents a special security that can be used for the purpose of posting collateral, thus facilitating other trades. If collateral in this sense is scarce in equilibrium, the value V_0 may well exceed the present value of δ. In the presence of a well-functioning competitive lease market in the security, V_0 equals the present value of the equilibrium lease payments resulting from lending out the security. The extent to which these lease payments exceed the dividends δ reflects the equilibrium value of holding an additional unit of inventory. $\diamond$

The preceding observations can be applied from the perspective of any other spot. Thus, in an arbitrage-free market, any two traded contracts (δ, V) and (δ, V') with a common dividend process δ (and therefore common terminal value) must have a common value process: $V = V'$. Buying the contract (δ, V) at spot $(F, t - 1)$ and selling it at each of its immediate successor spots generates the cash flow

$$- S1_{F \times \{t-1\}} + V1_{F \times \{t\}},$$

where $S \equiv V - \delta$. Therefore, if the contracts (δ^1, V^1) and (δ^2, V^2) are traded and $V_t^1 = V_t^2$ on some $F \in \mathcal{F}_{t-1}$, then $S_{t-1}^1 = S_{t-1}^2$ on F. As a corollary, if (δ^1, V^1) and (δ^2, V^2) are traded in an arbitrage-free market and $V_t^1 = V_t^2$ for all $t > 0$, then the two contracts are identical.

We have defined what it means for a contract to be traded in a given market. Conversely, trading in a given set of contracts implements a market, which can be succinctly defined as follows.

DEFINITION 1.3.4 The market **implemented** by a set of contracts is the smallest market (relative to inclusion) in which all contracts in the given set are traded.

Suppose $\mathcal{C}$ is a set of contracts and let X^0 be the set of all cash flows of the form (1.3.2) for every $(\delta, V) \in \mathcal{C}$ and spot (F, t). The set $\text{span}(X^0)$ of all finite linear combinations of elements of X^0 is a market in which all contracts in $\mathcal{C}$ are traded, and every market with this property must include X^0 and therefore $\text{span}(X^0)$. This shows that $X \equiv \text{span}(X^0)$ is the set implemented by $\mathcal{C}$. This construction also makes it clear that X is liquid. An element of $\text{span}(X^0)$ can be thought of as a contingent plan to buy or sell contracts at various spots, in other words, a trading strategy. Assuming X is arbitrage-free, a contract that is not in $\mathcal{C}$ but is traded in X is synthetic in $\mathcal{C}$, meaning that it is generated by a trading strategy in contracts in $\mathcal{C}$. Section 1.7 discusses trading strategies, synthetic contracts and associated budget equations more systematically. In Chapter 2, we will see that the minimum number of contracts implementing a complete market is equal to the maximum number of immediate successor spots to each spot.

1.4 Present-Value Functions

Taking a reference market X as given, in this section we discuss more systematically the dual notion of a present-value function. Recall that a **linear functional** on a vector space is a real-valued linear function whose domain is the entire vector space. Given a convention of what it means for a vector to be **positive**, a linear functional is said to be **positive** if it assigns a (strictly) positive value to every positive vector. For cash flows, positive and arbitrage are synonyms: The cash flow x is **positive** if and only if $0 \neq x \geq 0$.

DEFINITION 1.4.1 A (time-zero) **present-value function** (for the market X) is a positive linear functional Π on $\mathcal{L}$ such that $\Pi(x) \leq 0$ for all $x \in X$ and $\Pi(1_{\Omega \times \{0\}}) = 1$.

A present-value function Π specifies a time-zero value for every cash flow c, marketed or not. Given linearity, the requirement that Π is positive is equivalent to the monotonicity condition: $\Pi(b) > \Pi(a)$ for all cash flows a, b such that $b - a$ is an arbitrage. The essential restriction that Π is nonpositive on X is equivalent to $\Pi(x) = 0$ for all $x \in X$, since X is a linear subspace. (In extensions with trading constraints, X is no longer a linear subspace, but Definition 1.4.1 is still valid, and the present value of a traded cash flow can be strictly negative.) The requirement $\Pi(1_{\Omega \times \{0\}}) = 1$ is merely a normalization.

The properties of a present-value function Π combine to determine the value $\Pi(c)$ of a marketed cash flow c as the present value of c in the sense of Definition 1.2.3. Suppose the scalar α is such that $c - \alpha 1_{\Omega \times \{0\}} \in X$. The existence of Π implies that X is arbitrage-free and therefore α is unique. The fact that Π vanishes on X implies that $\Pi(c - \alpha 1_{\Omega \times \{0\}}) = 0$. The linearity of Π implies that $\Pi(c) = \alpha \Pi(1_{\Omega \times \{0\}})$. Finally, the normalization assumption results in $\Pi(c) = \alpha$. Another way of stating this conclusion is that if the contract (δ, V) is traded at time zero and Π is a present-value function, then $V_0 = \Pi(\delta)$.

If the market X is arbitrage-free and complete, letting $\Pi(c)$ equal the (unique) present value of c defines a function $\Pi \colon \mathcal{L} \to \mathbb{R}$ that is easily confirmed to be a present-value function. Therefore, every complete arbitrage-free market admits a unique present-value function. Conversely, every present-value function Π for X defines a unique complete market for which Π is a present-value function; it is the kernel of Π, that is, the set $\{c \in \mathcal{L} \mid \Pi(c) = 0\}$. If X is complete, then the kernel of Π is equal to X. If X is incomplete, then the kernel of Π is a proper superset of X and can be thought of as an arbitrage-free completion of X.

Recall that, since an adapted process x is an assignment of a scalar $x(F, t)$ to every spot (F, t), the set $\mathcal{L}$ of adapted processes can be identified with $\mathbb{R}^{1+N}$, where $1 + N$ is the total number of spots on the information tree. Using this identification, we endow $\mathcal{L}$ with the usual Euclidean inner product in $\mathbb{R}^{1+N}$, denoted

$$x \cdot y \equiv \sum_{\text{all spots } (F, t)} x(F, t) \, y(F, t).$$

DEFINITION 1.4.2 A **state-price process** is any adapted process p with $p_0 > 0$ such that

$$\Pi(c) \equiv \frac{p \cdot c}{p_0}, \quad c \in \mathcal{L},$$

defines a present-value function Π. In this case, we say that p **represents** Π. An **Arrow cash flow**[2] is a cash flow of the form $1_{F \times \{t\}}$, where (F, t) is a spot.

Every cash flow is a linear combination of the $1 + N$ Arrow cash flows, reflecting the identification of $\mathcal{L}$ and $\mathbb{R}^{1+N}$. If Π is a present-value function and we define $p(F, t) = \Pi(1_{F \times \{t\}})$ for every spot (F, t), then $\Pi(c) = p \cdot c$ for all $c \in \mathcal{L}$. Therefore, every present-value function can be represented by a state-price process. (In mathematical terms, the Arrow cash flows are a linear basis of $\mathcal{L}$ and p is the Riesz representation of Π.) If p is a state-price process representing the present-value function Π, then $p(F, t) = p_0 \Pi(1_{F \times \{t\}})$ for every spot (F, t), and therefore p is strictly positive, it is uniquely determined by Π up to a positive scaling factor, and it represents relative present value of Arrow cash flows: For all spots (F, t), (G, s),

$$\frac{p(F, t)}{p(G, s)} = \frac{\Pi(1_{F \times \{t\}})}{\Pi(1_{G \times \{s\}})}.$$

For a geometric interpretation of a state-price process p representing Π, note that, since $\Pi(x) = 0$ for all $x \in X$, p is orthogonal to X, and since Π is positive, p lies in $\mathbb{R}^{1+N}_{++}$ (the interior of the positive orthant of $\mathbb{R}^{1+N}$). The set of all state-price processes representing Π can be identified with a directed half line in $\mathbb{R}^{1+N}_{++}$. Inspection of Figure 1.2 suggests that such a direction exists if and only if $X \cap \mathbb{R}^{1+N}_{+} = \{0\}$, which is another way of saying that X is arbitrage-free. If, as in Figure 1.2, X is complete and therefore N-dimensional, there is only one orthogonal-to-X line within $\mathbb{R}^{1+N}$, and therefore the corresponding present-value function must be unique. If X is incomplete, the dimension of X is at most $N - 1$, leaving at least one dimension along which p can be rotated within $\mathbb{R}^{1+N}_{++}$

---

[2] The term, which is not entirely standard, is in recognition of Arrow (1953) (translated to English in Arrow (1964)) who, together with Debreu (1959), provided the modern conceptual framework for incorporating uncertainty in classical competitive equilibrium theory. The more common term "Arrow–Debreu security" refers to a claim to an Arrow cash flow. The (relative) time-zero prices of Arrow–Debreu securities are also known as Arrow–Debreu prices, corresponding to the notion of a state-price process here.

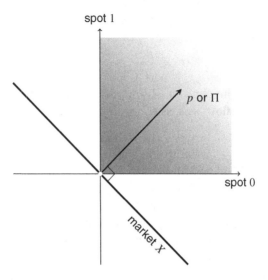

Figure 1.2 Example of a market and present-value function with
$N = 1$. The shaded region, including the axes but not the origin,
is the set of arbitrage cash flows. The market X does not cut into
the shaded region if and only if the half line defined by the
orthogonal vector p lies in the interior of the shaded region.

while maintaining orthogonality to X, which suggests that an incomplete
market admits multiple present-value functions. For an example, extend
Figure 1.2 by introducing a spot-2 orthogonal axis, while maintaining the
market X as a line. Rotating the orthogonal-to-X vector p around X while
staying within the interior of the positive orthant traces out all present-value
functions relative to X. A rigorous proof of these informal insights follows.

THEOREM 1.4.3 *A present-value function exists if and only if*[3] *X is
arbitrage-free; and is unique if and only if the market X is complete.*

---

[3] The equivalence of the lack of arbitrage opportunities and the existence of a present-value
function is an example of the so-called theorems of the alternative in convex analysis,
exposited in chapter 1 of Stoer and Witzgall (1970). In the current finite-dimensional
financial market context, the result is due to Ross (1978). It was extended to
infinite-dimensional spaces by Yan (1980) and Kreps (1981). In general, under infinitely
many states, the result requires the exclusion of a stronger notion of arbitrage
opportunities than mere positivity. A notable exception is the case of a market generated
by finitely many assets in discrete time, as shown by Dalang et al. (1990), with simplified
proofs given by Schachermayer (1992) and Kabanov and Kramkov (1994). Extensions to
models with continuous-time trading are reviewed by Delbaen and Schachermayer (2006).

PROOF We fix an enumeration $0, 1, \ldots, N$ of all spots, where $(\Omega, 0)$ is spot zero, and write $x = (x_0, x_1, \ldots, x_N)$ for the element of $\mathbb{R}^{1+N}$ corresponding to the adapted process x. In other words, $x_i \equiv x(F, t)$, where (F, t) is the ith spot. The conflicting notation x_t, where t is a time, is not used in this proof.

(Existence) The only-if part is immediate. Conversely, suppose X is arbitrage-free. For reasons explained in Remark 1.4.5, we will show the existence of a present-value function assuming only that X is a closed convex cone of cash flows. The idea is to define a state-price process p by separating the disjoint convex sets X and $\mathbb{R}^{1+N}_+ \setminus \{0\}$. This ensures that $p \cdot x \leq 0$ for all $x \in X$ and $p \geq 0$, but we need $p_n > 0$ for all n. To ensure strict positivity, we instead strictly separate X and the compact convex set

$$\Delta \equiv \left\{ x \in \mathbb{R}^{1+N}_+ \mid \sum_n x_n = 1 \right\}.$$

Let $p \equiv \bar{y} - \bar{x}$, where $(\bar{x}, \bar{y}) \in X \times \Delta$ minimizes $\|x - y\|$ over all $(x, y) \in X \times \Delta$. Assuming $(\bar{x}, \bar{y})$ exists, let us first confirm that p is a state-price process. Since $\bar{x}$ and $\bar{y}$ are the projection of each other on X and Δ, respectively, the first part of the Projection Theorem B.5.1 implies the support inequalities $p \cdot (x - \bar{x}) \leq 0$ for all $x \in X$ and $p \cdot (y - \bar{y}) \geq 0$ for all $y \in \Delta$. Since X is a cone, the first inequality implies $p \cdot x \leq p \cdot \bar{x} = 0$ for all $x \in X$. Given that $p \cdot \bar{x} = 0$, the second support inequality implies $p \cdot y \geq p \cdot \bar{y} = p \cdot p > 0$ for all $y \in \Delta$, and therefore $p_n > 0$ for all n, completing the proof that p is a state-price process. Finally, to show the existence of the pair $(\bar{x}, \bar{y})$, we apply the remaining parts of Theorem B.5.1 to conclude that the function mapping each point of Δ to its projection on X is well defined and continuous. Since norms are continuous, the function that maps each point of Δ to its least distance from X is also continuous (as well as convex) and therefore achieves a minimum over Δ by Proposition B.3.7 (or Proposition B.4.3).

(Uniqueness) Suppose Π is a present-value function and let M denote the set of marketed cash flows. We have already shown that X uniquely determines $\Pi(c)$ for all $c \in M$. If X is complete, then $M = \mathbb{R}^{1+N}$ and Π is uniquely determined. Suppose instead that X is incomplete, and fix any $\bar{c} \in \mathbb{R}^{1+N} \setminus M$. Taking an orthogonal projection, let $\bar{c} = m + z$, where $m \in M$ and $z \neq 0$ is orthogonal to M and therefore also to X and $(1, 0, \ldots, 0)$ (the vector corresponding to $1_{\Omega \times \{0\}}$). Let $p \in \mathbb{R}^{1+N}_{++}$ be the Riesz representation of Π, so $\Pi(c) = p \cdot c$ for all $c \in \mathbb{R}^{1+N}$. Consider any scalar $\alpha \neq 0$ such that $p^\alpha \equiv p + \alpha z \in \mathbb{R}^{1+N}_{++}$. Since z is orthogonal to X

and $(1, 0, \ldots, 0)$, $p^\alpha \cdot x = p \cdot x$ for all $x \in X$ and $p^\alpha \cdot (1, 0, \ldots, 0) = p_0 = 1$, and therefore $\Pi^\alpha(c) = p^\alpha \cdot c$ defines a present-value function, which is not equal to Π, since $\Pi^\alpha(\bar{c}) = \Pi(\bar{c}) + \alpha z \cdot z \neq \Pi(\bar{c})$. ☐

REMARK 1.4.4 The uniqueness argument just given shows that the value a present-value function assigns to a nonmarketed cash flow $\bar{c}$ is constrained only by the requirement that a present-value function is positive. Had we not required positivity of p^α, the parameter α could be chosen arbitrarily and $\Pi^\alpha(\bar{c})$ could take any value. ◇

REMARK 1.4.5 The preceding proof shows the existence of a present-value function assuming only that X is a closed convex cone, not necessarily a linear subspace. This generality is utilized in our later discussion of dominant choice. It also allows an easy extension of Theorem 1.4.3 to a **constrained market** X, which we define as a closed convex set of cash flows such that $0 \in X$ and there exists $\varepsilon > 0$ such that for all $x \in X$, $0 < \|x\| < \varepsilon$ implies $(\varepsilon / \|x\|) x \in X$. In words, every trade that is small in the sense that its norm is no more than ε can be scaled up so that its norm is ε. This allows for large position limits as well as short-sale constraints (since $x \in X$ need not imply $-x \in X$). We show that a *constrained market* X *is arbitrage-free if and only if a present-value function exists*. First note that X is arbitrage-free if and only if the cone $C \equiv \{sx \mid s \in \mathbb{R}_+, x \in X\}$ generated by X is arbitrage-free. Since C is a closed convex cone, the proof of Theorem 1.4.3 applies. That C is convex is immediate. That C is closed follows from the assumption that trades smaller than ε can be scaled up to have norm ε. The interested reader can prove this by confirming that C and X have the same (closed) intersection with $\{x \in \mathbb{R}^{1+N} \mid \|x\| \leq \varepsilon\}$. ◇

We have so far defined present-value functions from the perspective of time zero. A present-value function can be defined analogously from the perspective of every other spot (F, t). Suppose $X_{F,t} \subseteq \mathcal{L}_{F,t}$ is the set of traded cash flows from the perspective of spot (F, t), and the dynamic consistency assumption $X_{F,t} \subseteq X$ is satisfied (as in Proposition 1.2.4). Then the restriction of a time-zero present-value function Π to $\mathcal{L}_{F,t}$ is a positive linear functional on $\mathcal{L}_{F,t}$ satisfying $\Pi(x) \leq 0$ for all $x \in X_{F,t}$. After normalization so that $\Pi_{F,t}(1_{F \times \{t\}}) = 1$, we obtain the spot-$(F, t)$ **conditional** present-value function

$$\Pi_{F,t}(c) \equiv \frac{\Pi(c 1_{F \times \{t, \ldots, T\}})}{\Pi(1_{F \times \{t\}})}, \quad c \in \mathcal{L}. \tag{1.4.1}$$

Note that we have defined $\Pi_{F,t}$ over the entire domain $\mathcal{L}$, while only its restriction $\Pi_{F,t} : \mathcal{L}_{F,t} \to \mathbb{R}$ is meaningful as a spot-(F, t) present-value function. This is merely a convenience. The following proposition shows that for an arbitrage-free liquid market the preceding definition covers all spot-(F, t) present-value functions.

PROPOSITION 1.4.6 *Given an arbitrage-free liquid market X, suppose $\Phi: \mathcal{L}_{F,t} \to \mathbb{R}$ is a positive linear functional satisfying $\Phi(x) \leq 0$ for all $x \in X \cap \mathcal{L}_{F,t}$ and $\Phi(1_{F \times \{t\}}) = 1$. Then there exists a (time-zero) present-value function Π such that $\Phi(c) = \Pi_{F,t}(c)$ for all $c \in \mathcal{L}_{F,t}$.*

PROOF Fix any time-zero present-value function $\tilde{\Pi}$ and define the function $\Pi: \mathcal{L} \to \mathbb{R}$ by letting $\Pi(c) = \tilde{\Pi}(\tilde{c})$, where $\tilde{c}$ is the cash flow obtained from c after replacing the restriction of c on $F \times \{t, \ldots, T\}$ with a single payment at spot (F, t) equal to $\Phi(c) \equiv \Phi(c 1_{F \times \{t, \ldots, T\}})$, that is, $\tilde{c} \equiv c - c 1_{F \times \{t, \ldots, T\}} + \Phi(c) 1_{F \times \{t\}}$. Then Π is a positive linear functional that satisfies $\Pi(1_{\Omega \times \{0\}}) = 1$. We now show that Π vanishes on X and is therefore a present-value function whose conditional is Φ. Given any $x \in X$, the liquidity assumption allows us to find $y \in X \cap \mathcal{L}_{F,t}$ such that $(x - y) 1_{F \times \{t+1, \ldots, T\}} = 0$ and therefore $\Pi(x - y) = \tilde{\Pi}(x - y) = 0$. Since $y \in \mathcal{L}_{F,t}$ and $\Phi(y) = 0$, we also have $\Pi(y) = \tilde{\Pi}(0) = 0$. Therefore, $\Pi(x) = \Pi(x - y) + \Pi(y) = 0$. $\qquad \square$

An exercise in applying the conditioning formula (1.4.1) shows that if spot (F, t) has immediate successors $(F_0, t + 1), \ldots, (F_d, t + 1)$, then for every cash flow c,

$$\Pi_{F,t}(c) = c(F, t) + \Pi_{F,t}\left(\sum_{i=0}^{d} \Pi_{F_i, t+1}(c) 1_{F_i \times \{t+1\}} \right). \qquad (1.4.2)$$

These equations represent a recursion that determines the time-zero present value $\Pi(c)$, starting with the terminal values $\Pi_{F \times \{T\}}(c) = c(F, T)$ and proceeding backward on the information tree.

We conclude this section with some observations on the relationship between present-value functions and contract pricing.

DEFINITION 1.4.7 The positive linear functional $\Pi: \mathcal{L} \to \mathbb{R}$ is said to **price** the contract (δ, V) if for every spot (F, t),

$$V(F, t) = \Pi_{F,t}(\delta). \qquad (1.4.3)$$

PROPOSITION 1.4.8 *A present-value function prices all traded contracts. Conversely, a positive linear functional on $\mathcal{L}$ that prices all contracts implementing the market and assigns the value one to $1_{\Omega \times \{0\}}$ is a present-value function.*

PROOF To show the first claim, set to zero the present value of the cash flow (1.3.2) resulting from buying a traded contract at a given spot. To show the converse, let X^0 be the set of all cash flows of the form (1.3.2). As discussed in the last paragraph of Section 1.3, span(X^0) is the market implemented by the given contracts. A positive linear functional vanishes on span(X^0) if and only if it vanishes on X^0 if and only if it prices every contract defining X^0. □

Letting $S \equiv V - \delta$, equations (1.4.3) are equivalent to the backward recursion

$$S(F, t) = \Pi_{F,t}\left(V1_{F \times \{t+1\}}\right), \quad V_T = \delta_T.$$

This pricing equation follows either as an application of recursion (1.4.2) or by setting to zero the present value of the traded cash flow generated by buying the contract at spot (F, t) and selling it a period later. Conversely, given δ, V is uniquely determined by the backward recursion and must therefore satisfy (1.4.3).

1.5 Options and Dominant Choice

An option generalizes our earlier notion of a contract by allowing the owner to choose any dividend process within a specified set of cash flows O. For example, suppose O represents the possible cash flows resulting from a finite opportunity set of projects available to a firm. Shareholder owners of the firm may disagree on what is the best project to undertake. Assuming shareholders only care about the cash flow generated by a project and every cash flow in O can be sold in a complete market, then shareholders who would disagree absent a market can agree on a project that is assigned the highest price by the market. Extending this idea, we will see that a choice maximizes every present-value function if and only if it is dominant in that it is optimal given the market for every decision maker who is not averse to arbitrage cash flows. Moreover, assuming the market is liquid, a choice that is dominant at time zero remains dominant as uncertainty unfolds and therefore there is no incentive to deviate from an initially selected dominant choice. The existence of a dominant choice in turn leads to the pricing of a traded option. An important application of these arguments is to standardized traded financial options such as American call and put options.

We assume throughout that a market X is available and that O is a nonempty set of cash flows.

DEFINITION 1.5.1 A cash flow $\delta^* \in O$ is **dominant** in O if given any $\delta \in O$, there exists a trade $x \in X$ such that $\delta^* + x \geq \delta$.

A dominant choice is optimal for everyone who can trade in X and is not averse to an incremental arbitrage cash flow. An agent who would have preferred δ in the absence of a market prefers receiving δ^* and simultaneously entering a trade x such that $\delta^* + x \geq \delta$. While the trade x is agent specific, the dominance of δ^* is agent independent.

THEOREM 1.5.2 *Suppose the market X is arbitrage-free and O is a nonempty set of cash flows. Then $\delta^* \in O$ is dominant in O if and only if for every present-value function Π for X,*

$$\Pi(\delta^*) = \max\{\Pi(\delta) \mid \delta \in O\}. \tag{1.5.1}$$

PROOF Suppose δ^* is dominant. For all $\delta \in O$, we can write $\delta^* + x \geq \delta$ for some $x \in X$, and therefore $\Pi(\delta^*) = \Pi(\delta^* + x) \geq \Pi(\delta)$ for every present-value function Π.

For the converse, it is instructive, although formally redundant, to first consider a simple argument for a complete market X. Suppose $\delta^* \in O$ maximizes the unique present-value function Π over O. Given any $\delta \in O$, we have $\delta^* = \Pi(\delta^*) 1_{\Omega \times \{0\}} + x^*$ and $\delta = \Pi(\delta) 1_{\Omega \times \{0\}} + x$ for some $x^*, x \in X$, and therefore

$$\delta^* - x^* + x = \Pi(\delta^*) 1_{\Omega \times \{0\}} + x \geq \Pi(\delta) 1_{\Omega \times \{0\}} + x = \delta,$$

which proves the dominance of δ^*.

Suppose now that X is not necessarily complete and $\delta^* \in O$ is *not* dominant. We will show that there exists some present-value function that δ^* does *not* maximize. As in the proof of Theorem 1.4.3, we identify $\mathcal{L}$ with $\mathbb{R}^{1+N}$. Since δ^* is not dominant, there exists $\delta \in O$ such that $\{\delta^* - \delta + x \mid x \in X\} \cap \mathbb{R}^{1+N}_+ = \emptyset$. Then $\Delta \equiv \delta^* - \delta \notin X$ and the set $X^* \equiv \{x + \alpha \Delta \mid x \in X, \alpha \in \mathbb{R}_+\}$ is an arbitrage-free closed convex cone. By Remark 1.4.5, there exists $p \in \mathbb{R}^{1+N}_{++}$ such that $p \cdot x \leq 0$ for all $x \in X^*$. Such a vector p is a state-price process that satisfies $p \cdot \Delta \leq 0$. To ensure the latter is a strict inequality, we perturb p while maintaining the state-price property as follows. Let $\Delta = \bar{x} + z$, where $\bar{x} \in X$ and z is nonzero and orthogonal to X. Pick $\varepsilon > 0$ small enough so that $p - \varepsilon z$ is strictly positive and let $\Pi(x) \equiv (p - \varepsilon z) \cdot x$ for all $x \in \mathbb{R}^{1+N}$. For $x \in X$, $z \cdot x = 0$ and therefore $\Pi(x) = p \cdot x \leq 0$. Moreover,

$$\Pi(\Delta) = p \cdot \Delta - \varepsilon z \cdot \Delta \leq -\varepsilon z \cdot (\bar{x} + z) = -\varepsilon z \cdot z < 0.$$

Therefore Π is a present-value function such that $\Pi(\delta^*) < \Pi(\delta)$. $\qquad\square$

COROLLARY 1.5.3 *Suppose the market is arbitrage-free, the set of cash flows O is compact (for example, finite), and every element of O is marketed. Then a dominant choice in O exists.*

If a compact option contains nonmarketed cash flows, there may not be a dominant choice. For a simple example, assume $T = 1$, there is no uncertainty, $O = \{(0, 1), (1, 0)\}$ and $X = \{0\}$. Neither choice of O is dominant and each maximizes some present-value function.

Suppose now that the option holder does not have to commit to an initial option selection. Will there be an incentive to deviate from a previously made dominant selection in the face of new information? To address this issue we refine the definition of an option to incorporate the idea that the option holder can deviate from a previous cash-flow choice at any time. As before, let O represent a set of cash flows available to the option holder at time zero. For every $\delta \in O$ and spot (F, t), the set of cash flows in O that are equal to δ up to but not including spot (F, t) is

$$O_{F,t}(\delta) \equiv \big\{ \tilde{\delta} \in O \mid \tilde{\delta} = \delta \text{ on } F \times \{0, \ldots, t - 1\} \big\}.$$

Selecting $\delta \in O_{\Omega,0} \equiv O$ at time zero and switching to $\tilde{\delta} \in O_{F,t}(\delta)$ at spot (F, t) is equivalent to selecting $\delta + (\tilde{\delta} - \delta)1_{F \times \{t,\ldots,T\}}$ at time zero, which must therefore also belong to O if the option holder does not have to commit at time zero. This motivates the following formal definition.

DEFINITION 1.5.4 An **option** is a set $O \subseteq \mathcal{L}$ such that for every spot (F, t), $\delta \in O$ and $\tilde{\delta} \in O_{F,t}(\delta)$ implies $\delta + (\tilde{\delta} - \delta)1_{F \times \{t,\ldots,T\}} \in O$.

EXAMPLE 1.5.5 (American call) Fixing a **maturity** date $\bar{\tau} \in \{1, \ldots, T - 1\}$, let $\mathcal{T}$ denote the set of all stopping times that are valued in $\{0, \ldots, \bar{\tau}\} \cup \{\infty\}$. For $\tau \in \mathcal{T}$, let $1_{[\tau]}$ denote the process whose value at (ω, t) is one if $\tau(\omega) = t$ and zero otherwise. An **American option** with payoff process $D \in \mathcal{L}$ is the option

$$O \equiv \big\{ D1_{[\tau]} \mid \tau \in \mathcal{T} \big\}, \quad \text{where } D_\infty \equiv 0.$$

The option owner who selects the cash flow $D1_{[\tau]}$ is said to **exercise** the American option at time τ. The event $\{\tau = \infty\}$ corresponds to never exercising the option and thus receiving no payoff. In other words, along any path on the information tree, the option holder can choose at most one spot (F, t) where to collect $D(F, t)$, receiving zero if no spot is selected. An **American call** on some underlying traded contract with ex-dividend price process S is the American option with payoff process $D \equiv S - K$, where the scalar K is the **strike**. (Note that we have assumed $\bar{\tau} < T$ in order to avoid our earlier simplifying convention that $S_T = 0$.)

Suppose that the underlying contract, which we henceforth refer to as the **stock**, does *not* pay any dividends up to the maturity date $\bar{\tau}$. Suppose also that for every time prior to $\bar{\tau}$ there is a way to **save** K (units of account) and have K for sure at time $\bar{\tau}$ plus potentially some nonnegative interest. (Section 1.8 defines more formally a money-market account that can implement such savings, provided the interest rate process is nonnegative.) In this case, it is a dominant choice to *not* exercise the American call prior to maturity. To see why, suppose the option holder is considering exercising the option at some spot (F, t) with $t < \bar{\tau}$. The option holder is at least as well off keeping the option alive to maturity, shorting the stock and saving K. Exercising at spot (F, t) results in an inflow $S(F, t) - K$ and nothing thereafter. The alternative strategy also results in an inflow $S(F, t) - K$ at spot (F, t), but now the option holder still has the call option, a saved amount that is sufficient to pay the strike at maturity, and a short position on the underlying stock. At maturity, if the stock is worth more than K, the option holder can spend K from savings, exercise the call option and close out the short position. If on the other hand the stock price is less than K at maturity, the option holder ends up with $K - S_{\bar{\tau}}$ plus any additional interest on savings. Put together, relative to exercising at spot (F, t), the alternative strategy results in the additional time-$\bar{\tau}$ payoff of $\max\{K - S_{\bar{\tau}}, 0\}$ (which is the payoff of a European put option) plus any interest on the strike.

The argument fails if the stock pays sufficiently high dividends. Since dividend payouts reduce the stock value but do not affect the strike K, the benefit of collecting dividends can outweigh the cost of early exercise. This tension between immediate payoff and the benefit of waiting in order to condition actions on future information is typical of an American option. ◇

We have defined dominance from the perspective of time zero. Dominance at every other spot is defined similarly, relative to a dynamic market that assigns a set of traded cash flows $X_{F,t} \subseteq X \cap \mathcal{L}_{F,t}$ to every spot (F, t).

DEFINITION 1.5.6 A cash flow $\delta^* \in O$ is **dominant at spot** (F, t) if given any $\delta \in O_{F,t}(\delta^*)$, there exists a trade $x \in X_{F,t}$ such that $\delta^* + x \geq \delta$ on $F \times \{t, \ldots, T\}$.

We proceed under the assumption that the market is arbitrage-free and liquid and therefore, by Proposition 1.2.4, $X_{F,t} = X \cap \mathcal{L}_{F,t}$. The following proposition shows that ex-ante dominance implies ex-post dominance and therefore an option holder who selects a cash flow because it is dominant has no incentive to deviate from that choice because of new information.

PROPOSITION 1.5.7 (Dynamic consistency of dominance) *Given an arbitrage-free liquid market, a cash flow $\delta^* \in O$ that is dominant at time zero is also dominant at every other spot.*

PROOF Suppose δ^* is dominant at time zero and (F, t) is any other spot. Given $\delta \in O_{F,t}(\delta^*)$, let $\tilde{\delta} \equiv \delta^* + (\delta - \delta^*) 1_{F \times \{t,...,T\}} \in O$ and choose $\tilde{x} \in X$ such that $\delta^* + \tilde{x} \geq \tilde{\delta}$. By market liquidity, there exists a scalar α such that $x \equiv \tilde{x} 1_{F \times \{t,...,T\}} - \alpha 1_{F,t} \in X_{F,t} \subseteq X$, and therefore $\tilde{x} - \tilde{x} 1_{F \times \{t,...,T\}} + \alpha 1_{F,t} = \tilde{x} - x \in X$. Since outside the set $F \times \{t, \ldots, T\}$, $\delta^* + \tilde{x} \geq \tilde{\delta} = \delta^*$, we have $\tilde{x} - \tilde{x} 1_{F \times \{t,...,T\}} \geq 0$. Since X is arbitrage free, $\alpha \leq 0$. Therefore, on the set $F \times \{t, \ldots, T\}$, $\delta^* + x \geq \delta^* + \tilde{x} \geq \tilde{\delta} = \delta$, confirming the dominance of δ^* at spot (F, t). ☐

The dynamic consistency of dominance is also reflected in the structure of conditional present-value functions. Recall that, by Proposition 1.4.6, every spot-(F, t) present-value function can be expressed as the conditional $\Pi_{F,t}$ of a corresponding time-zero present-value function Π, as defined in (1.4.1). The argument of Theorem 1.5.2 applied to $X_{F,t} = X \cap \mathcal{L}_{F,t}$, therefore, shows that δ^* is dominant at spot (F, t) if and only if for every present-value function Π,

$$\Pi_{F,t}(\delta^*) = \max \left\{ \Pi_{F,t}(\delta) \mid \delta \in O_{F,t}(\delta^*) \right\}.$$

To show dynamic consistency, suppose that $\Pi_{F,t}(\delta) > \Pi_{F,t}(\delta^*)$ for some $\delta \in O_{F,t}(\delta^*)$ and present-value function Π. As in our earlier proof, let $\tilde{\delta} \equiv \delta^* + (\delta - \delta^*) 1_{F \times \{t,...,T\}} \in O$. Using equation (1.4.1) with $c = \tilde{\delta} - \delta^*$, it follows that $\Pi(\tilde{\delta}) > \Pi(\delta^*)$. Therefore, if δ^* is not dominant at (F, t) it is also not dominant at spot $(\Omega, 0)$. The argument applies from the perspective of any other spot, showing that if a cash flow is dominant at a spot, then it is also dominant at every successor of the given spot.

Dynamic consistency is the key to the dynamic programming[4] approach to present-value maximization. Associated with the maximization problem (1.5.1) is a **Bellman equation**, which recursively specifies, for every spot (F, t), a function $p_{F,t} \colon O \to \mathbb{R}$ with the property: $\delta^* = \delta$ on $F \times \{0, \ldots, t - 1\}$ implies $p_{F,t}(\delta^*) = p_{F,t}(\delta)$. In other words, just like the set $O_{F,t}(\delta)$, the value $p_{F,t}(\delta)$ depends on δ only through its history prior to spot (F, t). In particular, $p \equiv p_{\Omega,0}(\delta)$ does not depend on the argument δ at all. In order to focus on the main ideas, we assume throughout that all the maxima that follow exist, which, for example, is true if O is finite. The recursion starts with the terminal-spot functions

---

[4] Richard Bellman coined the term "dynamic programming" in 1950 at the RAND corporation. Dreyfus (2002) recounts some colorful context.

$$p_{F,T}(\delta^*) = \max\{\delta(F,T) \mid \delta \in O_{F,T}(\delta^*)\}, \quad \delta^* \in O, \quad (1.5.2)$$

and proceeds backward on the information tree. Suppose that at spot (F,t), with immediate successor spots $(F_0, t+1), \ldots, (F_d, t+1)$, the functions $p_{F_i, t+1}$ are already known. Then for all $\delta^* \in O$,

$$p_{F,t}(\delta^*) = \max_{\delta \in O_{F,t}(\delta^*)} \delta(F,t) + \Pi_{F,t}\left(\sum_{i=0}^{d} p_{F_i, t+1}(\delta) \, 1_{F_i \times \{t+1\}}\right). \quad (1.5.3)$$

Note that $p_{F_i, t+1}(\delta)$ depends on δ only through its history up to and including spot (F,t) and the values of δ prior to (F,t) are fixed by δ^*. The maximization in the Bellman equation is, therefore, over the scalar $\delta(F,t)$, which represents a spot-(F,t) payment that is consistent with prior selections by the option owner.

PROPOSITION 1.5.8 *The functions*

$$p_{F,t}(\delta^*) \equiv \max\{\Pi_{F,t}(\delta) \mid \delta \in O_{F,t}(\delta^*)\}, \quad \delta^* \in O,$$

satisfy the Bellman equations (1.5.3) *with terminal condition* (1.5.2).

PROOF Recursion (1.4.2) with $c = \delta \in O_{F,t}(\delta^*)$ implies that

$$\Pi_{F,t}(\delta) \le \delta(F,t) + \Pi_{F,t}\left(\sum_{i=0}^{d} p_{F_i, t+1}(\delta) \, 1_{F_i \times \{t+1\}}\right).$$

Maximizing over all $\delta \in O_{F,t}(\delta^*)$ shows (1.5.3) with inequality $\le$ in place of equality. We show equality by showing that the maximum is achieved by any $\delta \in O_{F,t}(\delta^*)$ such that $\Pi_{F,t}(\delta) = p_{F,t}(\delta^*)$. In other words, δ is dominant at spot (F,t) relative to the complete market with the given present-value function. By dynamic consistency, for all $i \in \{0, \ldots, d\}$, δ is also dominant at $(F_i, t+1)$ and therefore $p_{F_i, t+1}(\delta) = \Pi_{F_i, t+1}(\delta)$. Applying recursion (1.4.2) once again, we conclude that δ achieves the maximum in (1.5.3). $\qquad\square$

According to the dynamic programming approach, to maximize time-zero present value we apply the Bellman equation backward on the tree to find the functions $p_{F,t}$, which in particular gives the maximum time-zero present value as the constant value p of $p_{\Omega,0}(\delta)$. To find a $\delta^* \in O$ such that $p = \Pi(\delta^*)$, we start with the value δ_0^* of $\delta(\Omega, 0)$ that achieves the maximum of the Bellman equation at the time-zero spot. Given δ_0^*, we can then find values $\delta^*(F,1)$ that achieve the maximum of the Bellman equation at every time-one spot, thus defining δ_1^*, and so on up to the terminal date, thus constructing a δ^*. Let $V^*(F, t) \equiv p_{F,t}(\delta^*)$ and note that, by construction,

$$V^*(F, t) = \delta^*(F, t) + \Pi_{F,t}\left(V^* 1_{F \times \{t+1\}}\right), \quad V_T^* = \delta_T^*,$$

which is the backward recursive way of expressing the fact that Π prices the contract (δ^*, V^*) and therefore $p_0 = V_0^* = \Pi(\delta^*)$.

Given a small number of spots and option choices, the dynamic programming method can be easily implemented, since at each spot the possible values $p_{F,t}(\delta)$ can be listed and stored. In some applications it is simple to keep track of past option selections. For example, for an American option, all one needs to know at each spot is whether the option has been exercised or not. Even in this case, however, one quickly runs into the curse of dimensionality, since the number of spots increases exponentially with the number of periods. This issue motivates the modeling idea of a Markov state, which is introduced in the following chapter and is implicitly used in Exercise 1.9.9.

1.6 Option Pricing

Suppose that at time zero the option O can be bought or sold at a price or **premium** p. An option buyer makes a time-zero payment p to the option seller, who must then deliver the cash flow in O selected by the buyer. We assume throughout that there exists a dominant choice δ^* in O that is marketed in the arbitrage-free market X, with present value p^*. The option premium $p = p^*$ is consistent with the absence of arbitrage opportunities. Buying the option, exercising it to receive some $\delta \in O$ and entering a trade $x \in X$ generates the cash flow $-p^* 1_{\Omega \times \{0\}} + \delta + x$, which cannot be an arbitrage, since every present-value function assigns it a nonpositive value. Similarly, selling the option and entering any trade in X cannot guarantee an arbitrage—as long as the option buyer selects δ^*, the generated cash flow has zero present value and cannot be an arbitrage.

What if the option premium is not p^*? If $p < p^*$, then an arbitrageur can buy the option, exercise it to receive the dominant cash flow δ^*, while selling δ^* in the market X, netting the arbitrage cash flow $(p^* - p) 1_{\Omega \times \{0\}}$. Suppose now that $p > p^*$. The obvious first trade toward an arbitrage is to sell the option. If the arbitrageur knew in advance the option buyer's selection $\delta \in O$, then the arbitrage would be completed by buying δ^* and entering a trade $x \in X$ such that $\delta^* + x \geq \delta$. But the option buyer does not have to commit to the choice δ, so it is not clear what x should be. At spot (F, t) the arbitrageur observes δ up to spot (F, t), but the option buyer can continue with any choice that is consistent with this initial part of δ. Assuming $O_{F,t}(\delta)$ contains a dominant choice, we will show that the arbitrageur can hedge every possible choice by the option buyer, guaranteeing an arbitrage.

More precisely, we assume the arbitrageur has access to a dynamic market assigning a set of traded cash flows $X_{F,t} \subseteq X \cap \mathcal{L}_{F,t}$ to each spot (F,t). In order to hedge the short option position, the arbitrageur follows an O-**adapted strategy** h, which is an assignment to each nonterminal spot (F,t) of a function $h_{F,t} \colon O \to X_{F,t}$ such that $\delta = \tilde{\delta}$ on $F \times \{0, \ldots t\}$ implies $h_{F,t}(\delta) = h_{F,t}(\tilde{\delta})$. At spot (F,t), the arbitrageur observes that the option buyer selected δ up to (F,t) and enters the incremental trade $h_{F,t}(\delta) \in X_{F,t}$ in order to hedge all possible future choices by the option buyer. We will see that this is possible because of the assumption that $O_{F,t}(\delta)$ contains a dominant choice at spot (F,t). If δ and $\tilde{\delta}$ are identical up to spot (F,t), the arbitrageur does not know which cash flow the option buyer might continue with and therefore it must be the case that $h_{F,t}(\delta) = h_{F,t}(\tilde{\delta})$. To simplify the notation, for every $\delta \in O$ and time $t < T$, let the function $h_t(\delta) \colon \Omega \to X$ be defined by $h_t(\delta)(\omega) \equiv h_{F,t}(\delta)$ for all $\omega \in F$. The arbitrageur sells the option, receives the premium p at time zero and delivers whatever cash flow δ in O is selected by the option buyer, while following the O-adapted strategy h, resulting in the cash flow

$$p 1_{\Omega \times \{0\}} - \delta + \sum_{t=0}^{T-1} h_t(\delta). \tag{1.6.1}$$

The proof of the existence of an arbitrage of this form if $p > p^*$ follows.

PROPOSITION 1.6.1 *Suppose the market is arbitrage-free, δ^* is a marketed dominant cash flow in O with present value p^*, and for every $\delta \in O$ and spot (F,t), there exists a cash flow in $O_{F,t}(\delta)$ that is dominant at (F,t). If $p > p^*$, then there exists a O-adapted strategy h such that for all $\delta \in O$, the cash flow* (1.6.1) *is an arbitrage.*

PROOF Suppose $p > p^*$. Given $\delta \in O$, select $\delta^{(F,t)} \in O_{F,t}(\delta)$ to be dominant at spot (F,t) and let $\delta^{(t)} \equiv \sum_{i=1}^{n} \delta^{(F_i,t)} 1_{F_i \times \{0,\ldots,T\}}$, where $\{F_1, \ldots, F_n\}$ is the partition generating $\mathcal{F}_t$. The definition of an option implies that $\delta^{(t)} \in O_{F,t}(\delta)$ for every spot (F,t). Select $x \in X$ such that $\delta^* + x \geq \delta^{(1)}$ and let $h_0(\delta) \equiv -p^* 1_{\Omega \times \{0\}} + \delta^* + x$. Since $\delta^{(1)}$, and hence the choice of x, depends on δ only through the value δ_0, so does $h_0(\delta)$, as required by O-adaptedness. At time zero the arbitrageur sells the option, buys the cash flow δ^* and enters the trade x, which together with δ^*

dominates $\delta^{(1)}$. After paying $\delta_0 = \delta_0^{(1)}$ to the option buyer, the arbitrageur has generated the cash flow

$$c^{(0)} \equiv p1_{\Omega \times \{0\}} + h_0(\delta) - \delta 1_{\Omega \times \{0\}} \geq (p - p^*) 1_{\Omega \times \{0\}} + \delta^{(1)} 1_{\Omega \times \{1,...,T\}}.$$

Proceeding inductively, suppose after all transactions prior to time $t \in \{1, \ldots, T - 1\}$, the arbitrageur faces an overall cash flow

$$c^{(t-1)} \geq (p - p^*) 1_{\Omega \times \{0\}} + \delta^{(t)} 1_{\Omega \times \{t,...,T\}}. \tag{1.6.2}$$

Choose $h_{F,t}(\delta) \in X_{F,t}$ so that $\delta^{(t)} 1_{F \times \{t,.....T\}} + h_{F,t}(\delta) \geq \delta^{(t+1)} 1_{F \times \{t,.....T\}}$. The arbitrageur can identify $\delta^{(t+1)} 1_{F \times \{t,...,T\}}$, and therefore $h_{F,t}(\delta)$, having observed only $\delta 1_{F \times \{0,...,t\}}$. At time t the arbitrageur enters the trade $h_t(\delta)$ and pays out δ_t to the option holder, resulting in the cash flow

$$c^{(t)} \equiv c^{(t-1)} + h_t(\delta) - \delta 1_{\Omega \times \{t\}} \geq (p - p^*) 1_{\Omega \times \{0\}} + \delta^{(t+1)} 1_{\Omega \times \{t+1,...,T\}},$$

which is inequality (1.6.2) with t incremented by one. At time T the arbitrageur pays out δ_T to the option holder, resulting in the overall arbitrage cash flow $c^{(T-1)} - \delta 1_{\Omega \times \{T\}} \geq (p - p^*) 1_{\Omega \times \{0\}}$. The recursive construction of the cash flows $c^{(t)}$ implies that $c^{(T-1)} - \delta 1_{\Omega \times \{T\}}$ is equal to (1.6.1), and we have therefore produced an O-adapted strategy h such that (1.6.1) defines an arbitrage for all $\delta \in O$. $\qquad \square$

Under the assumptions of Proposition 1.6.1, we have shown that in an arbitrage-free market allowing time-zero trading of the option, the time-zero option price is the present value of the marketed dominant cash flow, even though an option seller cannot assume optimal exercise. An analogous discussion applies from the perspective of every other spot. Suppose, for simplicity, that every cash flow in O is marketed. If prior to spot (F, t) the option buyer has selected the cash flow $\delta \in O$, the arbitrage-free price $p_{F,t}(\delta)$ of the remaining option $O_{F,t}(\delta)$ is equal to the present value of a dominant cash flow in $O_{F,t}(\delta)$. If at spot (F, t) the option buyer selects a payment that is not consistent with any dominant cash flow in $O_{F,t}(\delta)$, then effectively the option buyer has gifted an arbitrage cash flow to the option seller. This creates an incentive to market complex options to inattentive or unsophisticated buyers with the expectation of suboptimal exercise.

1.7 Trading Strategies

As we saw in Section 1.2, the market implemented by given contracts can be described as the set of cash flows generated by trading strategies specifying contingent trades at every spot of the information tree. This section provides a more-systematic discussion of trading strategies and associated notation and terminology, which is tailored to the probabilistic methods introduced in Chapter 2.

Throughout this section, we take as given the contracts

$$\left(\delta^1, V^1\right), \ldots, \left(\delta^J, V^J\right), \tag{1.7.1}$$

with corresponding ex-dividend price processes $S^j = V^j - \delta^j$. Trading strategies in these contracts are time-indexed sequences of portfolios representing positions over each period. **Period** $t \in \{1, \ldots, T\}$ is the time interval whose beginning is time $t - 1$ and whose end is time t. A period-t portfolio $\left(\theta_t^1, \ldots, \theta_t^J\right)$ is formed at the beginning of period t, with θ_t^j representing a number of shares in contract j. By convention, we set $\theta_0^j = 0$. It is important to keep track of what quantities are known at the beginning of a period and what quantities are only known at the end of the period. To emphasize this distinction, we define a process x to be **predictable** if x_t is $\mathcal{F}_{t-1}$-measurable, for every time $t > 0$, and x_0 is constant. Thus if x is adapted, we can only claim that x_t is known at the end of period t, while if x is predictable, we know that x_t is revealed at the beginning of period t. Throughout this text, $\mathcal{P}$ denotes the set of all predictable processes and

$$\mathcal{P}_0 \equiv \{x \in \mathcal{P} \mid x_0 = 0\}.$$

DEFINITION 1.7.1 A **trading strategy** in contracts (1.7.1) is a J-dimensional row matrix whose entries are elements of $\mathcal{P}_0$. The trading strategy $\left(\theta^1, \ldots, \theta^J\right)$ **generates** the cash flow x, where

$$x_t = \sum_{j=1}^J \theta_t^j V_t^j - \theta_{t+1}^j S_t^j, \quad t < T; \quad x_T = \sum_{j=1}^J \theta_T^j V_T^j. \tag{1.7.2}$$

In the **budget equation** (1.7.2), x_t is the net of the time-t value $\sum_j \theta_t^j V_t^j$ of the period-t portfolio and the time-t cost $\sum_j \theta_{t+1}^j S_t^j$ of forming the period-$(t+1)$ portfolio. Since $\theta_0 = 0$, $x_0 = -\sum_j \theta_1^j S_0^j$ is the initial payment required to commence the strategy. The final payment x_T equals the portfolio's time-T liquidation value.

PROPOSITION 1.7.2 *The market implemented by contracts* (1.7.1) *is the set of all cash flows generated by trading strategies in these contracts.*

PROOF Let X be the market implemented by contracts (1.7.1) and let X' be the set of all cash flows generated by trading strategies in these contracts.

In the closing paragraph of Section 1.3 we saw that $X = \text{span}(X^0)$, where X^0 is the set of all $-V^j 1_{F \times \{t\}} + \delta^j 1_{F \times \{t,...,T\}}$, where (F, t) is a spot and $j \in \{1, ..., J\}$. One can easily check that $X^0 \subset X'$. Since X' is a linear subspace, this shows that $X \subseteq X'$. Conversely, suppose that $x \in X'$ is generated by a trading strategy $(\theta^1, ..., \theta^J)$. Note that $x = \sum_j x^j$, where $x_t^j = \theta_t^j V_t^j - \theta_{t+1}^j S_t^j$ for $t < T$ and $x_T^j = \theta_T^j V_T^j$. An induction in the number of trades of θ^j, defined as the number of spots where θ^j changes value, shows that $x^j \in \text{span}(X^0)$ for all j, and therefore $x \in X$. □

The budget equation (1.7.2) is more conveniently expressed using a process transform operator which directly corresponds to stochastic integrals in continuous-time extensions and facilitates the application of martingale methods introduced in Chapter 2. To state this form of the budget equation, we first introduce some more generally useful notation.

For any process x, the **lagged process** x_- is defined by

$$x_-(0) \equiv x(0), \quad x_-(t) \equiv x(t-1), \quad t = 1, ..., T.$$

The **increments process** of x is the process $\Delta x \equiv x - x_-$, that is,

$$\Delta x_0 \equiv 0, \quad \Delta x_t \equiv x_t - x_{t-1}, \quad t = 1, ..., T.$$

The **integral**[5] $x \bullet y$ of x with respect to the process y is the process

$$(x \bullet y)_0 \equiv 0, \quad (x \bullet y)_t \equiv \sum_{s=1}^{t} x_s \Delta y_s, \quad t = 1, ..., T.$$

We denote by $\mathbf{t}$ the process that counts time: $\mathbf{t}(t) \equiv t$. Thus

$$(x \bullet \mathbf{t})_0 = 0, \quad (x \bullet \mathbf{t})_t = \sum_{s=1}^{t} x_s, \quad t = 1, ..., T.$$

The **gain process** G^j of contract (δ^j, V^j) is defined by

$$G_t^j \equiv V_t^j + \sum_{s<t} \delta_s^j = S_t^j + \sum_{s \leq t} \delta_s^j, \quad t = 0, ..., T.$$

For times $t > s$, the increment $G_t^j - G_s^j$ represents the total gain resulting from purchasing contract j at time s and selling it at time t. If $(\theta^1, ..., \theta^J)$ is a trading strategy, then $\theta^j \bullet G^j$ represents total gains from trading contract j up to time t.

---

[5] The bullet notation for an integral is more commonly found in the more advanced literature on stochastic analysis. See, for example, Jacod and Shiryaev (2003).

PROPOSITION 1.7.3 *The trading strategy* $(\theta^1, \ldots, \theta^J)$ *generates the cash flow x if and only if*

$$\sum_{j=1}^{J} \theta^j V^j = -x_- \bullet \mathbf{t} + \sum_{j=1}^{J} \theta^j \bullet G^j, \quad \sum_{j=1}^{J} \theta_T^j V_T^j = x_T. \quad (1.7.3)$$

PROOF Let $W \equiv \sum_j \theta^j V^j$. Since $\Delta G_t^j = V_t^j - S_{t-1}^j$, the budget equation (1.7.2) can be written as $\Delta W_t = -x_{t-1} + \sum_j \theta_t^j \Delta G_t^j$, $W_T = x_T$, which is equivalent to (1.7.3). $\qquad\square$

Matrix notation helps us simplify expressions such as (1.7.3), so let us take a moment to introduce some associated conventions. For any set $\mathcal{Z}$, we write $\mathcal{Z}^{m \times n}$ for the set of m-by-n matrices whose entries are elements of $\mathcal{Z}$. For example, trading strategies are elements of $\mathcal{P}_0^{1 \times J}$. Unless explicitly specified otherwise, vectors of processes are assumed to be column vectors and we write $\mathcal{Z}^n$ rather than $\mathcal{Z}^{n \times 1}$. If $\mathcal{Z}$ is the set of all (adapted, predictable) processes, then $\mathcal{Z}^n$ is the set of n-**dimensional** (adapted, predictable) processes. If $\mathcal{Z}$ is a set of processes, we typically use superscripts to index vectors and matrices. For $x \in \mathcal{L}^{n \times m}$ and $y \in \mathcal{L}^{m \times l}$, the process $x \bullet y \in \mathcal{L}^{n \times l}$ is defined by the analog of the usual matrix multiplication formula:

$$(x \bullet y)^{ij} \equiv \sum_{k=1}^{m} x^{ik} \bullet y^{kj}.$$

With these conventions in place, we define as elements of $\mathcal{L}^J$

$$\delta \equiv \begin{pmatrix} \delta^1 \\ \vdots \\ \delta^J \end{pmatrix} \quad \text{and} \quad V \equiv \begin{pmatrix} V^1 \\ \vdots \\ V^J \end{pmatrix}, \quad (1.7.4)$$

as well as

$$S \equiv V - \delta \quad \text{and} \quad G \equiv S + \delta \bullet \mathbf{t} = V + \delta_- \bullet \mathbf{t}. \quad (1.7.5)$$

The **budget equation** (1.7.3) can be restated more succinctly as

$$\theta V = -x_- \bullet \mathbf{t} + \theta \bullet G, \quad \theta_T V_T = x_T. \quad (1.7.6)$$

Other versions of the budget equation are obtained by a change of the unit of account (also known as a change of numeraire). Suppose the strictly positive adapted process π represents a unit conversion factor at every spot. A cash flow or price process x expressed in the original unit of account becomes πx in the new unit of account. The gain process G in the new units becomes

$$G^{\pi} \equiv \pi V + (\pi \delta)_- \bullet \mathbf{t}.$$

A change of units should not affect the validity of a budget equation, resulting in the following version, whose proof is a simple exercise.

PROPOSITION 1.7.4 *For all* $\pi \in \mathcal{L}_{++}$, *the trading strategy* θ *generates the cash flow* x *if and only if*

$$\theta(\pi V) = -(\pi x)_- \bullet \mathbf{t} + \theta \bullet G^{\pi}, \quad \theta_T(\pi_T V_T) = \pi_T x_T.$$

Using the vector notation (1.7.4), we write (δ, V) to refer to the J contracts (1.7.1). A trading strategy θ in (δ, V) defines a synthetic contract, which can be thought of as a share in a fund following strategy θ.

DEFINITION 1.7.5 A trading strategy $\theta \in \mathcal{P}_0^{1 \times J}$, generating the cash flow x, defines the **synthetic contract** $(\delta^{\theta}, V^{\theta})$, where

$$(\delta_0^{\theta}, V_0^{\theta}) \equiv (0, \theta_1 V_0),$$
$$(\delta_t^{\theta}, V_t^{\theta}) \equiv (x_t, \theta_t V_t), \quad t = 1, \ldots, T.$$

A contract is **synthetic** in (δ, V) if it is of the form $(\delta^{\theta}, V^{\theta})$ for some trading strategy θ.

Note that the ex-dividend price process $S^{\theta} \equiv V^{\theta} - \delta^{\theta}$ is given by

$$S_{t-1}^{\theta} = \theta_t S_{t-1}, \quad t = 1, \ldots, T; \quad S_T^{\theta} = 0,$$

and the gain process of the contract $(\delta^{\theta}, V^{\theta})$ is given by

$$G^{\theta} \equiv V^{\theta} + \delta_-^{\theta} \bullet \mathbf{t} = V_0^{\theta} + \theta \bullet G,$$

as can be seen by rearranging the budget equation (1.7.6).

Consider now the trading strategies $\theta_1, \ldots, \theta_m$ in the original contracts (1.7.1) and let α be a trading strategy in the synthetic contracts $(\delta^{\theta_i}, V^{\theta_i})$, generating the cash flow x. One can think of α as a trading strategy in m funds, where fund i follows trading strategy θ_i. The same cash flow x can be generated by the trading strategy $\theta \equiv \sum_{i=1}^{m} \alpha^i \theta_i$ in the original contracts (δ, V). This argument justifies the following observations, where $X(\delta, V)$ denotes the market implemented by (δ, V).

PROPOSITION 1.7.6 *The market implemented by any synthetic contracts in* (δ, V) *is a subset of* $X(\delta, V)$. *The market implemented by* (δ, V) *and any number of synthetic contracts in* (δ, V) *is* $X(\delta, V)$.

Dividend processes of synthetic contracts correspond to the cash flows that are marketed in $X(\delta, V)$.

PROPOSITION 1.7.7 *A cash flow c is marketed in $X(\delta, V)$ if and only if there exists a trading strategy θ in (δ, V) such that*

$$c_t = \delta_t^\theta, \quad t = 1, \ldots, T. \tag{1.7.7}$$

In particular, the market $X(\delta, V)$ is complete if and only if every cash flow c with $c_0 = 0$ is the dividend process of some contract that is synthetic in (δ, V).

PROOF Note that c is marketed in $X(\delta, V)$ if and only if there exists $x \in X(\delta, V)$ such that $c_t = x_t$ for $t > 0$. By Proposition 1.7.2 and the definition of a synthetic contract, the latter condition is equivalent to the existence of a trading strategy θ in (δ, V) satisfying (1.7.7) $\qquad\square$

Finally, we relate synthetic contracts to traded contracts. Note that in order to conclude that a traded contract is synthetic, the assumption that the market is arbitrage-free is essential (see Exercise 1.9.8).

PROPOSITION 1.7.8 *Every contract that is synthetic in (δ, V) is traded in $X(\delta, V)$. Conversely, if the market $X(\delta, V)$ is arbitrage-free, every contract that is traded in $X(\delta, V)$ is synthetic in (δ, V).*

PROOF Buying a contract (δ^*, V^*) that is synthetic in (δ, V) at any spot generates a cash flow in $X(\delta^*, V^*)$, which, by Proposition 1.7.6, is a subset of $X(\delta, V)$. Therefore, a synthetic contract in (δ, V) is traded in $X(\delta, V)$.

Conversely, suppose that (δ^*, V^*) is traded in $X(\delta, V)$ and let x be the cash flow generated by buying (δ^*, V^*) at time zero. Since $x \in X(\delta, V)$, there exists a trading strategy θ in (δ, V) that generates x. Since the time-zero dividend of every contract is assumed to be zero by convention, it follows that $\delta^* = \delta^\theta$. Assuming the market is arbitrage-free, it must also be the case that $V^* = V^\theta$. $\qquad\square$

1.8 Money-Market Account and Returns

A special type of contract we call a money-market account implements single-period default-free borrowing and lending: A unit of account invested at time $t - 1$ pays $1 + r_t$ at time t, where the interest rate r_t is determined at time $t - 1$ and is therefore $\mathcal{F}_{t-1}$-measurable. In practice, a loan can be made default-free by posting sufficient collateral, an important aspect of real-world markets that is not modeled here. The rate r_t is often referred to in the literature as the (period-t) risk-free rate, although it can vary randomly from period to period. The rate r_t applies to a loan over a single period and is therefore a short-term interest rate. The predictable process r (with the convention $r_0 = 0$) is a short-term interest-rate process, a term we abbreviate to short-rate process.

More precisely, we adopt the following terminology.

DEFINITION 1.8.1 A **money-market account (MMA)** is a contract (δ^0, V^0), with ex-dividend price process $S^0 \equiv V^0 - \delta^0$, such that for some $r \in \mathcal{P}_0$,

$$S^0_{t-1} = 1 \quad \text{and} \quad V^0_t = 1 + r_t, \quad t = 1, \dots, T.$$

The predictable process r is the account's (interest) **rate process.** Given a reference market X, a process $r \in \mathcal{P}_0$ is a **short-rate process** if r is the rate process of an MMA that is traded in X.

The following observations can be verified by the reader. We call two contracts **equivalent** if each is synthetic in the other.

PROPOSITION 1.8.2 *Suppose the market is arbitrage-free and r is a short-rate process. Then r is unique and $1 + r$ is strictly positive. Moreover, every traded contract (δ^0, V^0) whose value process V^0 is predictable and strictly positive is equivalent to an MMA and satisfies*

$$\frac{V^0_t}{S^0_{t-1}} = 1 + r_t, \quad S^0 \equiv V^0 - \delta^0, \quad t = 1, \dots, T.$$

In applications where the market is implemented by a given finite set of contracts, it is common to assume that one of these contracts is an MMA. In all such applications, we label the contracts generating the market as

$$(\delta^0, V^0), (\delta^1, V^1), \dots, (\delta^J, V^J), \tag{1.8.1}$$

where contract zero is the MMA, with rate process r and gain process G^0. Since $\delta^0_0 = r_0 = 0$, we have

$$V^0 = 1 + r, \quad \delta^0_- = r_-, \quad \delta^0_T = V^0_T, \quad G^0 = 1 + r \bullet \mathbf{t}. \tag{1.8.2}$$

Of course, the results of Section 1.7 apply to this case after a simple relabeling of the contracts, which affects the form of the budget equation. We adopt the matrix notation (1.7.4) and (1.7.5), where (δ, V) and S and G refer to contracts $1, \dots, J$ and exclude contract zero. A trading strategy in the $1 + J$ contracts (1.8.1) is denoted as

$$\left(\theta^0, \theta\right) \in \mathcal{P} \times \mathcal{P}^{1 \times J},$$

where θ_t^0 represents the ex-dividend value in the MMA at the beginning of period $t > 0$, and θ is a trading strategy in the remaining J contracts. Adapting the budget equation (1.7.6) to this notation, it follows that the trading strategy $\left(\theta^0, \theta\right)$ generates cash flow x if and only if

$$\theta^0 V^0 + \theta V = \left(\theta^0 r - x_-\right) \bullet \mathbf{t} + \theta \bullet G,$$
$$\theta_T^0 V_T^0 + \theta_T V_T = x_T.$$

In applications where portfolio values can be assumed to stay positive it is common to focus on returns rather than prices. The **return process** R^j associated with contract j is defined by

$$R_0^j \equiv 1, \quad R_t^j \equiv \frac{V_t^j}{S_{t-1}^j}, \quad t = 1, \ldots, T, \tag{1.8.3}$$

provided S_{t-1}^j is nowhere zero, which we assume for the remainder of this section. We refer to R_t^j as the period-t return of contract j. Note that if r is the market's short-rate process, then

$$R^0 \equiv 1 + r.$$

The return process R^θ of a trading strategy $\left(\theta^0, \theta\right)$ is defined as the return process of the corresponding synthetic contract, which we denote by $\left(\delta^\theta, V^\theta\right)$. Letting $S^\theta \equiv V^\theta - \delta^\theta$, the return process R^θ is well defined provided the denominator in the following definition is nowhere zero:

$$R_0^\theta \equiv 1, \quad R_t^\theta \equiv \frac{V_t^\theta}{S_{t-1}^\theta}, \quad t = 1, \ldots, T.$$

The **excess return** process of the trading strategy $\left(\theta^0, \theta\right)$ is $R^\theta - R^0$.

Portfolio returns can be more parsimoniously represented in terms of returns and portfolio weights, provided of course all relevant returns are well defined. Consider any trading strategy $\left(\theta^0, \theta\right)$ such that the time-$(t - 1)$ ex-dividend portfolio value

$$S_{t-1}^\theta = \theta_t^0 + \theta_t S_{t-1}$$

is nowhere zero. Associated with $\left(\theta^0, \theta\right)$ is a row vector

$$\psi = \left(\psi^1, \ldots, \psi^J\right) \in \mathcal{P}_0^{1 \times J},$$

defined by

$$\psi_0^j = 0, \quad \psi_t^j \equiv \frac{\theta_t^j S_{t-1}^j}{S_{t-1}^\theta}, \quad t = 1, \ldots, T. \tag{1.8.4}$$

At the beginning period t, which is time $t - 1$, ψ_t^j represents the proportion of the portfolio's ex-dividend value S_{t-1}^θ that is allocated to contract j. The vector ψ_t omits the proportion allocated to the MMA, which can be computed as $\psi_t^0 \equiv 1 - \sum_{j=1}^J \psi_t^j$. We will refer to ψ_t, which can be any element of $L(\mathcal{F}_{t-1})^{1 \times J}$, as a (period-$t$) **portfolio allocation**, and to ψ, which can be any element of $\mathcal{P}_0^{1 \times J}$, as a **portfolio allocation policy**. The period-t return R_t^θ can be computed entirely in terms of ψ_t and contract returns, which we therefore also denote, abusing notation, by R_t^ψ:

$$R_t^\theta \equiv R_t^\psi \equiv R_t^0 + \sum_{j=1}^J \psi_t^j \left(R_t^j - R_t^0\right). \tag{1.8.5}$$

In the following chapters we will discuss notions of optimal portfolio allocations.

1.9 Exercises

EXERCISE 1.9.1 (a) (Forward pricing) Assume that there is a single period ($T = 1$) and therefore the filtration consists of spot zero and N time-one spots. In an arbitrage-free market X, you can trade a **stock** with ex-dividend price process (S_0, S_1) and a dividend yield y, where $1 + y > 0$. (Note that S_0 and y are scalars and S_1 is a random variable or element of $\mathbb{R}^N$.) Buying a share of the stock at time zero generates the cash flow $(-S_0, S_1 (1 + y)) \in X$ and selling (or shorting) the stock generates the cash flow $(S_0, -S_1 (1 + y)) \in X$. You can also trade a money-market account (**MMA**) with interest rate (or dividend yield) r. Investing a unit of account in the MMA generates the cash flow $(-1, 1 + r) \in X$ and borrowing a unit of account from the MMA generates the cash flow $(1, -(1 + r)) \in X$. Note that, since the market is arbitrage-free, $1 + r > 0$.

A **forward contract** for delivery of the stock at time one is a contract whose time-zero price is by definition zero and whose time-one payoff takes the form $S_1 - K$ for a scalar K, which is the contract's **delivery price**. Entering a **long** contract generates the cash flow $(0, S_1 - K)$ and entering a **short** contract generates the cash flow $(0, K - S_1)$. The forward contract is **traded** in X if these cash flows are in X, in which case K defines the stock's **forward price** F. Notice that the delivery price K is part of the forward contract's specification. The unique value of K such that $(0, S_1 - K) \in X$ defines F. How can you create a synthetic forward contract using the stock and the MMA? What is the implied relationship between S_0 and F? What is an explicit arbitrage, assuming the forward is traded, if the claimed relationship between S_0 and F is violated?

(b) (Leasing, convenience yield) Assume that the stock dividend yield is zero and instead y represents a lease rate of one share of the stock. That is, one can borrow a share at time zero and return $1 + y$ shares at time one, or one can deliver one share at time zero and receive back $1 + y$ shares at time one. In both cases, there is no possibility of default (which in practice can be implemented by posting appropriate collateral). In an ideal frictionless market, one would expect that $y = 0$, since the stock pays no dividends and therefore lending it out for a period should not have any consequences. In realistic situations, which typically take us beyond this text's formalism, there can be reasons for the lease rate to be positive in a competitive lease market, despite the fact that the stock pays no cash dividends over the term of the lease (or if there is a cash dividend yield, y can be strictly higher than that yield). For example, what we call a "stock" can be a US treasury bond that can be used for the purpose of posting collateral in other transactions. In this case, the lease rate y is referred to as a "convenience yield." The term also applies in situations where what we call a "stock" is some commodity, say heating oil, and there is an immediacy value of carrying inventory, as in having quick availability of heating oil in the winter months. In such contexts, how should your formal analysis of part (a) be modified, if at all, and how should it be interpreted?

EXERCISE 1.9.2 Complete Exercise 1.9.1 and assume the setting of part (a). A trader entering a long forward position has the right to receive a positive payoff on the event $\{S_1 > K\}$ and the obligation to pay the negative payoff on the event $\{S_1 < K\}$. A (European) **call option** (on the stock), or just **call**, with **strike** K is defined by the same payoff but without the obligation. The call's payoff is therefore $(S_1 - K)^+$ (where $x^+ \equiv \max\{x, 0\}$). Starting with a short forward contract and removing the obligation part results in the definition of the payoff $(K - S_1)^+$ of a (European) **put option** (on the stock), or just **put**, with **strike** K. A call or put option with payoff V is traded in X if $(-p, V) \in X$ for a necessarily unique scalar p, which defines the option's **price** or **premium**. The cash flow $(-p, V)$ is generated by **buying** or going **long** the option and the cash flow $(p, -V)$ is generated by **selling** or going **short** or **writing** the option. Given a commitment to pay S_1 (resp. receive S_1) at time one, buying the call (resp. put) is a form of insurance, paying a premium p at time zero in return for the right to pay K rather than S_1 on the event $\{S_1 > K\}$ (resp. receive K rather than S_1 on the event $\{S_1 < K\}$).

(a) (Put–call parity) Suppose the call and the put, both with strike K, are traded, with respective premia p_{call} and p_{put}. (The strike need not equal the stock's forward price.) What is the relationship, known as put–call parity, between $p_{\text{call}} - p_{\text{put}}$ and the stock price? Suppose the relationship is violated. Provide an explicit arbitrage resulting from trading the options, the stock and the MMA.

(b) (Hard-to-short stock) As in Exercise 1.9.1(b), assume that the stock dividend yield is zero but the stock can be leased at a rate y. Sometimes demand for borrowing stock shares for the purpose of shorting them is high relative to the shares available for lending, pushing the lease rate y up. (For example, aggressive sentiment-based trading by groups of traders can temporarily push prices to seemingly unreasonable levels relative to fundamentals, causing high demand for shorting by other traders who want to take advantage of the presumed reversion of the stock price to more reasonable levels. The activity is risky, since shorting requires the posting of collateral and further price increases can force the closing of short positions at a loss, further fueling an upward trend.) The short seller must borrow a share and pay a fee yS_1 at time one. Conversely, a long position over the period can lease out the position and collect this fee. How should your formal analysis of part (a) be modified, if at all, and how should it be interpreted?

EXERCISE 1.9.3 (Simple binomial pricing[6]) This exercise introduces the simplest version of the binomial pricing model, which will be extended in other exercises. Specialize the setting of Exercise 1.9.1(a) by further assuming there are only two states ($N = 2$) corresponding to the time-one stock price S_1 taking the values $S_0(1 + u)$ and $S_0(1 + d)$, where $u > d > -1$. For simplicity, assume the stock pays no dividends ($y = 0$). The market X is implemented by trading in the stock and the MMA. What are necessary and sufficient conditions on the parameters r, u, and

[6] The binomial model introduced in this exercise and further developed in Exercises 1.9.9 and 2.9.6 originates in Sharpe (1978), Cox et al. (1979) and Rendleman and Bartter (1979).

d for *X* to be arbitrage-free? Proceeding under the assumption that *X* is arbitrage-free, show that *X* is complete and compute the corresponding present-value function. Review the definition of a (European) call option in Exercise 1.9.2. What is the premium of a call according to this present-value function? Compute a replicating portfolio, that is, a portfolio in the stock and the MMA whose time-one payoff is the same as that of the call option. Finally, confirm that the time-zero value of the replicating portfolio is consistent with your earlier call premium calculation.

EXERCISE 1.9.4 Take as given a reference arbitrage-free market *X*.

(a) Suppose the market *X* is complete. Explain why for every cash flow *c*, there is a unique scalar $\Pi(c)$ such that $c - \Pi(c) 1_{\Omega \times \{0\}} \in X$. Then show that the function $\Pi : \mathcal{L} \to \mathbb{R}$ so defined is a present-value function.

(b) Without assuming market completeness, show that a cash flow is marketed if and only if it is assigned the same value by every present-value function. As a corollary, show that a cash flow is traded if and only if it is assigned the value zero by every present-value function.

EXERCISE 1.9.5 Show the present-value backward recursion (1.4.2) using the conditional present-value definition (1.4.1).

EXERCISE 1.9.6 Fix a reference complete market with corresponding present-value function Π. For every spot (F,t), Π defines a spot-(F,t) conditional present-value function $\Pi_{F,t}$ by equation (1.4.1). Consider an American option with payout process *D* as defined in Example 1.5.5, whose notation we use here. For every stopping time τ, let $\delta^\tau \equiv D1_{[\tau]}$ and define the adapted process V^τ by letting, for every spot (F,t), $V^\tau(F,t) \equiv \Pi_{F,t}(\delta^\tau)$. Note that, by construction, Π prices the contract (δ^τ, V^τ). Define the adapted process V^* by letting, for every spot (F,t),

$$V^*(F,t) \equiv \max \left\{ V^\tau(F,t) \mid \tau \in \mathcal{T} \right\}.$$

Show that V^* uniquely solves the backward recursion

$$V^*(F,t) = \max \left\{ D(F,t), \Pi_{F,t}\left(V^* 1_{F \times \{t+1\}} \right) \right\}, \quad V_T^* = D_T^+.$$

Label each spot (F,t) **red** if $V^*(F,t) = D(F,t)$ and **green** otherwise. A state ω corresponds to a path from spot zero to a terminal spot. Define $\tau^*(\omega)$ to be the time of the first red spot along the path ω, with the convention $\tau^*(\omega) = \infty$ if ω is a sequence of green spots only. Show that the stopping time τ^* so defined is dominant (in the sense that $D1_{[\tau^*]}$ is a dominant cash flow choice for the option).

EXERCISE 1.9.7 (a) Show that

$$x \bullet (y \bullet z) = (xy) \bullet z, \quad x, y, z \in \mathcal{L}.$$

(This is sometimes called the associative property of stochastic integrals and applies in more general stochastic settings.)

(b) Given contracts $(\delta, V) \in \mathcal{L}^{J \times 2}$, the trading strategies $\theta_1, \dots, \theta_m$ define corresponding synthetic contracts

$$\left(\delta^{\theta_i}, V^{\theta_i} \right), \quad i = 1, \dots, m.$$

Let $\alpha = \left(\alpha^1, \dots, \alpha^m \right)$ be a trading strategy in the m synthetic contracts, generating cash flow x. Use part (a) and the budget equation

$$\sum_{i=1}^{m} \alpha^i V^{\theta_i} = -x_- \bullet \mathbf{t} + \sum_{i=1}^{m} \alpha^i \bullet G^{\theta_i},$$

$$\sum_{i=1}^{m} \alpha^i_T V^{\theta_i}_T = x_T,$$

to verify that the cash flow x is also generated by the trading strategy $\theta \equiv \sum_{i=1}^{m} \alpha^i \theta_i$ in the original contracts (δ, V).

EXERCISE 1.9.8 Give an example of contracts $(\delta, V) \in \mathcal{L}^{J \times 2}$ and a contract $\left(\delta^*, V^* \right)$ that is *not* a synthetic contract in (δ, V), yet it is traded in $X(\delta, V)$.

EXERCISE 1.9.9 (Multiperiod binomial replication) This exercise extends the replication argument of Exercise 1.9.3 to the multiperiod case, using the notation $U \equiv 1 + u$ and $D \equiv 1 + d$, where $U > D > 0$. The assumption of a zero dividend yield is also relaxed.

Suppose $\Omega = \{0, 1\}^T$ and the filtration $\{\mathcal{F}_t\}$ is generated by the process b, where $b_0 \equiv 0$ and $b_t(\omega) \equiv \omega_t$ for every time $t > 0$ and state $\omega = (\omega_1, \dots, \omega_T) \in \Omega$. The process Z is specified by a given initial value $Z_0 \in (0, \infty)$ and the recursion

$$\frac{Z_t}{Z_{t-1}} \equiv b_t U + (1 - b_t) D, \quad t = 1, \dots, T.$$

Note that the process Z also generates the filtration $\{\mathcal{F}_t\}$, since $(Z_1, \dots, Z_t)$ and $(b_1, \dots, b_t)$ are mutually uniquely determined path by path.

(a) The **recombining tree** is the graph whose **nodes** are all possible values of Z_t / Z_0, and whose arrows connect a value z of Z_{t-1}/Z_0 to the corresponding possible values zU and zD of Z_t/Z_0. For example, here is the recombining tree for $T = 3$.

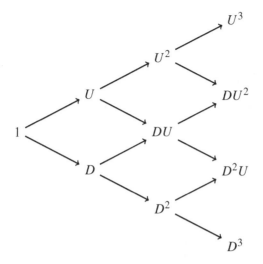

Note that spots correspond to paths on the recombining tree. How many nodes are there and how many spots? How do these numbers increase with T? What is their order of magnitude for $T = 100$?

(b) Assume that the market is arbitrage-free and is implemented by two contracts: An MMA $\left(\delta^0, V^0\right)$ with a constant rate process $r > -1$, and a **stock** (δ, V) with ex-dividend price process $S \equiv V - \delta$, specified in terms of a constant **dividend yield** $y > -1$ by

$$S_{t-1} \equiv Z_{t-1} \quad \text{and} \quad V_t \equiv (1 + y) Z_t, \quad t = 1, \ldots, T.$$

Explain why $U(1 + y) > 1 + r > D(1 + y)$.

(c) Consider a contract $\left(\delta^*, V^*\right)$ that pays no dividends prior to T, that is, $\delta_-^* = 0$. Assume that the contract's price can be expressed as $V_t^* = f_t(Z_t)$ for some functions $f_t \colon \mathcal{Z}_t \to (0, \infty)$, where $\mathcal{Z}_t$ denotes the set of all possible values of Z_t. Let (θ^0, θ) be a trading strategy in the contracts $\left(\delta^0, V^0\right)$ and (δ, V), defining the synthetic contract $\left(\delta^\theta, V^\theta\right)$. Show that if $\left(\delta^*, V^*\right) = \left(\delta^\theta, V^\theta\right)$, then

$$\theta_t^0 = g_t(Z_{t-1}) \quad \text{and} \quad \theta_t = h_t(Z_{t-1}), \quad t = 1, \ldots, T,$$

for functions $g_t, h_t \colon \mathcal{Z}_{t-1} \to \mathbb{R}$, for which you should be able to give explicit formulas in terms of f_t and the model parameters. Finally, provide a recursive algorithm for computing f_t.

2

Probabilistic Methods in Arbitrage Pricing

The arbitrage-pricing theory of Chapter 1 postulates an exhaustive set of possible states but makes no use of any probabilities over these states. This chapter introduces probabilistic representations of valuation rules that are consistent with a given arbitrage-free market. These are useful in developing theoretical and computational methodology and essential in empirical applications.

2.1 Probability Basics

In this section we review some essential probabilistic concepts and notation. To the Chapter 1 primitives of a finite state space Ω and a filtration on this state space, we add a reference **probability** (measure) on the subsets of Ω, that is, a function $P \colon 2^\Omega \to [0, 1]$ such that $P(\Omega) = 1$ and for all events A and B,

$$A \cap B = \emptyset \quad \Longrightarrow \quad P(A \cup B) = P(A) + P(B).$$

We assume that P has **full support**: $P(A) > 0$ if $A \neq \emptyset$.

The **expectation** or **mean** of a random variable x (under P) is

$$\mathbb{E}[x] \equiv \sum_{\omega \in \Omega} x(\omega) \, P(\{\omega\}) = \sum_{\alpha \in \{x(\omega) \,|\, \omega \in \Omega\}} \alpha P(\{x = \alpha\}).$$

The function $\mathbb{E} \colon \mathbb{R}^\Omega \to \mathbb{R}$ so defined is the **expectation operator** relative to P; it is a linear functional on $\mathbb{R}^\Omega$ that is positive ($\mathbb{E}[x] > 0$ if $0 \neq x \geq 0$) and satisfies $\mathbb{E}[1_A] = P(A)$ for every event A. We often omit excessive parentheses, as in $\mathbb{E}x = \mathbb{E}[x]$ and $P[x \leq \alpha] = P(\{x \leq \alpha\})$. We write $\hat{x}$ for the demeaned version of a random variable x:

$$\hat{x} \equiv x - \mathbb{E}x.$$

The **covariance** of two random variables x, y is the scalar

$$\operatorname{cov}[x, y] \equiv \mathbb{E}[\hat{x}\hat{y}] = \mathbb{E}[xy] - \mathbb{E}x\mathbb{E}y. \tag{2.1.1}$$

We can then define the **variance** $\text{var}[x] \equiv \text{cov}[x, x]$, the **standard deviation** $\text{stdev}[x] \equiv \sqrt{\text{var}[x]}$, and provided x and y have positive variance, the **correlation coefficient**

$$\text{corr}[x, y] \equiv \frac{\text{cov}[x, y]}{\text{stdev}[x]\,\text{stdev}[y]}.$$

The random variables x and y are **uncorrelated** if $\text{cov}[x, y] = 0$. Two uncorrelated random variables can be nontrivially determined by the same random source. For example, if $\Omega = \{-1, 0, 1\}$ and each state is assigned the same probability, then the identity random variable $x(\omega) = \omega$ is uncorrelated with its square $x^2(\omega) = |\omega|$. If $f(x)$ and $g(y)$ are uncorrelated for all functions $f, g \colon \mathbb{R} \to \mathbb{R}$, then the random variables x, y are said to be **stochastically independent**, or just **independent** where there is no risk of confusion with other types of independence (like linear independence). By virtue of Proposition 1.1.4, the independence of x and y is really a property of the algebras $\sigma(x)$ and $\sigma(y)$. Define two algebras $\mathcal{A}$ and $\mathcal{B}$ to be (stochastically) **independent** if every random variable in $L(\mathcal{A})$ is uncorrelated with every random variable in $L(\mathcal{B})$. Then x and y are independent if and only if $\sigma(x)$ and $\sigma(y)$ are independent.

PROPOSITION 2.1.1 *For all algebras $\mathcal{A}$ and $\mathcal{B}$ and corresponding partitions $\mathcal{A}^0$ and $\mathcal{B}^0$, the following are equivalent conditions.*

(1) $\mathcal{A}$ *and* $\mathcal{B}$ *are independent.*
(2) $P(A \cap B) = P(A)\,P(B)$ *for all* $A \in \mathcal{A}$ *and* $B \in \mathcal{B}$.
(3) $P(A \cap B) = P(A)\,P(B)$ *for all* $A \in \mathcal{A}^0$ *and* $B \in \mathcal{B}^0$.

PROOF Since the covariance operator is linear in each of its arguments and every random variable is a linear combination of indicator functions, $\mathcal{A}$ and $\mathcal{B}$ are independent if and only if 1_A and 1_B are uncorrelated for all $A \in \mathcal{A}^0$ and $B \in \mathcal{B}^0$. Since $\text{cov}(1_A, 1_B) = P(A \cap B) - P(A)\,P(B)$, this proves the equivalence of the first and third conditions. The same argument with $\mathcal{A}, \mathcal{B}$ in place of $\mathcal{A}^0, \mathcal{B}^0$ shows the equivalence between the first two conditions. $\square$

More generally, the random variables $x_1, \ldots, x_n$ are said to be (stochastically) **independent** if for each $i \in \{1, \ldots, n\}$, the algebras $\sigma(x_i)$ and $\sigma(\{x_j \mid j \neq i\})$ are independent.[1]

---

[1] This is not the same as pairwise independence. For example, suppose x_1 and x_2 are independent random variables, each taking the values $+1$ and -1 with equal probability. (As in Example 1.1.1 with $T = 2$, $x_i(\omega) = \omega_i$ and equal probability for each state.) Then

PROPOSITION 2.1.2 *For all random variables* $x_1, \ldots, x_n$, *the following are equivalent conditions.*

(1) $x_1, \ldots, x_n$ *are stochastically independent.*
(2) $\sigma(x_1, \ldots, x_{i-1})$ *and* $\sigma(x_i)$ *are independent for* $i = 2, \ldots, n$.
(3) $\mathbb{E} \prod_{i=1}^{n} f_i(x_i) = \prod_{i=1}^{n} \mathbb{E} f_i(x_i)$ *for all* $f_1, \ldots, f_n : \mathbb{R} \to \mathbb{R}$.
(4) $P\left[\bigcap_{i=1}^{n} \{x_i \in S_i\}\right] = \prod_{i=1}^{n} P[x_i \in S_i]$ *for all* $S_1, \ldots, S_n \subseteq \mathbb{R}$.

PROOF That (1) implies (2) is immediate. We show (3) assuming (2) by induction in n. For $n = 2$, the claim is valid by definition. Suppose it is true for $n - 1$ variables. For all $f_i : \mathbb{R} \to \mathbb{R}$, (2) implies that $\prod_{i=1}^{n-1} f_i(x_i)$ and $f_n(x_n)$ are uncorrelated and therefore the expectation of their product equals the product of their expectation, which together with the inductive hypothesis gives (3). To show that (3) implies (4), set $f_i(x_i) = 1_{\{x_i \in S_i\}}$. Finally, we show (1) assuming (4) holds. The partition generating $\sigma(x_1)$ is the set of all nonempty events of the form $A = \{x_1 = \alpha_1\}$, $\alpha_1 \in \mathbb{R}$, and the partition generating $\sigma(x_2, \ldots, x_n)$ is the set of the nonempty events of the form $B = \{x_2 = \alpha_2, \ldots, x_n = \alpha_n\}$, $\alpha_2, \ldots, \alpha_n \in \mathbb{R}$. For such A and B, (4) implies $P(A \cap B) = \prod_{i=1}^{n} P[x_i = \alpha_i]$ and $P(B) = \prod_{i=2}^{n} P[x_i = \alpha_i]$. Therefore $P(A \cap B) = P(A) P(B)$. By Proposition 2.1.1, this proves that $\sigma(x_1)$ is stochastically independent of $\sigma(x_2, \ldots, x_n)$. The same argument applies for any permutation of the $x_1, \ldots, x_n$, completing the proof. □

Uncorrelatedness and stochastic independence can equivalently be thought of as orthogonality conditions in the space $\hat{L} = \{\hat{x} \mid x \in \mathbb{R}^{\Omega}\}$ of zero-mean random variables, with the covariance inner product $\langle \hat{x} \mid \hat{y} \rangle = \text{cov}[x, y]$ and induced norm $\|\hat{x}\| = \sqrt{\langle \hat{x} \mid \hat{x} \rangle} = \text{stdev}[x]$. The random variables x, y are uncorrelated if and only if $\hat{x}$ and $\hat{y}$ are orthogonal in $\hat{L}$. Similarly, the algebras $\mathcal{A}$ and $\mathcal{B}$ are independent if and only if $\hat{L}(\mathcal{A}) \equiv \{\hat{x} \mid x \in L(\mathcal{A})\}$ and $\hat{L}(\mathcal{B})$ are orthogonal in $\hat{L}$. The Cauchy–Schwarz inequality in this context states that $|\text{corr}[x, y]| \leq 1$ with equality holding if and only if $\hat{x} = \alpha \hat{y}$ or $\hat{y} = \alpha \hat{x}$ for some scalar α.

The conditional probability $P(\cdot \mid B) : 2^{\Omega} \to [0, 1]$ given a positive probability event B is defined by **Bayes' rule**:

$$P(A \mid B) = \frac{P(A \cap B)}{P(B)}.$$

---

the random variables $x_1, x_2, x_1 x_2$ are not independent even though any two of them are independent.

The expectation operator corresponding to $P(\cdot \mid B)$ is denoted by $\mathbb{E}[\cdot \mid B]$ and is easily seen to satisfy

$$\mathbb{E}[x \mid B] = \frac{\mathbb{E}[x 1_B]}{P(B)}, \quad \text{for all } x \in \mathbb{R}^\Omega. \tag{2.1.2}$$

The **conditional expectation operator** (under P) given the algebra $\mathcal{B}$ generated by the partition $\{B_1, \ldots, B_n\}$ is defined by

$$\mathbb{E}[x \mid \mathcal{B}] = \sum_{i=1}^{n} \mathbb{E}[x \mid B_i] 1_{B_i}, \quad \text{for all } x \in \mathbb{R}^\Omega. \tag{2.1.3}$$

The following proposition shows that the random variable $\mathbb{E}[x \mid \mathcal{B}]$ is the best estimate of x given information $\mathcal{B}$, in the sense of minimizing expected squared error.

PROPOSITION 2.1.3 *For all algebras $\mathcal{B}$, and random variables x, y, the following are equivalent conditions, provided y is $\mathcal{B}$-measurable.*

(1) $y = \mathbb{E}[x \mid \mathcal{B}]$.
(2) $\mathbb{E}[(x-y)^2] \leq \mathbb{E}[(x-z)^2]$ *for all* $z \in L(\mathcal{B})$.
(3) $\mathbb{E}[yz] = \mathbb{E}[xz]$ *for all* $z \in L(\mathcal{B})$.
(4) $\mathbb{E}y = \mathbb{E}x$ *and* corr$[x-y, z] = 0$ *for all* $z \in L(\mathcal{B})$.
(5) $\mathbb{E}[y 1_B] = \mathbb{E}[x 1_B]$ *for all* $B \in \mathcal{B}$.

PROOF Condition (2) is a norm-minimization problem in the vector space of random variables with the inner product $\langle x \mid y \rangle = \mathbb{E}[xy]$. The $y \in L(\mathcal{B})$ satisfying condition (2) is the projection of x onto $L(\mathcal{B})$. By the orthogonal projection theorem (Corollary B.5.2), such a $y \in L(\mathcal{B})$ is equivalently characterized by the orthogonality condition $\langle x - y \mid z \rangle = 0$ for all $z \in L(\mathcal{B})$, which is condition (3). The equivalence of (3) and (4) is immediate from the definitions. The equivalence of (3) and (5) is also straightforward, given that every $z \in L(\mathcal{B})$ is a linear combination of indicator functions of events in $\mathcal{B}$. Finally, let $\mathcal{B}^0 \equiv \{B_1, \ldots, B_n\}$ denote the partition generating $\mathcal{B}$. Since the indicator of any $B \in \mathcal{B}$ is the sum of random variables of the form 1_{B_i}, condition (5) is equivalent to its version resulting after replacing $\mathcal{B}$ with $\mathcal{B}^0$, whose equivalence to condition (1) follows easily from the definitions. $\qquad \square$

REMARK 2.1.4 In infinite state-space extensions of the theory, definition (2.1.3) is not generally meaningful and a version of the orthogonality condition of Proposition 2.1.3 forms the basis for the usual textbook definition of a conditional expectation. The uniqueness claim in the last proposition relies on the full-support assumption on P. In general, any two random variables y and y' that are conditional expectations of x given $\mathcal{B}$ must satisfy $P[y = y'] = 1$ but can take arbitrary values on any $B \in \mathcal{B}$ such that $P(B) = 0$. We will have no need for this generality here, but the technicality becomes unavoidable in infinite state-space extensions of the theory. $\diamond$

COROLLARY 2.1.5 *The algebras $\mathcal{A}$ and $\mathcal{B}$ are stochastically independent if and only if $\mathbb{E}[x \mid \mathcal{B}] = \mathbb{E}x$ for all $x \in L(\mathcal{A})$.*

PROOF Recall that $\mathcal{A}$ and $\mathcal{B}$ are independent if and only if $\hat{L}(\mathcal{A})$ and $\hat{L}(\mathcal{B})$ are orthogonal in $\hat{L}$ under the covariance inner product, a condition that is in turn equivalent to the requirement that the projection of every $\hat{x} \in \hat{L}(\mathcal{A})$ onto $\hat{L}(\mathcal{B})$ is zero. By Proposition 2.1.3, the last condition can be restated as $\mathbb{E}[\hat{x} \mid \mathcal{B}] = 0$ for all $\hat{x} \in \hat{L}(\mathcal{A})$. $\qquad\square$

The important **law of iterated expectations** states that

$$\mathcal{B} \subseteq \mathcal{A} \quad \Longrightarrow \quad \mathbb{E}\big[\mathbb{E}[x \mid \mathcal{A}] \mid \mathcal{B}\big] = \mathbb{E}[x \mid \mathcal{B}], \qquad (2.1.4)$$

for every random variable x and algebras $\mathcal{A}$ and $\mathcal{B}$. Indeed if $\mathcal{B} \subseteq \mathcal{A}$, $L(\mathcal{B})$ is a linear subspace of $L(\mathcal{A})$, and therefore projecting x on $L(\mathcal{B})$ is equivalent to first projecting x on $L(\mathcal{A})$ and then further projecting on $L(\mathcal{B})$, which translates to $\mathbb{E}[x \mid \mathcal{B}] = \mathbb{E}\big[\mathbb{E}[x \mid \mathcal{A}] \mid \mathcal{B}\big]$.

Another important property of conditional expectations, stated below, is an immediate consequence of identities (2.1.2) and (2.1.3) in the current context. The following less direct proof is instructive, however, in illustrating how the orthogonality characterization of expectations can be used in ways that also apply in infinite state-space extensions of the theory.

PROPOSITION 2.1.6 *For every random x and algebra $\mathcal{B}$, if $b \in L(\mathcal{B})$, then $\mathbb{E}[bx \mid \mathcal{B}] = b\mathbb{E}[x \mid \mathcal{B}]$.*

PROOF Fix any $b \in L(\mathcal{B})$ and let $y = b\mathbb{E}[x \mid \mathcal{B}]$. We use the characterization of condition (3) of Proposition 2.1.3 twice, first to justify the middle equality in

$$\mathbb{E}[(bx)\,z] = \mathbb{E}[x\,(bz)] = \mathbb{E}\big[\mathbb{E}[x \mid \mathcal{B}]\,(bz)\big] = \mathbb{E}[yz] \text{ for all } z \in L(\mathcal{B}),$$

and then to conclude that since $y \in L(\mathcal{B})$, the above condition implies $y = \mathbb{E}[bx \mid \mathcal{B}]$. $\qquad\square$

The **conditional expectation** of a random variable x **given the random variables** $y_1, \dots, y_n$ is defined by

$$\mathbb{E}[x \mid y_1, \dots, y_n] = \mathbb{E}\big[x \mid \sigma(y_1, \dots, y_n)\big],$$

which is a function of $(y_1, \ldots, y_n)$ in the sense of Proposition 1.1.4. Thus $\mathbb{E}[x \mid y_1, \ldots, y_n]$ is the projection of x onto $L(\sigma(y_1, \ldots, y_n))$, which is the linear space of all random variables of the form $f(y_1, \ldots, y_n)$ for arbitrary $f: \mathbb{R}^n \to \mathbb{R}$. This contrasts with a simple linear regression,[2] where x is projected on the smaller subspace of all random variables of the same form but with f restricted to be linear.

As in Chapter 1, throughout this chapter we take as given an underlying filtration $(\mathcal{F}_0, \mathcal{F}_1, \ldots, \mathcal{F}_T)$ on Ω, abbreviated to $\{\mathcal{F}_t\}$, with $\mathcal{F}_0 = \{\emptyset, \Omega\}$ and $\mathcal{F}_T = 2^\Omega$. We write $L_t \equiv L(\mathcal{F}_t)$ for the set of all $\mathcal{F}_t$-measurable random variables, $\mathbb{E}_t[x]$ or $\mathbb{E}_t x$ for the conditional expectation $\mathbb{E}[x \mid \mathcal{F}_t]$, and cov_t for the conditional covariance given $\mathcal{F}_t$, defined as in (2.1.1) but with $\mathbb{E}_t$ in place of $\mathbb{E}$. Conditional variances, standard deviations and correlation coefficients given $\mathcal{F}_t$ are defined and denoted analogously.

A **martingale** (under P) is any adapted process M such that

$$M_t = \mathbb{E}_t M_u \quad \text{for all } u > t.$$

By the law of iterated expectations (2.1.4), an adapted process M is a martingale if and only if $\mathbb{E}_{t-1}[\Delta M_t] = 0$ for all $t > 0$ if and only if $M_t = \mathbb{E}_t M_T$ for all t. We let $\mathcal{M}$ denote the set of all martingales and $\mathcal{M}_0 \equiv \{M \in \mathcal{M} \mid M_0 = 0\}$, which is the set of all **zero-mean martingales** since a martingale M is in $\mathcal{M}_0$ if and only if $\mathbb{E} M_t = 0$ for all t.

The following simple observation has far-reaching implications for finance interpretations as well as martingale theory in general, including the extension of the theory of stochastic integrals in more-general settings.

PROPOSITION 2.1.7 *For every martingale M and predictable process ϕ, the integral $\phi \bullet M$ is a martingale.*[3]

PROOF If $\phi_t \in L_{t-1}$, then $\mathbb{E}_{t-1}[\Delta(\phi \bullet M)_t] = \mathbb{E}_{t-1}[\phi_t \Delta M_t] = \phi_t \mathbb{E}_{t-1}[\Delta M_t] = 0$. $\square$

EXAMPLE 2.1.8 In the context of Example 1.1.1, assume that P represents a fair coin: $P(\{\omega\}) = 2^{-T}$ for all $\omega \in \Omega$. A gambler betting on a coin toss gains a dollar if the outcome is heads and loses a dollar otherwise. Contingent on state ω, a gambler betting on the first t coin tosses gains (or loses if negative) $B_t(\omega) = \omega_1 + \omega_2 + \cdots + \omega_t$. Letting $B_0 = 0$, this defines an adapted process B whose increments $\Delta B_1, \ldots, \Delta B_T$ are

---

[2] A prominent case where the two projections coincide is when $(x, y_1, \ldots, y_n)$ has a Gaussian distribution, the definition of which requires an infinite state space.

[3] Note that it is not enough to assume that ϕ is adapted for $\phi \bullet M$ to be a martingale, a fact that is the main motivation behind the introduction of the notion of a predictable process.

stochastically independent and generate the underlying filtration. Since the algebra $\mathcal{F}_{t-1} = \sigma(\Delta B_1, \ldots, \Delta B_{t-1})$ is independent of (the algebra generated by) ΔB_t, $\mathbb{E}_{t-1}\Delta B_t = \mathbb{E}\Delta B_t = 0$ and therefore $B \in \mathcal{M}_0$. A gambler who starts betting at time t and quits at a later time u can expect zero total gains: $\mathbb{E}_t[B_u - B_t] = 0$. A natural question is whether the gambler could beat the odds by following some clever strategy, wagering $\phi_t \in L_t$ dollars on the tth coin toss at time $t - 1$. With the convention $\phi_0 = 0$, such a strategy defines an element ϕ of $\mathcal{P}_0$ and the process $M = \phi \bullet B$ defines the corresponding cumulative gains. By Proposition 2.1.7, M is also a zero-mean martingale. Therefore, a gambler following strategy ϕ from time t up to a later time u must again expect zero total gains: $\mathbb{E}_t[M_u - M_t] = 0$. In this sense, the gambler cannot beat the odds. If, on the other hand, the gambler were allowed to bet indefinitely, then the following doubling strategy would beat the odds: Bet one dollar, if heads stop, otherwise bet two dollars on a second coin toss, if heads stop, otherwise bet four dollars on a third coin toss, and so on. If allowed to play indefinitely, eventually heads is bound to show up, resulting in a gain of one dollar. Of course, losses can be staggering in the meantime and any finite borrowing limit is enough to rule out the strategy. Here this type of doubling strategy is ruled out by the finite number of periods, but it is something one has to technically deal with in infinite-horizon or continuous-time models. $\diamond$

Every adapted process x has a unique **Doob decomposition**:

$$x = x_0 + x^p + M, \quad x^p \in \mathcal{P}_0, \quad M \in \mathcal{M}_0. \qquad (2.1.5)$$

To see why, note that (2.1.5) implies $\Delta x = \Delta x^p + \Delta M$ and therefore,

$$\Delta x_t^p = \mathbb{E}_{t-1}\Delta x_t, \quad t = 1, \ldots, T.$$

Conversely, if x^p is recursively defined by the last equation and $x_0^p = 0$, then $x^p \in \mathcal{P}_0$ and $x - x^p$ is a martingale. The process x^p is known as the **compensator** of x.

2.2 Beta Pricing and Frontier Returns

In Section 1.8 we defined excess returns as the difference between the returns of a traded contract and that of a traded (default-free) money-market account (MMA). In this section we show that in an arbitrage-free market expected excess returns are proportional to the return's covariance with a traded return that is characterized by the property of minimizing variance given its expected value. The resulting expression for expected returns is known as a beta-pricing equation.

Beta pricing is a characterization of single-period traded cash flows; it is not about the specific contracts implementing the market and a version of the theory can be formulated with or without a traded MMA. For expositional simplicity, however, we adopt the setting of Section 1.8. We take as given an arbitrage-free market that is implemented by the $1 + J$ contracts (1.8.1), where $S_{t-1}^j \neq 0$ everywhere (meaning at all states) and therefore returns are well defined in (1.8.3). Contract zero is an MMA with rate process $r \in \mathcal{P}_0$. The corresponding period-t return is

$$R_t^0 = 1 + r_t \in L_{t-1}.$$

A period-t portfolio allocation $\psi_t = (\psi_t^1, \ldots, \psi_t^J) \in L_{t-1}^{1 \times J}$ results in the portfolio return

$$R_t^\psi \equiv R_t^0 + \sum_{j=1}^J \psi_t^j (R_t^j - R_t^0).$$

The set of period-t **traded returns** is

$$\mathcal{R}_t \equiv \left\{ R_t^\psi \mid \psi_t \in L_{t-1}^{1 \times J} \right\}.$$

A key observation is that returns can be mixed: For all $R_t^1, R_t^2 \in \mathcal{R}_t$ and $\alpha_t \in L_{t-1}$, $(1 - \alpha_t) R_t^1 + \alpha_t R_t^2 \in \mathcal{R}_t$. To avoid trivialities, we assume throughout that for every time t, there exists some $R_t \in \mathcal{R}_t$ such that $\mathrm{var}_{t-1}[R_t] > 0$ everywhere.

DEFINITION 2.2.1 The traded return $R_t^* \in \mathcal{R}_t$ is a (minimum variance) **frontier return** if for all $R_t \in \mathcal{R}_t$,

$$\left\{ \mathbb{E}_{t-1} R_t = \mathbb{E}_{t-1} R_t^* \right\} \subseteq \left\{ \mathrm{var}_{t-1}[R_t] \geq \mathrm{var}_{t-1}[R_t^*] \right\}.$$

The property of being a frontier return can also be defined spot by spot: $R_t^* \in \mathcal{R}_t$ is a frontier return at spot $(F, t - 1)$ if for all $R_t \in \mathcal{R}_t$,

$$\mathbb{E}[R_t \mid F] = \mathbb{E}[R_t^* \mid F] \implies \mathrm{var}[R_t \mid F] \geq \mathrm{var}[R_t^* \mid F].$$

Since for all $F \in \mathcal{F}_{t-1}$, $R_t, R_t^* \in \mathcal{R}_t$ implies $R_t 1_F + R_t^* 1_{F^c} \in \mathcal{R}_t$, it follows that R_t^* is a frontier return if and only if it is a frontier return at every time-$(t - 1)$ spot. Similarly, the following characterization of frontier returns, as well as the entire discussion of the remainder of this section, applies separately spot by spot and all the arguments are encapsulated by the special case of a single period ($T = 1$).

LEMMA 2.2.2 $R_t^* \in \mathcal{R}_t$ *is a frontier return if and only if for all* $R_t \in \mathcal{R}_t$,
$\mathbb{E}_{t-1}[R_t - R_t^*] = 0$ *implies* $\mathrm{cov}_{t-1}[R_t - R_t^*, R_t^*] = 0$.

PROOF Fix any spot $(F, t-1)$ and let (F_i, t), $i = 0, 1, \ldots, d$, denote its
immediate successor spots. On $\mathbb{R}^{1+d}$, we use the inner product

$$\langle x \mid y \rangle \equiv \sum_{i=0}^{d} x_i y_i P[F_i \mid F].$$

Let $\mathcal{R}_{F,t} \equiv \{ (R(F_0, t), \ldots, R(F_d, t)) \mid R \in \mathcal{R}_t \}$ and

$$z^* \equiv \left(R^*(F_0, t), \ldots, R^*(F_d, t) \right),$$

and define the linear manifold

$$M \equiv \left\{ z \in \mathcal{R}_{F,t} \mid \sum_{i=0}^{d} (z_i - z_i^*) P[F_i \mid F] = 0 \right\}.$$

Since R_t^* is a frontier return, $\|z\| \geq \|z^*\|$ for all $z \in M$. In other
words, z^* is the orthogonal projection of the zero vector onto M. By
the orthogonal projection theorem (Corollary B.5.2), z^* is characterized
by the orthogonality condition $\langle z - z^* \mid z^* \rangle = 0$ for all $z \in M$, which
can be restated as, for all $R_t \in \mathcal{R}_t$, $\mathbb{E}[R_t - R_t^* \mid F] = 0$ implies
$\mathbb{E}[(R_t - R_t^*)R_t^* \mid F] = 0$. The lemma's claim follows. $\qquad\square$

We can use Lemma 2.2.2 to determine all **frontier allocations**, that is,
all portfolio allocations generating frontier returns. Let

$$1 + \mu_t^i \equiv \mathbb{E}_{t-1} R_t^i \quad \text{and} \quad \Sigma_t^{ij} \equiv \mathrm{cov}_{t-1}[R_t^i, R_t^j], \quad i, j = 1, \ldots, J,$$

and define the column vector $\mu_t \equiv (\mu_t^1, \ldots, \mu_t^J)'$ and the $J \times J$ symmetric
matrix $\Sigma_t \equiv [\Sigma_t^{ij}]$, which is positive semidefinite (at every time-$(t-1)$
spot). For simplicity, we assume that Σ_t is full rank and therefore positive
definite. The assumption of a full-rank Σ_t is easily seen to be equivalent to
the assumption that the contracts are **everywhere nonredundant**, meaning
that conditionally on any spot $(F, t-1)$, there is no $j \in \{0, 1, \ldots, J\}$ such
that the return R_t^j on F can be generated by an allocation in the remaining
contracts. By Lemma 2.2.2, an allocation ψ_t^* generates a frontier return R_t^*
if and only if for every period-t allocation ψ_t,

$$(\psi - \psi_t^*)(\mu_t - r_t) = 0 \quad \text{implies} \quad (\psi - \psi_t^*) \Sigma_t \psi_t^{*'} = 0.$$

In other words, whenever $\psi_t - \psi_t^*$ is orthogonal to $\mu_t - r_t$, it is also orthogonal to $\Sigma_t \psi_t^{*'}$ (all conditionally on each time-$(t-1)$ spot). Therefore,
$\mu_t - r_t$ is collinear to $\Sigma_t \psi_t^{*'}$ in the sense that there exists $\alpha_t \in L_{t-1}$ such

that $\alpha_t(\mu_t - r_t) = \Sigma_t \psi_t^{*\prime}$. Rearranging, we can parametrically describe all frontier allocations by

$$\psi_t^* = \alpha_t(\mu_t - r_t)' \Sigma_t^{-1}, \quad \alpha_t \in L_{t-1}. \tag{2.2.1}$$

Note that if ψ_t^* is any nonzero frontier allocation, then every other frontier allocation takes the form $\alpha_t \psi_t^*$ for some $\alpha_t \in L_{t-1}$, a fact known as **two-fund separation**. The term reflects the idea that every frontier return can be achieved by allocating a value proportion α_t in a fund allocated according to ψ_t^* and the rest in a fund that is an MMA. As we will see shortly, two-fund separation is valid even without the assumption that the contracts are everywhere nonredundant.

The frontier returns other than the MMA return are exactly the returns relative to which beta pricing is possible.

PROPOSITION 2.2.3 *For all $R_t^* \in \mathcal{R}_t$ such that $\mathrm{var}_{t-1}\big[R_t^*\big] > 0$ everywhere, the following two conditions are equivalent.*

(1) R_t^* *is a frontier return.*
(2) *For all $R_t \in \mathcal{R}_t$,*

$$\mathbb{E}_{t-1}R_t - R_t^0 = \frac{\mathrm{cov}_{t-1}\big[R_t^*, R_t\big]}{\mathrm{var}_{t-1}\big[R_t^*\big]} \left(\mathbb{E}_{t-1}R_t^* - R_t^0\right) \tag{2.2.2}$$

and for some $R_t \in \mathcal{R}_t$, $\mathbb{E}_{t-1}R_t \neq R_t^0$ everywhere.

PROOF (1 $\implies$ 2) Suppose $R_t^* \in \mathcal{R}_t$ is a frontier return and therefore $0 = \mathrm{var}_{t-1}\big[R_t^0\big] \geq \mathrm{var}_{t-1}\big[R_t^*\big]$ on the event $\{\mathbb{E}_{t-1}R_t^* = R_t^0\}$. Since $\mathrm{var}_{t-1}\big[R_t^*\big] > 0$ everywhere, the event $\{\mathbb{E}_{t-1}R_t^* = R_t^0\}$ is empty.

Given any $R_t \in \mathcal{R}_t$, define $\tilde{R}_t \in \mathcal{R}_t$ by letting

$$\tilde{R}_t \equiv R_t^* + R_t - R_t^0 \text{ on } \{\mathbb{E}_{t-1}R_t = R_t^0\}, \text{ and}$$

$$\tilde{R}_t \equiv R_t^0 + \frac{\mathbb{E}_{t-1}R_t^* - R_t^0}{\mathbb{E}_{t-1}R_t - R_t^0}(R_t - R_t^0) \text{ on } \{\mathbb{E}_{t-1}R_t \neq R_t^0\}.$$

By construction, $\mathbb{E}_{t-1}\big[\tilde{R}_t - R_t^*\big] = 0$ and therefore, by Lemma 2.2.2, $\mathrm{cov}_{t-1}\big[\tilde{R}_t - R_t^*, R_t^*\big] = 0$, which expands to equation (2.2.2).

(2 $\implies$ 1) Conversely, suppose the second condition holds, which clearly implies that $\mathbb{E}_{t-1}R_t^* \neq R_t^0$ everywhere. We show that R_t^* is a frontier return by verifying the orthogonality condition of Lemma 2.2.2. Consider any $R_t \in \mathcal{R}_t$ such that $\mathbb{E}_{t-1}\big[R_t - R_t^*\big] = 0$. Applying the beta-pricing equation and canceling out the term $\mathbb{E}_{t-1}R_t^* - R_t^0$ on each side, we conclude that $\mathrm{cov}_{t-1}\big[R_t - R_t^*, R_t^*\big] = 0$. $\qquad \square$

The so-called beta-pricing equation (2.2.2) has been of considerable interest in empirical work, since it suggests that the slope coefficient of a linear regression (commonly denoted by β) can be used to explain expected excess returns. In practice, we can only identify a frontier return with error. We therefore have to consider the beta-pricing equation relative to some proxy return $R_t^p = R_t^* + \varepsilon_t$, where the error ε_t is judged to be small. It is an instructive exercise to show that an arbitrarily small value of $\mathbb{E}_{t-1}\varepsilon_t^2$ is consistent with the existence of a traded return whose beta with respect to R_t^p is arbitrarily different from its beta with respect to R_t^*. The basic idea is that leverage (that is, borrowing from the MMA to invest in a risky portfolio) can arbitrarily amplify any discrepancy between the two betas. This pitfall can be avoided if we instead focus on excess returns normalized by their standard deviation.

The time-$(t-1)$ **Sharpe ratio** of a period-t return R_t is the ratio

$$S_{t-1}[R_t] \equiv \frac{\mathbb{E}_{t-1}R_t - R_t^0}{\text{stdev}_{t-1}[R_t]},$$

provided the denominator is nowhere zero. We adopt the convention that $S_{t-1}[R_t] \equiv 0$ on the event $\{\text{var}_{t-1}[R_t] = 0\}$. The beta-pricing equation (2.2.2) can be restated in terms of Sharpe ratios as

$$S_{t-1}[R_t] = \text{corr}_{t-1}[R_t^*, R_t] S_{t-1}[R_t^*]. \tag{2.2.3}$$

This version of the beta-pricing equation is robust to replacing R_t^* with a highly correlated proxy R_t^p, in the following sense.[4]

PROPOSITION 2.2.4 *For all $R_t^*, R_t^p, R_t \in \mathcal{R}_t$, equation (2.2.3) implies its approximate version:*

$$S_{t-1}[R_t] = \text{corr}_{t-1}[R_t^p, R_t] S_{t-1}[R_t^p] + \epsilon_{t-1}, \tag{2.2.4}$$

where

$$|\epsilon_{t-1}| \leq \sqrt{2}\,|S_{t-1}[R_t^*]|\,\sqrt{1 - \text{corr}_{t-1}[R_t^*, R_t^p]}.$$

PROOF We adopt the notation of the proof of Lemma 2.2.2, since the argument relates to a single step of the information tree following a given spot $(F, t-1)$. Analogously to z^*, let

---

[4] While this section's results are standard textbook material, to my knowledge, Proposition 2.2.4 first appeared in Skiadas (2009).

$$z \equiv \left(R(F_0,t),\ldots,R(F_d,t)\right) \quad \text{and} \quad z^p \equiv \left(R^p(F_0,t),\ldots,R^p(F_d,t)\right).$$

We write $\mathcal{S}[z]$ for the value of $\mathcal{S}_{t-1}[R_t]$ on F, and analogously for $\mathcal{S}[z^*]$ and $\mathcal{S}[z^p]$. The vectors z, z^*, z^p can be thought of as random variables on $\{0, 1, \ldots, d\}$. For any such random variable x, we use the notation

$$\tilde{x} \equiv \frac{x - \sum_i x_i P[F_i \mid F]}{\text{stdev}[x]},$$

where the standard deviation is relative to the probability assigning mass $P[F_i \mid F]$ to state i. Note that $\langle \tilde{x} \mid \tilde{x} \rangle = 1$ and $\langle \tilde{x} \mid \tilde{y} \rangle = \text{corr}[x, y]$. Let $1 - \delta \equiv \langle \tilde{z}^* \mid \tilde{z}^p \rangle$, which is the value of $\text{corr}_{t-1}[R_t^*, R_t^p]$ on F, and let $\epsilon \equiv \mathcal{S}[z] - \langle \tilde{z}^p \mid \tilde{z} \rangle \mathcal{S}[z^p]$, which is the value of ϵ_{t-1} on F, as defined by (2.2.4). Condition (2.2.3) implies that $\mathcal{S}[z] = \langle \tilde{z}^* \mid \tilde{z} \rangle \mathcal{S}[z^*]$ and $\mathcal{S}[z^p] = (1 - \delta) \mathcal{S}[z^*]$. Therefore,

$$\frac{\epsilon^2}{\mathcal{S}[z^*]^2} = \langle \tilde{z}^* - (1 - \delta)\tilde{z}^p \mid \tilde{z} \rangle^2$$

$$\leq \langle \tilde{z}^* - (1 - \delta)\tilde{z}^p \mid \tilde{z}^* - (1 - \delta)\tilde{z}^p \rangle,$$

where the last inequality follows by the Cauchy–Schwarz inequality. Expanding the last term, we find that it equals $2\delta - \delta^2$. Therefore $\epsilon^2 \leq \mathcal{S}[z^*]^2 \, 2\delta$, which is the claimed bound. $\qquad \square$

The set of frontier traded returns other than the MMA return are the set of the traded returns of maximum absolute Sharpe ratio. We state this claim more formally below, using the convention that indeterminate Sharpe ratios are assigned the value zero.

PROPOSITION 2.2.5 *A traded return R_t^* is a frontier return if and only if $|\mathcal{S}_{t-1}[R_t^*]| \geq |\mathcal{S}_{t-1}[R_t]|$ for all $R_t \in \mathcal{R}_t$.*

PROOF The "only if" part is a corollary of Proposition 2.2.3, as can easily be seen by taking absolute values on both sides of equation (2.2.3) and using the fact that the absolute correlation is less than one. (Alternatively, one can show the claim more directly from the definition of frontier returns.) The converse is immediate from the definitions. $\qquad \square$

Absolute Sharpe ratios are invariant to positions in the MMA. For any $\alpha_t \in L_{t-1}$, a mix of α_t in a portfolio allocation ψ_t and $1 - \alpha_t$ in the MMA results in a return that has the same absolute Sharpe ratio as R_t^ψ:

$$\left| \mathcal{S}_{t-1}\left[R_t^0 + \alpha_t\left(R_t^\psi - R_t^0\right)\right] \right| = \left| \mathcal{S}_{t-1}\left[R_t^\psi\right] \right|.$$

Moreover, as we vary α_t we trace out all (conditional) mean-standard deviation pairs of traded returns that are consistent with the given absolute Sharpe-ratio value. In particular, if $R_t^{\psi} = R_t^*$ is a frontier return other than the risk-free return R_t^0, varying α_t traces out all traded returns with maximum absolute Sharpe ratio, that is, all frontier returns. We therefore obtain once again the two-fund separation result introduced earlier, but without the nonredundancy assumption. If the contracts are assumed to be everywhere nonredundant, we can use expression (2.2.1) to compute the maximum squared Sharpe ratio as

$$\mathcal{S}_{t-1}\big[R_t^*\big]^2 = (\mu_t - r_t)' \, \Sigma_t^{-1} \, (\mu_t - r_t).$$

The frontier returns with a positive (conditional) Sharpe ratio are known as (conditionally) **mean-variance efficient**, since besides minimizing variance given an expected value, they also achieve a maximum expected return given the variance of the return (all conditionally on beginning-of-period information). Since the seminal contribution of Markowitz (1952), mean-variance efficiency has played a prominent role as a criterion for portfolio choice. The criterion is simplistic in that it is myopic (only focuses on single-period returns) and uses variance as a measure of portfolio risk (treating high and low returns symmetrically). A more-sophisticated theory of portfolio choice, albeit with its own limitations, is developed in Chapter 3. Mean-variance efficiency resurfaces as a building block of optimal portfolios in the more sophisticated theory for a useful class of return dynamics and high trading frequency. So the concept is in theory more robust than this section's discussion suggests. Its empirical implementation, however, has its own serious limitations, which are beyond the scope of this text. Ultimately, the theory of mean-variance efficiency serves mainly as a parsimonious model that highlights the benefits of portfolio diversification.

2.3 State-Price Densities

A state-price process was defined in Section 1.2 as a representation of a present-value function. A state-price density process is essentially the same object, but with its values expressed as a density relative to a given reference probability. This simple construct opens the door for the use of probabilistic methods.

We continue to take as given a market X in the usual stochastic setting, consisting of a filtration $\{\mathcal{F}_t\}$ on the finite state space Ω, where $\mathcal{F}_0 = \{\emptyset, \Omega\}$ and $\mathcal{F}_T = 2^\Omega$, and a full-support probability P on $\mathcal{F}_T$. A state-price density can be defined as an adapted process π such that a state-price process p is well defined by letting $p(F, t) = \pi(F, t) P(F)$ for every spot (F, t). The following equivalent definition bypasses reference to a state-price process and extends directly to infinite state-space settings.

DEFINITION 2.3.1 A **state-price density process** or **SPD** (relative to the probability P) is any π in $\mathcal{L}_{++}$ such that a present-value function $\Pi : \mathcal{L} \to \mathbb{R}$ is well defined by

$$\Pi(c) = \frac{1}{\pi_0} \mathbb{E} \left[\sum_{t=0}^{T} \pi_t c_t \right], \quad c \in \mathcal{L}.$$

In this case, π is said to **represent** Π.

PROPOSITION 2.3.2 *Every present-value function admits an SPD representation, which is unique up to positive scaling.*

PROOF Given a present-value function Π, let π denote its (unique) Riesz representation in $\mathcal{L}$ with the inner product $\langle x \mid y \rangle \equiv \mathbb{E} \sum_t x_t y_t$. The positivity of Π implies that $\pi \in \mathcal{L}_{++}$, and therefore π is an SPD representing Π. Conversely, if π is an SPD representing Π, then π/π_0 is the Riesz representation of Π. □

Recall that a time-zero present-value function Π defines a spot-(F, t) conditional present-value function $\Pi_{F,t}$ by equation (1.4.1). If the SPD π represents Π, then

$$\Pi_{F,t}(c) = \frac{1}{\pi(F,t)} \mathbb{E} \left[\sum_{u=t}^{T} \pi_u c_u \mid F \right]. \tag{2.3.1}$$

This link between conditional valuation and conditional expectation turns out to be methodologically quite useful.

In Proposition 1.4.8 we saw that if Π is a present-value function and (δ, V) is a traded contract, then Π prices the contract:

$$V(F,t) = \Pi_{F,t}(\delta) \quad \text{for every spot } (F, t).$$

If π is an SPD representing Π, then the same pricing condition can be expressed as

$$V_t = \frac{1}{\pi_t} \mathbb{E}_t \left[\sum_{u=t}^{T} \pi_u \delta_u \right], \quad t = 0, \ldots, T-1. \quad (2.3.2)$$

In this case we say that π **prices** the contract (δ, V). The following variant of the argument used in Proposition 1.4.8 shows the pricing condition (2.3.2) directly in a way that extends to infinite state-space settings, where expression (2.3.1) is not generally meaningful. Given any time t and $F \in \mathcal{F}_t$, set to zero the present value of the cash flow (1.3.2) generated by buying the contract at time t on the event F and holding it to time T to find

$$\mathbb{E}[\pi_t V_t 1_F] = \mathbb{E}\left[\left(\sum_{u=t}^{T} \pi_u \delta_u \right) 1_F \right].$$

Equation (2.3.2) follows after applying Proposition 2.1.3.

The second part of Proposition 1.4.8 can also be restated in terms of SPDs, as follows.

PROPOSITION 2.3.3 *A process* $\pi \in \mathcal{L}_{++}$ *is an SPD for the market implemented by a set of contracts if and only if it prices every contract in the given set.*

The following link of pricing to martingales plays a central methodological role, especially in technically more-advanced incarnations of the theory.

PROPOSITION 2.3.4 *A process* $\pi \in \mathcal{L}_{++}$ *prices the contract* (δ, V) *if and only if* $G^\pi \equiv \pi V + (\pi \delta)_- \bullet t$ *is a martingale.*

PROOF Multiply equation (2.3.2) by π_t and add $\sum_{u=0}^{t-1} \pi_u \delta_u$ on both sides to find $G_t^\pi = \mathbb{E}_t[G_T^\pi]$. This calculation can be reversed. $\square$

REMARK 2.3.5 The preceding martingale condition characterizes an SPD for a market implemented by given contracts. Suppose that the market X is implemented by contracts (δ^j, V^j), $j = 1, \ldots, J$, and let $\delta = (\delta^1, \ldots, \delta^J)'$ and $V = (V^1, \ldots, V^J)'$. Write G for the corresponding column vector of gain processes. Then, by the last two propositions, $\pi \in \mathcal{L}_{++}$ is an SPD if and only if G^π is a martingale, meaning that each $G^{j\pi} = \pi V^j + (\pi \delta^j)_- \bullet t$ is a martingale. For the synthetic contract generated by a trading strategy θ, we have $G^{\theta\pi} = \pi_0 V_0^\theta + \theta \bullet G^\pi$, which is a martingale if G^π is a martingale, since θ is predictable. This observation merely reflects the fact that an SPD prices every traded contract, and every synthetic contract is traded. $\diamond$

Yet another useful way of expressing the fact that an SPD π prices the contract (δ, V) is in the recursive form

$$S_{t-1} = \mathbb{E}_{t-1}\left[\frac{\pi_t}{\pi_{t-1}}V_t\right], \quad t = 1, \ldots, T, \tag{2.3.3}$$

where $S \equiv V - \delta$. The equivalence of this condition to (2.3.2) follows from the law of iterated expectations (2.1.4). Alternatively, equation (2.3.3) can be rearranged to $\mathbb{E}_{t-1}\left[\Delta G_t^\pi\right] = 0$, which is one of the equivalent ways of stating the martingale property of G^π.

If $S_{t-1} \neq 0$ everywhere, then the return $R_t = V_t/S_{t-1}$ is well defined and recursion (2.3.3) implies

$$1 = \mathbb{E}_{t-1}\left[\frac{\pi_t}{\pi_{t-1}}R_t\right], \quad t = 1, \ldots, T. \tag{2.3.4}$$

Conversely, the validity of the pricing restriction (2.3.4) for every return R of a traded contract implies that π prices an arbitrary traded contract (δ, V), provided that at every nonterminal spot there exists at least some traded contract with nonzero ex-dividend price. If the event $F \equiv \{S_{t-1} = 0\}$ is empty, then clearly (2.3.4) with $R_t = V_t/S_{t-1}$ implies (2.3.3). What if F is nonempty? Equation (2.3.4) cannot be applied directly to the contract (δ, V). Instead, apply (2.3.4) twice—once to any traded contract with a well-defined return on F, and once to the return of a portfolio that holds this same contract as well as the contract (δ, V) whose ex-dividend price vanishes on F. Solving the resulting pair of equations gives the desired pricing condition (2.3.3).

Applying the pricing equation (2.3.4) to a traded MMA with rate process r results in an expression that gives r as a function of an SPD:

$$\frac{1}{1+r_t} = \mathbb{E}_{t-1}\left[\frac{\pi_t}{\pi_{t-1}}\right], \quad t = 1, \ldots, T. \tag{2.3.5}$$

This equation can be rearranged to

$$r_t = -\mathbb{E}_{t-1}\left[\frac{\Delta\pi_t}{\pi_{t-1}}\right] + \varepsilon_t, \quad \text{where } \varepsilon_t \equiv \frac{r_t^2}{1+r_t}.$$

If a period in the model represents a sufficiently short time interval, then ε_t is numerically negligible and the period-t risk-free interest rate r_t is approximately equal to minus the expected growth rate of the SPD π_t over the period, conditionally on the beginning-of-period information. In Section 2.8 we will see that this approximation becomes an exact relationship in continuous time, where r_t represents a continuously compounded interest rate.

Suppose now that π is an SPD, an MMA is traded and therefore the short-rate process r is given in terms of π as just described. The contract (δ, V) is said to be priced risk neutrally if $S_{t-1} = \mathbb{E}_{t-1} V_t / (1 + r_t)$, a relationship that is valid for an MMA. For a general contract, the ex-dividend price S_{t-1} must be adjusted relative to the risk-neutral valuation to reflect the uncertainty of the end-of-period payoff V_t. The appropriate risk adjustment can be written precisely by rearranging equation (2.3.3) to

$$S_{t-1} - \frac{\mathbb{E}_{t-1} V_t}{1 + r_t} = \mathrm{cov}_{t-1}\left[\frac{\pi_t}{\pi_{t-1}}, V_t\right] = \mathrm{cov}_{t-1}\left[\frac{\Delta \pi_t}{\pi_{t-1}}, V_t\right]. \quad (2.3.6)$$

In words, the risk adjustment relative to the risk-neutral valuation is equal to the covariance of the end-of-period value with the SPD growth rate conditionally on the beginning-of-period information.

Assuming the return $R_t = V_t / S_{t-1}$ is well defined, the preceding relationship is often expressed as a restriction on the excess return $R_t - R_t^0$, where $R^0 \equiv 1 + r$. Using the definition of covariance and the pricing restrictions (2.3.4) and (2.3.5), we find

$$\mathbb{E}_{t-1} R_t - R_t^0 = -R_t^0 \, \mathrm{cov}_{t-1}\left[\frac{\pi_t}{\pi_{t-1}}, R_t\right]. \quad (2.3.7)$$

To make the connection to the beta-pricing pricing notion of Section 2.2, we need to project π_t / π_{t-1} to the space of traded returns. In the context of Section 2.2, we can write

$$\frac{\pi_t}{\pi_{t-1}} = R_t^* + \epsilon_t,$$

where R_t^* is a period-t traded return and $\mathbb{E}_{t-1}[\epsilon_t R_t] = 0$ for every period-t traded return R_t (and therefore $\mathbb{E}_{t-1}\epsilon_t = 0$). The existence of a (unique) such decomposition follows by a simple projection argument conditionally at each time-$(t-1)$ spot, using the inner product of the proof of Lemma 2.2.2. Substituting into equation (2.3.7) and dividing the resulting equation by its special case with R^* in place of R (excluding a degenerate case), we obtain the beta-pricing equation

$$\mathbb{E}_{t-1} R_t - R_t^0 = \frac{\mathrm{cov}_{t-1}\left[R_t^*, R_t\right]}{\mathrm{var}_{t-1}\left[R_t^*\right]} \left(\mathbb{E}_{t-1} R_t^* - R_t^0\right).$$

The return R_t^* maximizes correlation with π_t / π_{t-1} conditionally on time-$(t-1)$ information among all period-t traded returns; a claim that reduces to the Cauchy–Schwarz inequality in the geometry used to define R_t^* and ϵ_t as the components of an orthogonal projection of π_t / π_{t-1}.

Another noteworthy consequence of (2.3.7) is the inequality

$$\mathcal{S}_{t-1}\big[R_t\big]^2 \le (1+r_t)^2 \, \mathrm{var}_{t-1}\left[\frac{\pi_t}{\pi_{t-1}}\right]. \tag{2.3.8}$$

To show it, write the cov_{t-1} of equation (2.3.7) in terms of corr_{t-1} and use the fact that the latter is valued in $[-1,1]$. Given the short rate r_t, the conditional variance of an SPD growth rate places an upper bound on the maximum conditional squared Sharpe ratio of a traded return (an expression for which was derived at the end of Section 2.2). This fact is known as the **Hansen–Jagannathan bound.**[5]

2.4 Equivalent Martingale Measures

Equivalent martingale measures give us another methodologically useful way of representing present-value functions. Prior to introducing the concept, let us fix, for the entire section, a reference arbitrage-free market X and underlying full-support probability P. Suppose that π is an SPD relative to these primitives, as defined in the last section. Suppose also, for now, that an MMA is traded in X, with corresponding short-rate process r. Consider any nonterminal spot $(F, t-1)$, with immediate successor spots (F_i, t), $i = 0, \ldots, d$, and define the positive scalars $Q(F_i \mid F)$ by

$$\frac{\pi(F_i, t)}{\pi(F, t-1)} = \frac{1}{1 + r(F, t)} \frac{Q(F_i \mid F)}{P(F_i \mid F)}, \quad i = 0, \ldots, d. \tag{2.4.1}$$

Pricing of the MMA, as in equation (2.3.5), gives

$$\frac{1}{1 + r(F, t)} = \sum_{i=0}^{d} \frac{\pi(F_i, t)}{\pi(F, t-1)} P(F_i \mid F), \tag{2.4.2}$$

and therefore $\sum_i Q(F_i \mid F) = 1$. In other words, $Q(F_i \mid F)$ can be thought of as a positive transition probability from spot $(F, t-1)$ to spot (F_i, t). For every path on the information tree, which corresponds to a state revealed at the terminal date T, these transition probabilities can be multiplied through, thus defining a full-support probability Q, which is the equivalent martingale measure (EMM) corresponding to π. In fact, this construction applies without the assumption that an MMA is traded by taking equation (2.4.2) to be the definition of $r(F, t)$.

We write $\mathbb{E}^Q$ for the expectation operator relative to Q, simplifying the conditional expectation notation $\mathbb{E}^Q\big[x \mid \mathcal{F}_t\big]$ to $\mathbb{E}_t^Q[x]$ or $\mathbb{E}_t^Q x$. Omitting

---

[5] Named after the contribution of Hansen and Jagannathan (1991).

the superscript implies the default reference probability: $\mathbb{E} = \mathbb{E}^P$. Identity (2.4.1) can be used to price any traded contract (δ, V), with $S = V - \delta$, resulting in

$$S_{t-1} = \frac{\mathbb{E}^Q_{t-1} V_t}{1 + r_t}, \quad t = 1, \dots, T. \tag{2.4.3}$$

This pricing relationship is often referred to as **risk-neutral pricing**,[6] although the term is misleading in terms of economic content; the risk associated with the payoff V_t is priced in equation (2.4.3) just as it would by the SPD π. In fact, Q can be thought of as representing a pure (conditional) price of risk over a single period in the sense that if at spot $(F, t - 1)$ one enters a forward contract to exchange a time-t fixed payment $f_{i,t}$ for a contingent unit of account that is paid if and only if spot (F_i, t) materializes, then $f_{i,t} = Q(F_i \mid F)$. In other words, $Q(F_i \mid F)$ is a spot-$(F, t - 1)$ forward price of a unit payment at spot (F_i, t). The following example shows the role of the conditional expectation $\mathbb{E}^Q_{t-1}$ as a one-period-ahead forward pricing operator more formally.

EXAMPLE 2.4.1 Apply the pricing equation (2.4.3) for an assumed traded contract $(\delta, V) = (D - f, D - f)$, where D is an adapted process and f is a predictable process, with $f_0 = D_0 = 0$. For every $t > 0$, we interpret D_t as the time-t value of some asset and f_t as the time-$(t - 1)$ forward price of the asset for time-t delivery. Buying the contract at time $t - 1$ and selling it at time t results in the net cash flow x, where $x_u = 0$ for $u \neq t$ and $x_t = D_t - f_t$. From the perspective of time $t - 1$, the trade is equivalent to entering a forward contract for time-t delivery of the asset at the forward price f_t, whose value is determined at time $t - 1$. The pricing equation (2.4.3) in this case reduces to $f_t = \mathbb{E}^Q_{t-1} D_t, t = 1, \dots, T$. Note that this argument does not apply in general if we extend the delivery of the asset to two or more periods, since the interest r can vary stochastically from period to period. $\diamond$

The development so far relies critically on the finite information tree and the use of a state-price density. We now introduce an equivalent but more-direct EMM definition that extends readily to infinite state-space versions of the theory. $\mathcal{Q}$ denotes the set of all full-support probability measures on 2^Ω.

---

[6] The idea of risk-neutral pricing already appears in Arrow (1971) and Drèze (1971), and it is exploited in option pricing by Cox and Ross (1976). The term "equivalent martingale measure" is due to Harrison and Kreps (1979).

DEFINITION 2.4.2 A **discount process** is a predictable strictly positive process ρ such that $\rho_0 = 1$. An **equivalent martingale measure**[7] (**EMM**) is a probability $Q \in \mathcal{Q}$ relative to which there exists a predictable SPD. An **EMM-discount pair** is a $(Q, \rho) \in \mathcal{Q} \times \mathcal{P}$ such that ρ is both a discount process and an SPD relative to Q. The present-value function **represented** by this pair is defined by

$$\Pi(c) = \mathbb{E}^Q \left[\sum_{t=0}^{T} \rho_t c_t \right], \quad c \in \mathcal{L}.$$

The results in Section 2.3 on pricing using state-price densities all apply to an EMM-discount pair (Q, ρ), with simplifications resulting from the fact that ρ is predictable. For example, if an MMA with rate process r is traded, then the pricing equation (2.3.5) with P replaced by Q and π replaced by ρ results in $1/(1 + r_t) = \rho_t/\rho_{t-1}$ and therefore

$$r = -\frac{\Delta\rho}{\rho} \quad \text{and} \quad \rho_t = \prod_{u=0}^{t} \frac{1}{1 + r_u}, \quad t = 0, \ldots, T. \tag{2.4.4}$$

In general, given a discount process ρ, we refer to the $r \in \mathcal{P}_0$ defined by the first equation in (2.4.4) as the **rate process implied** by ρ, and given $r \in \mathcal{P}_0$, we refer to the ρ defined by the second equation in (2.4.4) as the **discount process implied** by r. The preceding discussion shows the following claim.

PROPOSITION 2.4.3 *If r is the market's short-rate process and (Q, ρ) is an EMM-discount pair, then r is the rate process implied by ρ.*

Whether an MMA is traded or not, an EMM-discount pair (Q, ρ) prices a traded contract according to (2.4.3), where r is the rate process implied by ρ, as can be seen by formally replacing P by Q and π by ρ in equation (2.3.3). The other pricing relationships in Section 2.3 can similarly be stated in terms of an EMM-discount pair.

Let us now establish a correspondence between SPDs relative to P and EMM-discount pairs, defined by the requirement that they represent the same present-value function. Not surprisingly, this correspondence is the same as that of equations (2.4.1) and (2.4.2), thus establishing

---

[7] The terms "equivalent" and "martingale measure" have their origin in probability theory. A probability (measure) Q is said to be **equivalent** to P if $Q(A) = 0 \iff P(A) = 0$ for every event A. In our setting, for all $P \in \mathcal{Q}$, the set of all equivalent-to-P probabilities on 2^{Ω} is $\mathcal{Q}$. A **martingale measure** is a probability relative to which a given set of processes is a set of martingales. In the current context, such a set is that of the properly discounted gain processes of all traded contracts, as discussed at the end of this section.

the equivalence of the EMM notion of Definition 2.4.2 and that of this section's opening discussion. As preparation, we review two generally useful probabilistic constructs, starting with a change-of-measure formula.

Associated with every probability $Q \in \mathcal{Q}$ is the **density**[8] dQ/dP, which is the strictly positive unit-mean random variable

$$\frac{dQ}{dP}(\omega) = \frac{Q(\{\omega\})}{P(\{\omega\})}, \quad \omega \in \Omega,$$

and the **conditional density process**

$$\xi_t = \mathbb{E}_t\left[\frac{dQ}{dP}\right], \quad t = 0, 1, \ldots, T.$$

Since $\mathcal{F}_T = 2^\Omega$, $\xi_T = dQ/dP$ and therefore ξ determines Q and conversely. By the law of iterated expectations (2.1.4), ξ is a unit-mean strictly positive martingale (under P).

LEMMA 2.4.4 *For every $Q \in \mathcal{Q}$ and adapted process x,*

$$\mathbb{E}^Q[x_t] = \mathbb{E}[\xi_t x_t], \quad t = 0, \ldots, T.$$

PROOF For every random variable z, we have

$$\mathbb{E}^Q z = \sum_{\omega \in \Omega} z(\omega)\, Q(\{\omega\}) = \sum_{\omega \in \Omega} \frac{Q(\{\omega\})}{P(\{\omega\})} z(\omega)\, P(\{\omega\}) = \mathbb{E}\left[\frac{dQ}{dP} z\right].$$

The law of iterated expectations and the fact that $x_t \in L_t$ can now be used to argue that

$$\mathbb{E}^Q x_t = \mathbb{E}\mathbb{E}_t\left[\frac{dQ}{dP} x_t\right] = \mathbb{E}\left[\mathbb{E}_t\left[\frac{dQ}{dP}\right] x_t\right] = \mathbb{E}[\xi_t x_t].$$

$\square$

REMARK 2.4.5 For every spot (F, t), Lemma 2.4.4 with $x_t = 1_F$ implies that $Q(F) = \mathbb{E}^Q 1_F = \mathbb{E}[\xi_t 1_F] = \xi(F, t)\, P(F)$. The value $\xi(F, t)$ that ξ_t takes on the event F is, therefore, the likelihood ratio $Q(F)/P(F)$, which is the ratio of the Q-probability of the path on the information tree leading from time zero to spot (F, t) to the P-probability of the same path. The random variable $\xi_T = dQ/dP$ has the same interpretation for terminal spots. $\diamond$

---

[8] Also known as the **Radon–Nikodym derivative** of Q with respect to P.

REMARK 2.4.6 A $Q \in \mathcal{Q}$ defines the density dQ/dP, which is strictly positive and satisfies $\mathbb{E}[dQ/dP] = 1$. Conversely, a strictly positive random variable Z such that $\mathbb{E}Z = 1$ defines a unique $Q \in \mathcal{Q}$ such that $Z = dQ/dP$; it is given by $Q(F) = \mathbb{E}[Z1_F]$ for every event F. $\diamond$

The second probabilistic construction we need to relate SPDs to EMMs is the following multiplicative version of the Doob decomposition. $\mathcal{P}_{++}$ denotes the set of all strictly positive predictable processes, and $\mathcal{M}$ denotes the set of all martingales under P.

LEMMA 2.4.7 *Every strictly positive adapted process* π *admits a unique decomposition of the form*

$$\pi = \pi_0 \rho \xi, \quad \rho \in \mathcal{P}_{++}, \quad \xi \in \mathcal{M}, \quad \xi_0 = 1. \tag{2.4.5}$$

PROOF Given $\pi \in \mathcal{L}_{++}$, let $\rho \in \mathcal{P}_{++}$ be defined recursively by

$$\rho_0 = 1 \quad \text{and} \quad \frac{\rho_t}{\rho_{t-1}} = \frac{\mathbb{E}_{t-1}\pi_t}{\pi_{t-1}}, \quad t = 1, \dots, T. \tag{2.4.6}$$

Define the process ξ so that $\pi = \pi_0 \rho \xi$ and therefore $\xi_0 = 1$. The second equality in (2.4.6) is equivalent to $\xi_{t-1} = \mathbb{E}_{t-1}\xi_t$. This proves the existence of decomposition (2.4.5) and also its uniqueness, since the martingale property of ξ implies that ρ must be given by (2.4.6). $\qquad \square$

Lemmas 2.4.4 and 2.4.7 together reveal the relationship between an SPD π representing a present-value function Π and a corresponding EMM-discount pair (Q, ρ) also representing Π: If $dQ/dP = \xi_T$ and ρ and ξ are as in decomposition (2.4.5), then for every cash flow c,

$$\Pi(c) = \frac{1}{\pi_0}\mathbb{E}\left[\sum_{t=0}^{T} \pi_t c_t\right] = \mathbb{E}\left[\sum_{t=0}^{T} \rho_t \xi_t c_t\right] = \mathbb{E}^Q\left[\sum_{t=0}^{T} \rho_t c_t\right].$$

Here is a more detailed statement of this conclusion.

PROPOSITION 2.4.8 *Suppose* π *is an SPD representing the present-value function* Π. *Let* ρ *be defined by* (2.4.6), *and let*

$$Q(F) = \mathbb{E}[\xi_T 1_F], \ F \in \mathcal{F}_T, \quad \text{where} \quad \xi_T \equiv \frac{1}{\rho_T}\frac{\pi_T}{\pi_0}.$$

Then (Q, ρ) *is an EMM-discount pair representing* Π. *Conversely, suppose* (Q, ρ) *is an EMM-discount pair representing the present-value function* Π. *Let* ξ *denote the conditional density process of* Q *relative to* P, *and let* π_0 *be an arbitrary positive scalar. Then* $\pi = \pi_0 \rho \xi$ *is an SPD representing* Π.

COROLLARY 2.4.9 *For every present-value function* Π, *there exists a unique EMM-discount pair representing* Π.

The relationship $\pi = \pi_0 \rho \xi$ of Proposition 2.4.8 connects the EMM notion of Definition 2.4.2 to the earlier construction of equations (2.4.1) and (2.4.2). To see how, consider any nonterminal spot $(F, t-1)$ with immediate successor spots (F_i, t), $i = 0, \ldots, d$. Bayes' rule and Remark 2.4.5 imply

$$\frac{Q(F_i \mid F)}{P(F_i \mid F)} = \frac{Q(F_i)/Q(F)}{P(F_i)/P(F)} = \frac{\xi(F_i, t)}{\xi(F, t-1)}, \quad i = 0, \ldots, d. \quad (2.4.7)$$

Therefore, if r is the rate process implied by ρ, then equation (2.4.1) applied to all nonterminal spots is the same as $\pi/\pi_- = (\rho\xi)/(\rho\xi)_-$, which is an equivalent recursive expression of the condition $\pi = \pi_0 \rho \xi$.

At the heart of the connection between pricing in terms of an SPD and pricing in terms of an EMM is a probabilistic change-of-measure formula, such as Lemma 2.4.4. We now review two other versions of the change-of-measure formula, each offering another insight on the relationship between pricing in terms of SPDs and EMMs.

We have encountered the pricing of risk in terms of covariances in equation (2.3.6), and in terms of an expectation under an EMM in equation (2.4.3). The connection between the two is made directly by the following conditional change-of-measure formula, which on a finite information tree is an immediate consequence of equations (2.4.7). The result applies in more general stochastic settings, however, where an expression like (2.4.7) is not meaningful. The following alternative argument applies to such settings and offers a chance to practice formal properties of conditional expectations.

LEMMA 2.4.10 *For every* $Q \in \mathcal{Q}$, $V \in \mathcal{L}$ *and time* $t > 0$,

$$\mathbb{E}^{Q}_{t-1} V_t = \mathbb{E}_{t-1}\left[\frac{\xi_t}{\xi_{t-1}} V_t \right] = \mathbb{E}_{t-1} V_t + \mathrm{cov}_{t-1}\left[\frac{\xi_t}{\xi_{t-1}}, V_t \right], \quad (2.4.8)$$

where ξ *is the conditional density process of* Q *relative to* P.

PROOF Let $y = \mathbb{E}_{t-1}[\xi_t V_t]/\xi_{t-1}$. For all $z \in L_{t-1}$, Lemma 2.4.4, the law of iterated expectations, Proposition 2.1.6 and the fact that $yz \in L_{t-1}$ imply

$$\mathbb{E}^{Q}\left[yz\right] = \mathbb{E}\left[\xi_{t} yz\right] = \mathbb{E}\left[\mathbb{E}_{t-1}\left[\xi_{t} yz\right]\right] = \mathbb{E}\left[\mathbb{E}_{t-1}\left[\xi_{t}\right] yz\right] = \mathbb{E}\left[\xi_{t-1} yz\right]$$
$$= \mathbb{E}\left[\mathbb{E}_{t-1}\left[\xi_{t} V_{t}\right] z\right] = \mathbb{E}\left[\mathbb{E}_{t-1}\left[\xi_{t} V_{t} z\right]\right] = \mathbb{E}\left[\xi_{t} V_{t} z\right] = \mathbb{E}^{Q}\left[V_{t} z\right].$$

By Proposition 2.1.3, it follows that $y = \mathbb{E}_{t-1}^{Q} V_{t}$. The second equation in (2.4.8) follows from the definition of the conditional covariance and the fact that $\mathbb{E}_{t-1}\xi_{t}/\xi_{t-1} = 1$. □

Now insert the SPD factorization $\pi = \pi_{0}\rho\xi$, where ρ is a discount process with implied rate process r and ξ is a strictly positive unit-mean martingale, in the pricing equation (2.3.6) to find

$$S_{t-1} = \frac{1}{1+r_{t}}\left(\mathbb{E}_{t-1} V_{t} + \operatorname{cov}_{t-1}\left[\frac{\xi_{t}}{\xi_{t-1}}, V_{t}\right]\right).$$

This is the same as equation (2.4.3) if Q is the probability defined by $\xi_{T} = dQ/dP$, since the term in parentheses is equal to $\mathbb{E}_{t-1}^{Q} V_{t}$ by Lemma 2.4.10.

Yet another way in which state pricing is expressed is through the martingale property of properly discounted gain processes. The connection between EMMs and SPDs in those terms is made through a probabilistic result on the martingale property after a change of measure reviewed below. For simplicity, let us assume that the market X is implemented by $1 + J$ contracts, where contract zero is an MMA, as described in Section 1.8, whose notation we adopt here. Consider any $\pi \in \mathcal{L}_{++}$ with decomposition (2.4.5), where ξ is the conditional density process of $Q \in \mathcal{Q}$. As an SPD, π prices the MMA if and only if the discount process ρ and the short-rate process r are related by (2.4.4), a condition we henceforth assume. By Proposition 2.3.3, π is an SPD if and only if it prices the remaining J contracts with underlying probability P, if and only if ρ prices the same J contracts with underlying probability Q. Therefore, by Proposition 2.3.4, π is an SPD if and only if $G^{\pi} \equiv \pi V + (\pi\delta)_{-} \bullet \mathbf{t}$ is a martingale (under P); and Q is an EMM if and only if[9] $G^{\rho} \equiv \rho V + (\rho\delta)_{-} \bullet \mathbf{t}$ is a Q-**martingale**; that is, a martingale under Q. The fact that π is an SPD if and only if Q is an EMM corresponds to the purely probabilistic fact that G^{π} is a P-martingale if and only if G^{ρ} is a Q-martingale, which can be shown directly as a corollary of Lemma 2.4.10.

---

[9] It is this equivalence that justifies the term "martingale measure" for Q.

EXAMPLE 2.4.11 (American call) This example uses some basic facts from martingale theory to provide a dual argument showing the American call late exercise result of Example 1.5.5. Throughout this example, we restrict the time horizon to be the American call maturity $\bar{\tau}$ (rather than T), and we let $\bar{\mathcal{T}}$ denote the set of every stopping time valued in $\{0, \ldots, \bar{\tau}\}$. From Section 2.1, recall that every adapted process has a unique Doob decomposition, whose predictable part defines its compensator. A **submartingale** is an adapted process x with an **increasing** compensator x^p, meaning $\Delta x_t^p \geq 0$ or, equivalently, $\mathbb{E}_{t-1} \Delta x_t \geq 0$ (at every state and time $t > 0$). We will use two basic facts about a submartingale x.

Fact 1. $\tau \in \bar{\mathcal{T}}$ implies $\mathbb{E} x_{\bar{\tau}} \geq \mathbb{E} x_\tau$.

PROOF Define the martingale $M \equiv x - x^p$ and given any $\tau \in \bar{\mathcal{T}}$, the process $\phi_t = 1_{\{\tau < t \leq \bar{\tau}\}}$ (with $1_\emptyset \equiv 0$). Since ϕ is predictable, $\phi \bullet M$ is a zero-mean martingale and $0 = \mathbb{E}(\phi \bullet M)_T = \mathbb{E}[M_{\bar{\tau}} - M_\tau]$. Therefore $\mathbb{E} M_{\bar{\tau}} = \mathbb{E} M_\tau$ and, since x^p is increasing, $\mathbb{E} x_{\bar{\tau}} \geq \mathbb{E}[M_{\bar{\tau}} + x_\tau^p] = \mathbb{E} x_\tau$. $\square$

Fact 2. Suppose the function $f \colon \mathbb{R} \to \mathbb{R}$ is increasing and convex. Then $f(x)$ is also a submartingale.

PROOF The submartingale property $\mathbb{E}_{t-1} \Delta x_t \geq 0$ and the assumption that f is increasing imply $f(\mathbb{E}_{t-1} x_t) \geq f(x_{t-1})$. The claim is therefore a corollary of **Jensen's inequality**

$$\mathbb{E}_{t-1} f(x_t) \geq f(\mathbb{E}_{t-1} x_t).$$

Using only the convexity of f, note that the right derivative of f at $\mathbb{E}_{t-1} x_t$ defines $d_{t-1} \in L_{t-1}$ as the slope of a supporting line of the graph of f at $\mathbb{E}_{t-1} x_t$ and therefore $f(x_t) \geq f(\mathbb{E}_{t-1} x_t) + d_{t-1}(x_t - \mathbb{E}_{t-1} x_t)$. Applying $\mathbb{E}_{t-1}$ on both sides results in Jensen's inequality. $\square$

Consider now the American call of Example 1.5.5 with the underlying stock paying no dividends up to $\bar{\tau}$. Assume also that there is a traded MMA with a nonnegative interest-rate process, which means that the corresponding discount process ρ is decreasing ($s > t$ implies $\rho_s \leq \rho_t$). For every EMM Q, we will show that $\bar{\tau}$ maximizes the corresponding present value (using the notation $x^+ \equiv \max\{x, 0\}$):

$$\mathbb{E}^Q \left[\rho_{\bar{\tau}} (S_{\bar{\tau}} - K)^+ \right] = \max_{\tau \in \bar{\mathcal{T}}} \mathbb{E}^Q \left[\rho_\tau (S_\tau - K)^+ \right].$$

By Theorem 1.5.2, this proves that *not* exercising the call prior to maturity is a dominant choice. Fixing the reference EMM Q, let us select the underlying probability P to equal Q. The no-dividends assumption implies that ρS is a martingale (up to the assumed time-horizon $\bar{\tau}$), and since ρK is a predictable decreasing process, $\rho S - \rho K$ is a submartingale. The function that takes the positive part is convex and increasing. By Fact 2, $x \equiv \rho (S - K)^+$ is a submartingale, and by Fact 1, $\mathbb{E}x_{\bar{\tau}} \geq \mathbb{E}x_{\tau}$. ◇

2.5 Predictable Representations

In linear algebra it is often convenient to introduce a linear basis and refer to vectors by their representation relative to the reference basis. In what is essentially the same idea applied at each node of the information tree, in this section, we introduce processes representing risk sources that span all uncertainty. We show how every adapted process can be represented in terms of its exposure to these risk sources, known as volatility, and we show how a present-value function implies a relationship between expected excess returns and return volatility. The resulting scaffolding is especially useful as we pass from high-frequency models to the continuous-time limit. Unless otherwise indicated, expectations, variances, covariances and the martingale property are all relative to a given underlying full-support probability P. Recall that $\mathcal{M}_0$ denotes the set of zero-mean martingales and $\mathcal{P}_0$ denotes the set of predictable processes that take the value zero at time zero.

The essential idea behind predictable representations is easiest to see in the one-period case, with a single time-zero spot and $1 + d$ time-one spots. A random variable in this case is any vector in $\mathbb{R}^{1+d}$ and the linear space of zero-mean random variables has dimension d. Let the column vector $b = (b^1, \ldots, b^d)'$ list the elements of an orthonormal basis of the space of all zero-mean random variables with the covariance inner product. By construction, the b^i are uncorrelated to each other and they each have zero mean and unit variance. In this simple case, an $M \in \mathcal{M}_0$ can be represented as $M_1 = \Delta M_1 = \sigma b$ for some row vector $\sigma \in \mathbb{R}^{1 \times d}$, necessarily given by $\sigma = \mathbb{E}\left[\Delta M_1 b'\right]$.

The same construction can be applied over the single period following a nonterminal spot, for any time-horizon T. For simplicity, we assume that every nonterminal spot has exactly $1 + d$ immediate successor spots. Focusing on any nonterminal spot $(F, t - 1)$ and its immediate successor spots (F_i, t), $i = 0, \ldots, d$, we apply the earlier single-period construction conditionally on F. Adding the subscript F, t, we end up with the vector $b_{F,t} = \left(b^1_{F,t}, \ldots, b^d_{F,t}\right)'$, where the $b^i_{F,t}$ are random variables taking the value zero outside F and such that

$$\mathbb{E}\big[b^i_{F,t}\,|\,F\big] = 0 \quad\text{and}\quad \mathbb{E}\big[b^i_{F,t}b^j_{F,t}\,|\,F\big] = \begin{cases} 1 & \text{if } i = j, \\ 0 & \text{if } i \neq j. \end{cases} \tag{2.5.1}$$

For all $M \in \mathcal{M}_0$, $\Delta M_t 1_F = \sigma_{F,t} b_{F,t}$, where $\sigma_{F,t} = \mathbb{E}\big[\Delta M_t b'_{F,t}\,|\,F\big]$.

The bases $b_{F,t}$ are conveniently stitched together by defining the process $B = (B^1, \ldots, B^d)'$, where $B^i_0 = 0$ and $\Delta B^i_t 1_F = b^i_{F,t}$ for every nonterminal spot $(F, t-1)$. Condition (2.5.1) is equivalent to the requirement that $B \in \mathcal{M}^d_0$ and

$$\mathbb{E}_{t-1}\big[\Delta B^i_t \Delta B^j_t\big] = \begin{cases} 1 & \text{if } i = j, \\ 0 & \text{if } i \neq j, \end{cases} \qquad t = 1, \ldots, T. \tag{2.5.2}$$

The spot-by-spot representations of the increments of a martingale $M \in \mathcal{M}_0$ become

$$M = \sigma \bullet B, \quad \sigma \in \mathcal{P}^{1 \times d}_0, \tag{2.5.3}$$

where $\sigma_t = \mathbb{E}_{t-1}\big[\Delta M_t \Delta B'_t\big]$.

We call a $B \in \mathcal{M}^d_0$ satisfying (2.5.2) and such that every $M \in \mathcal{M}_0$ admits a representation of the form (2.5.3) a **dynamically orthonormal basis**.

PROPOSITION 2.5.1 *The underlying filtration is generated by a dynamically orthonormal basis $B \in \mathcal{M}^d_0$ if and only if every nonterminal spot has $1 + d$ immediate successors. In this case every adapted process x can be uniquely represented in the form*

$$x = x_0 + \mu \bullet \mathbf{t} + \sigma \bullet B, \quad \mu \in \mathcal{P}_0, \quad \sigma \in \mathcal{P}^{1 \times d}_0. \tag{2.5.4}$$

The processes μ and σ can be computed in terms of x by

$$\mu_t = \mathbb{E}_{t-1}\big[\Delta x_t\big] \quad\text{and}\quad \sigma_t = \mathbb{E}_{t-1}\big[\Delta x_t \Delta B'_t\big], \quad t = 1, \ldots, T.$$

PROOF Suppose every nonterminal spot has $1 + d$ immediate successors. We have already shown the existence of a dynamically orthonormal basis $B \in \mathcal{M}^d_0$. Representation (2.5.4) follows from the Doob decomposition (2.1.5) of x with $\mu = \Delta x^p$. To show that B generates the underlying filtration, we use induction in t to show that the realization of $B_1, \ldots, B_t$ reveals the realized time-t spot. For $t = 0$, the claim is vacuously true. Suppose it is true for $t - 1$ and ω is the realized state, corresponding to spots $(F, t-1)$ and (G, t), where $\omega \in G \subseteq F$. The inductive hypothesis means that $(F, t-1)$ is the unique spot that is consistent with $B_s(\omega)$ for $s < t$. Let x be the adapted process that takes the value zero everywhere except for $x(G, t) = 1$. By representation (2.5.4) of x, we can write $\Delta x_t = \mu_t + \sigma_t \Delta B_t$, where μ_t and σ_t take a constant value on F, revealed by $B_s(\omega)$ for $s < t$. If in addition we know $B(G, t)$, then the value $x(G, t)$ is also revealed, and hence the identity of the spot (G, t). This completes the inductive step. The necessity of the condition that each nonterminal spot has exactly $d + 1$ immediate successor spots follows by a dimensionality argument that is left to the reader. □

We henceforth fix a reference dynamically orthonormal basis $B \in \mathcal{M}_0^d$ generating the underlying filtration. We refer to (2.5.4) as the **predictable representation** or the **dynamics** of x (relative to B). The processes μ and σ are, respectively, the **drift** and **volatility** of x.

Turning our attention back to an arbitrage-free market X, consider a state-price density π (relative to P) and corresponding decomposition $\pi = \pi_0 \rho \xi$, where ρ is a (predictable) discount process and ξ is the conditional density process of the EMM Q associated with π. We are interested in the predictable representation of π, how it relates to the dynamics of ρ and ξ, and finally how it can be used to price a traded contract.

We saw in Section 2.4 that if there is a traded MMA then the short-rate process is $r = -\Delta\rho/\rho$. Let $r \in \mathcal{P}_0$ be defined by this equation, whether an MMA is traded or not. Simple algebra shows that

$$\frac{\Delta\pi}{\pi_-} = \frac{\rho}{\rho_-} \left(\frac{\Delta\rho}{\rho} + \frac{\Delta\xi}{\xi_-} \right) = -\frac{1}{1+r} \left(r - \frac{\Delta\xi}{\xi_-} \right).$$

The last term leads us to the process $(1/\xi_-) \bullet \xi$, which is a zero-mean martingale. Define the process η by the corresponding unique predictable representation:

$$\frac{1}{\xi_-} \bullet \xi = -\eta' \bullet B, \quad \eta \in \mathcal{P}_0^d. \tag{2.5.5}$$

The state-price dynamics can then be expressed as

$$\frac{\Delta\pi}{\pi_-} = -\frac{1}{1+r} \left(r + \eta' \Delta B \right). \tag{2.5.6}$$

By Proposition 2.5.1 and Lemma 2.4.10,

$$\eta_t = -\mathbb{E}_{t-1} \left[\frac{\xi_t}{\xi_{t-1}} \Delta B_t \right] = -\mathbb{E}_{t-1}^Q [\Delta B_t]. \tag{2.5.7}$$

Equivalently,

$$B + \eta \bullet \mathbf{t} \text{ is a } Q\text{-martingale.} \tag{2.5.8}$$

Conversely, the process η determines the probability Q, since representation (2.5.5) can be restated as the recursion

$$\frac{\xi_t}{\xi_{t-1}} = 1 - \eta_t' \Delta B_t, \quad \xi_0 = 1, \tag{2.5.9}$$

which can be multiplied through to obtain

$$\xi_t = \prod_{s=0}^{t} \left(1 - \eta_s' \Delta B_s\right), \quad t = 0, 1, \ldots, T. \qquad (2.5.10)$$

In particular, η determines $\xi_T = dQ/dP$ and hence Q. Not every $\eta \in \mathcal{P}_0^d$ defines a $Q \in \mathcal{Q}$ in this manner, however, since ξ must be strictly positive. In fact, the preceding correspondence between η and Q defines a bijection between $\mathcal{Q}$ and the set

$$\mathcal{H} = \left\{ \eta \in \mathcal{P}_0^d \mid 1 - \eta' \Delta B \in \mathcal{L}_{++} \right\}.$$

In equations (2.4.4), we saw that every rate process $r \in \mathcal{P}_0$ implies a unique discount process ρ (Definition 2.4.2) and vice versa. The preceding construction shows that every $Q \in \mathcal{Q}$ defines a unique $\eta \in \mathcal{H}$ by letting $\eta_t = -\mathbb{E}_{t-1}^{Q}[\Delta B_t]$, and conversely, every $\eta \in \mathcal{H}$ defines a $Q \in \mathcal{Q}$ where $dQ/dP = \xi_T$ is defined by (2.5.10). With $Q \in \mathcal{Q}$ and $\eta \in \mathcal{H}$ so related, we call $\eta \in \mathcal{H}$ a **market-price-of-risk** process if Q is an EMM.

In Sections 2.3 and 2.4, we saw a number of equivalent expressions of what it means for an SPD π or corresponding EMM-discount pair (Q, ρ) to price a given contract (δ, V). One of those is that $G^\rho \equiv \rho V + (\rho \delta)_- \bullet \mathbf{t}$ is a Q-martingale. Let us now formulate the same condition in terms of the corresponding pair $(\eta, r) \in \mathcal{H} \times \mathcal{P}_0$ and the gain-process predictable representation

$$G \equiv V + \delta_- \bullet \mathbf{t} \equiv V_0 + \mu^G \bullet \mathbf{t} + \sigma^G \bullet B. \qquad (2.5.11)$$

A direct calculation of the increment ΔG^ρ shows that

$$G^\rho = \left(\rho\left(\mu^G - rS_- - \sigma^G \eta\right)\right) \bullet \mathbf{t} + \left(\rho\sigma^G\right) \bullet (B + \eta \bullet \mathbf{t}).$$

By (2.5.8), G^ρ is a Q-martingale if and only if the first term, which is the drift of G^ρ under Q, is zero. Summarizing, we have shown that (Q, ρ) prices the contract (δ, V) if and only if

$$\mu^G = rS_- + \sigma^G \eta. \qquad (2.5.12)$$

This argument has the advantage of generalizing to continuous-time formulations, where we cannot refer to a single-period recursion. For a more-direct connection, substitute $\mu_t^G = \mathbb{E}_{t-1}[V_t] - S_{t-1}$ and $\sigma_t^G = \mathbb{E}_{t-1}\left[V_t \Delta B_t'\right]$ into (2.5.12), solve for S_{t-1} and use (2.5.9) and the change-of-measure formula (2.4.8) to recover the risk-neutral pricing recursion (2.4.3).

Assuming S_{t-1} is nonzero everywhere, pricing condition (2.5.12) is often expressed in terms of the return dynamics

$$\frac{\Delta G_t}{S_{t-1}} = \mu_t^R + \sigma_t^R \Delta B_t, \quad \mu^R \in \mathcal{P}_0, \quad \sigma^R \in \mathcal{P}_0^{1 \times d}.$$

In this case, equation (2.5.12) can be restated as

$$\mu^R - r = \sigma^R \eta,$$

which can more directly be viewed as a corollary of the expected excess returns expression (2.3.7) and the SPD dynamics (2.5.6). Thus conditional expected excess returns relative to the risk-free rate over a single period are explained by exposure to the risk sources ΔB_t as measured by the return volatility σ^R. This type of pricing is also known as *factor pricing*, with ΔB_t being the (risk) *factors*, the volatility representing conditional *factor loadings*, and η_t representing the *factor prices*.

In applications it is common to assume that the market is implemented by exogenously specified contracts. Let us adopt the setting of Section 1.8, where X is implemented by $1 + J$ contracts, with contract zero being an MMA defining the short-rate process r and implied discount process ρ. Modifying our earlier notation, let $\delta = \left(\delta^1, \ldots, \delta^J \right)'$ and define $V, G \in \mathcal{L}^J$ analogously, with G having dynamics (2.5.11), where $\mu^G \in \mathcal{P}_0^J$ and $\sigma^G \in \mathcal{P}_0^{J \times d}$. Given $Q \in \mathcal{Q}$ and corresponding $\eta \in \mathcal{H}$, Q is an EMM if and only if it prices contracts 1 through J, a condition that is equivalent to $\eta \in \mathcal{H}$ satisfying equation (2.5.12).

REMARK 2.5.2 The existence of some $\eta \in \mathcal{P}_0$ satisfying equation (2.5.12) does not exclude all arbitrage opportunities (the positivity part of the definition of $\mathcal{H}$ is essential), and is instead equivalent to the weaker condition of no MMA arbitrage:

$$\text{For all } \theta \in \mathcal{P}_0^{1 \times J}, \quad \theta \sigma^G = 0 \implies \theta \left(\mu^G - r S_- \right) = 0. \tag{2.5.13}$$

An orthogonal decomposition at each spot gives

$$\mu^G - r S_- = \sigma^G \eta + \varepsilon', \quad \varepsilon \sigma^G = 0, \tag{2.5.14}$$

for some $\eta \in \mathcal{P}_0^d$ and $\varepsilon \in \mathcal{P}_0^{1 \times J}$. Applying (2.5.13) with $\theta = \varepsilon$ gives $\varepsilon \left(\mu^G - r S_- \right) = 0$ and therefore $\varepsilon \left(\sigma^G \eta + \varepsilon' \right) = 0$. Since $\varepsilon \sigma^G \eta = 0$, we have $\varepsilon \varepsilon' = 0$. This proves that $\varepsilon = 0$ and (2.5.14) reduces to (2.5.12). The converse claim is immediate. $\diamond$

For a typical arbitrage-pricing application, consider a market maker who can trade in the $1 + J$ contracts implementing the arbitrage-free market

X, and sells contract (δ^*, V^*) to a customer at time zero. We assume that (δ^*, V^*) is synthetic in X, that is, there exists a trading strategy (θ^0, θ) such that $(\delta^*, V^*) = (\delta^\theta, V^\theta)$. The customer could in principle bypass the market maker by following trading strategy (θ^0, θ), but the idea is that the customer is less suited to carry out the necessary transactions. At time zero, the market maker charges V_0^* for the contract plus whatever fee competition for market-making activity allows and hedges the sale by purchasing the synthetic contract $(\delta^\theta, V^\theta)$, resulting in a net cash flow that is just the time-zero fee charged. The hedge is perfect within the market maker's model. (Of course, there is always the risk that the model does not comply with reality.)

The market maker's pricing model consists of the short-rate process r and the gain or return dynamics specified earlier, which imply a market-price-of-risk process η and associated EMM Q. This data can be used to recursively compute V^* as the present-value process of δ^*. The requirement that $(\delta^*, V^*) = (\delta^\theta, V^\theta)$ dictates, for all $t > 0$, the MMA position θ_t^0 in terms of the positions θ_t in the remaining J contracts by matching either cum-dividend values, $V_t^* = \theta_t^0 (1 + r_t) + \theta_t V_t$, or ex-dividend values, $S_{t-1}^* = \theta_t^0 + \theta_t S_{t-1}$, where $S^* \equiv V^* - \delta^*$ and $S \equiv V - \delta$. The MMA position is therefore given by

$$\theta_t^0 = S_{t-1}^* - \theta_t S_{t-1} = \frac{V_t^* - \theta_t V_t}{1 + r_t}, \quad t > 0; \quad \theta_0^0 = 0. \qquad (2.5.15)$$

The trading strategy θ in the remaining J contracts is computed to eliminate volatility period by period. To see why, define the gain-process predictable representation

$$G^* \equiv V^* + \delta_-^* \bullet \mathbf{t} \equiv V_0^* + \mu^* \bullet \mathbf{t} + \sigma^* \bullet B \qquad (2.5.16)$$

and use the budget equation for θ to find

$$V^* - \theta V = \text{ predictable term } + \left(\sigma^* - \theta \sigma^G \right) \bullet B.$$

By (2.5.15), $V^* - \theta V$ is predictable and therefore

$$\sigma^* = \theta \sigma^G, \qquad (2.5.17)$$

which must have a solution in θ in order for (δ^*, V^*) to be synthetic in X. At time $t - 1$, the market maker enters the trade $\theta_t - \theta_{t-1}$ in the J contracts, with the MMA acting as the clearing account, whose new balance is dictated by the budget equation.

Assuming the market is arbitrage-free, a corollary of this discussion is that every synthetic contract is uniquely generated by a trading strategy if and only if the rows of the $J \times d$ matrix $\sigma^G(\omega, t)$ are linearly independent for all $(\omega, t) \in \Omega \times \{1, \ldots, T\}$. In this case, let us call the contracts $(\delta^1, V^1), \ldots, (\delta^J, V^J)$ **dynamically independent**. By Proposition 1.7.7, another corollary is that the market is complete if and only if the rank of $\sigma^G(\omega, t)$ is d for all $(\omega, t) \in \Omega \times \{1, \ldots, T\}$. In particular, the MMA (δ^0, V^0) together with the dynamically independent contracts $(\delta^1, V^1), \ldots, (\delta^J, V^J)$ implement a complete market if and only if $J = d$ and $\sigma^G(\omega, t)$ is invertible for all $(\omega, t) \in \Omega \times \{1, \ldots, T\}$. In this case, for every $j \in \{1, \ldots, J\}$, the contract (δ^j, V^j) is necessarily **everywhere risky**, meaning that for every time $t > 0$ and every nonempty event $F \in \mathcal{F}_t$, the random variable $V_t^j 1_F$ is not $\mathcal{F}_{t-1}$-measurable. The number of contracts required to implement a complete market is dictated by the underlying filtration.

THEOREM 2.5.3 *Suppose that every nonterminal spot has $1 + d$ immediate successors. An arbitrage-free market is complete if and only if it can be implemented by a money-market account and d dynamically independent everywhere risky contracts.*

PROOF The "if" part follows from our earlier discussion. Conversely, suppose the market X is complete and arbitrage-free, and let (Q, ρ) be the corresponding unique EMM-discount pair implying the short-rate process r and market-price-of-risk process η. Market completeness implies that a contract is traded (in X) if and only if it is priced by (Q, ρ). Define the MMA (δ^0, V^0) through r in (1.8.2) and the everywhere risky contracts $(\delta, V) \in \mathcal{L}^{d \times 2}$ by letting $\delta_- \equiv 0$ and $G^\rho = \rho V \equiv B + \eta \bullet t$. By (2.5.8), G^ρ is a Q-martingale and therefore, by Proposition 2.3.4, (Q, ρ) prices contracts (δ, V), which are therefore traded. The market implemented by the contracts (δ^0, V^0) and (δ, V) is included in X and is complete and is therefore equal to X. □

2.6 Independent Increments and the Markov Property

Assuming every spot has at least two immediate successors, the number of spots rises exponentially with the number of periods, rendering intractable any method that requires spot-by-spot computation. Applications typically address this issue by relying on a so-called Markovian structure, where the entire history is summarized by a relatively low-dimensional state vector.

We outline the basic idea in the setting of Section 2.5, where the underlying filtration is generated by the d-dimensional dynamically orthonormal basis B, relative to the full-support probability P. We further assume, throughout this section, that B has **independent increments** (relative to P): For every time $t > 0$ and function $f : \mathbb{R}^d \to \mathbb{R}$,

$$\mathbb{E}_{t-1}\big[f(\Delta B_t)\big] = \mathbb{E}\big[f(\Delta B_t)\big]. \tag{2.6.1}$$

Note that by Proposition 1.1.4 and Corollary 2.1.5, B has independent increments if and only if for every time $t > 0$, the algebras $\sigma(\Delta B_t)$ and $\mathcal{F}_{t-1} = \sigma(\Delta B_1, \ldots, \Delta B_{t-1})$ are stochastically independent. While condition (2.6.1) is all we need to apply the independent-increments property in this section, Proposition 2.1.2 explains the terminology: B has independent increments if and only if the random variables $\Delta B_1, \ldots, \Delta B_T$ are stochastically independent. Example 2.1.8 gives a prototypical independent increments process. As in that example, an independent increments process B such that $\mathbb{E}\Delta B_t = 0$ for all $t > 0$ is necessarily a martingale, but an arbitrary martingale need not have independent increments.

A time-t spot corresponds to a specific realization of one of the $(1+d)^t$ possible histories of B up to that spot. The idea is to hypothesize that there is a k-dimensional random variable Z_t whose value summarizes all that is relevant of the history of B up to t, where k is a computationally manageable positive integer. Fixing an initial value $Z_0 \in \mathbb{R}^k$, we assume that Z is the k-dimensional adapted process defined recursively by

$$\Delta Z_t = a_t(Z_{t-1}) + b_t(Z_{t-1})\,\Delta B_t, \quad t = 1, \ldots, T, \tag{2.6.2}$$

for given functions $a_t : \mathbb{R}^k \to \mathbb{R}^k$ and $b_t : \mathbb{R}^k \to \mathbb{R}^{k \times d}$, and is therefore a **Markov process** (relative to P): For every time $t > 0$ and function $f : \mathbb{R}^k \to \mathbb{R}$,

$$\mathbb{E}_{t-1} f(Z_t) = \mathbb{E}\big[f(Z_t) \mid Z_{t-1}\big]. \tag{2.6.3}$$

The **Markov property** (2.6.3), which is equivalent to the $\sigma(Z_{t-1})$-measurability of $\mathbb{E}_{t-1} f(Z_t)$, formalizes the idea that every statistic of Z_t can be calculated knowing only Z_{t-1} rather than the entire path of B up to time $t - 1$. The claim that Z is a Markov process relative to P is a corollary of the following more general result.

LEMMA 2.6.1 *Suppose B has independent increments relative to P, and $Q \in \mathcal{Q}$ has a conditional density process $\xi_t = \mathbb{E}_t\big[dQ/dP\big]$ that satisfies, for some $\eta_t : \mathbb{R}^k \to \mathbb{R}^d$,*

$$\frac{\Delta \xi_t}{\xi_{t-1}} = -\eta_t(Z_{t-1})'\,\Delta B_t, \quad t = 1, \ldots T; \quad \xi_0 = 1.$$

Then Z is a Markov process relative to Q.

PROOF Consider any time $t > 0$, vector $z \in \mathbb{R}^k$ and function $f : \mathbb{R}^k \to \mathbb{R}$. Using the conditional change of measure formula of Lemma 2.4.10 and recursion (2.6.2) for Z, note that on the event $\{Z_{t-1} = z\}$ the conditional expectation $\mathbb{E}^Q_{t-1}[f(Z_t)]$ is equal to

$$\mathbb{E}_{t-1}\left[\left(1 - \eta(z)'_t \Delta B_t\right) f(z + a_t(z) + b_t(z) \Delta B_t)\right].$$

Since ΔB_t is independent of $\mathcal{F}_{t-1}$, the above quantity is constant on $\{Z_{t-1} = z\}$ and therefore so is $\mathbb{E}^Q_{t-1}[f(Z_t)]$. This proves that $\mathbb{E}^Q_{t-1}[f(Z_t)]$ is $\sigma(Z_{t-1})$-measurable. □

Let $\mathcal{Z}_t \equiv \{Z_t(\omega) \mid \omega \in \Omega\}$ denote the set of all possible time-t Markov states. For $x \in \mathcal{L}$, we abuse notation and write $x_t = x_t(Z_t)$ to mean that there exists a function $x_t : \mathcal{Z}_t \to \mathbb{R}$ such that $x_t(\omega) = x_t(Z_t(\omega))$ for all $\omega \in \Omega$, or, equivalently, that x_t is $\sigma(Z_t)$-measurable. Similarly, if x is a predictable process, we write $x_t = x_t(Z_{t-1})$ to express the condition that there exists a function $x_t : \mathcal{Z}_{t-1} \to \mathbb{R}$ such that $x_t(\omega) = x_t(Z_{t-1}(\omega))$ for all $\omega \in \Omega$ (a condition that is equivalent to the $\sigma(Z_{t-1})$-measurability of x_t). Even though x_t denotes two separate mathematical objects, the meaning is clear from the context.

With these conventions, suppose that the market is arbitrage-free and is implemented by the MMA and J contracts, as specified in Section 2.5, with the additional restriction that for every time $t > 0$,

$$\left(r_t, \delta_t, V_t, \mu_t^G, \sigma_t^G\right) = \left(r_t(Z_{t-1}), \delta_t(Z_t), V_t(Z_t), \mu_t^G(Z_{t-1}), \sigma_t^G(Z_{t-1})\right),$$

and therefore $S_t \equiv V_t - \delta_t = S_t(Z_t)$. A market-price-of-risk process $\eta \in \mathcal{H}$, defining an EMM Q, is specified as a predictable solution to equation (2.5.12) such that $1 - \eta'\Delta B$ is strictly positive. It is not hard to see that we can select $\eta \in \mathcal{H}$ to satisfy $\eta_t = \eta_t(Z_{t-1})$ for all $t > 0$. Assuming such a selection, it follows from Lemma 2.6.1 that Z is a Markov process relative to Q, as well as P.

As in the last section, consider the pricing and replication of a traded contract (δ^*, V^*), with the added assumption that $\delta_t^* = \delta_t^*(Z_t)$ for all t. Pricing the contract recursively using the EMM Q and the Markov property of Z relative to Q gives

$$V^*_{t-1} = \delta^*_{t-1}(Z_{t-1}) + \frac{\mathbb{E}^Q\left[V_t^*(Z_t) \mid Z_{t-1}\right]}{1 + r_t(Z_{t-1})} = V^*_{t-1}(Z_{t-1}).$$

Since $V_T^* = \delta_T^*(Z_T)$, it follows that $V^* = V_t^*(Z_t)$ for all t. Using the notation (2.5.16) for the corresponding gain process, we also have $\mu_t^* = \mu_t^*(Z_{t-1})$ and $\sigma_t^* = \sigma_t^*(Z_{t-1})$. A trading strategy that replicates the contract (δ^*, V^*) can be selected to have an analogous Markovian structure, thanks to equations (2.5.17) and (2.5.15).

This type of Markovian scaffolding extends to our earlier discussion of option pricing. The following example illustrates the basic idea.

EXAMPLE 2.6.2 (American call) Assuming the above Markovian structure, we revisit the American call of Examples 1.5.5 and 2.4.11, but with the underlying stock potentially paying dividends, so that early exercise at some spot can be dominant. For simplicity, we assume a complete market with EMM Q. The option's value process and an associated dominant exercise time can be determined by solving the backward recursion

$$V_t^*(z) = \max\left\{ S_t(z) - K, \frac{\mathbb{E}^Q\left[V_{t+1}^*(Z_{t+1}) \mid Z_t = z\right]}{1 + r_{t+1}(z)} \right\}, \quad V_{\bar{\tau}+1}^*(z) \equiv 0,$$

where z ranges over all possible values of the Markov state. The idea is that at time t and Markov state z, assuming the option has not been already exercised, it is optimal (in the sense of present-value maximization) to exercise the option and collect $S_t(z) - K$ if the maximum is achieved by the first term and it is optimal to keep the option alive if the maximum is achieved by the second term. Define the stopping time

$$\tau^* \equiv \min\left\{ t \mid V_t^*(Z_t) = (S_t(Z_t) - K)^+ \right\}$$

and for any stopping time $\tau \in \bar{\mathcal{T}}$, let V^τ be the present-value process of the dividend process $\delta_t^\tau \equiv (S_t - K)^+ 1_{\{\tau = t\}}$. It is left as an exercise to verify (using a backward-in-time inductive argument) that $V_t^\tau \leq V_t^*$, with equality if $\tau = \tau^*$. This confirms that τ^* maximizes present value and V^* represents the option's value process in the sense of Section 1.5. ◇

2.7 A Glimpse of the Continuous-Time Theory

We have so far taken the unit of time to coincide with a period, which implies that the time horizon T is the same as the number of periods N. For this section, we fix the time horizon T in a given time unit, which we call a year, and discuss approximations and associated simplifications that arise as the number of periods N becomes very large. Continuous-time models of interest are idealized representations of such large discrete models in the sense that the quantitative predictions of the discrete and continuous models are very close provided N is large enough. Theoretical support for this claim is provided by limit theorems as N goes to infinity. The rigorous details of both the continuous-time limiting model and arguments of convergence are highly technical and well beyond the scope of this presentation. Moreover, infinite models bring into scope set-theoretic esoteric aspects of the theory that are far removed from applications. An introduction founded on this chapter's finite information tree analysis should help demystify some of the continuous-time tools, as well as provide a basis for prioritizing aspects of the continuous-time theory that are most relevant for economic theory.

While one can embed an arbitrary sequence of discrete risks into a continuous-time model, tractability benefits derive from the assumption that uncertainty evolves as a sequence of risks that are in a sense small or infinitesimal in the continuous-time limit. The fundamental building blocks for constructing such high-frequency sequences of small risks are two types of stochastic process: Brownian motion and the Poisson process. The incremental change of either type of process over an infinitesimal time interval represents a small risk, but for opposing reasons. For Brownian motion, a change is certain to happen, but the magnitude of the change is infinitesimal. For a Poisson process, the change is fixed at a unit amount, but the probability of change is infinitesimal. Mixing of Brownian motions and Poisson processes can be used to construct an arbitrary Lévy process,[10] which can be thought of as a representation of an arbitrary sequence of independent identically distributed small risks in high frequency. Lévy processes generalize to classes of so-called semimartingales,[11] where the conditional moments of each small risk can be time varying and path dependent, presenting a rich palette for formulating statistically testable stochastic models. This being a pedagogical introduction, we focus on the one-dimensional Brownian case.

The set of **times** for this section is $\{0, h, 2h, 3h, \ldots, Nh\}$, where $h \equiv T/N$ represents the time length of each period in years. The idea is to fix T and consider versions of the model as N goes to infinity and h goes to zero. We proceed informally with a given value of N, but with the understanding that N is big in the following sense. Suppose $h = 10^{-6}$ years and we are only interested in computing quantities to a precision of six decimal places. We therefore think of h as small but not negligible and quantities of higher order, like $h^{1.5} = 10^{-9}$ and $h^2 = 10^{-12}$, as negligible. On the other hand

---

[10] Cinlar (2010) provides a broad introduction to probability theory that includes Lévy processes. Applebaum (2004) provides an overview of Lévy processes with a focus on stochastic integration.

[11] See Jacod and Shiryaev (2003) and Jacod and Protter (2012).

$\sqrt{h} = 10^{-3}$ is 1000 times bigger than h and definitely not negligible. Note that as h goes to zero, $\sqrt{h}$ becomes infinitely larger than h.

The underlying filtration $\{\mathcal{F}_{nh} \mid n = 0, 1, \ldots, N\}$ is defined analogously to Example 1.1.1, where information available at time $t = nh$ can be thought of as being the outcome of n consecutive coin tosses. Equivalently, we assume that the underlying filtration is generated by a process B, where $B_0 \equiv 0$, and over every period the increment of B can take one of two possible values. As in Section 2.5, we fix an underlying full-support probability P, relative to which B is a zero-mean martingale. Moreover, we assume that $v \equiv \mathbb{E}_t\big[(B_{t+h} - B_t)^2\big]$ is the same for all $t = nh < T$, and B is normalized so that $\mathbb{E}\big[B_T^2\big] = T$. Since $B_T = \sum_n \Delta B_{nh}$, where $\Delta B_{nh} \equiv B_{nh} - B_{(n-1)h}$,

$$\mathbb{E}\big[B_T^2\big] = \sum_{n=1}^{N} \mathbb{E}\big[(\Delta B_{nh})^2\big] + 2 \sum_{m<n} \mathbb{E}\big[\Delta B_{mh} \Delta B_{nh}\big].$$

For $m < n$, $\mathbb{E}\big[\Delta B_{mh} \Delta B_{nh}\big] = \mathbb{E}\big[\Delta B_{mh} \mathbb{E}_{mh} \Delta B_{nh}\big] = 0$. Therefore $T = \mathbb{E}\big[B_T^2\big] = Nv$ or $v = T/N = h$.

Another assumption we make requires that, loosely speaking, the increment $B_{t+h} - B_t$ is small, so that in the limit the paths of B are continuous. For example, if the two scenarios following each spot are equally likely,[12] the fact that $B_{t+h} - B_t$ has zero conditional mean and conditional variance equal to h implies that $B_{t+h} - B_t$ takes the possible values $\pm\sqrt{h}$, and therefore $(B_{t+h} - B_t)^2 = h$. More generally, we assume that for every smooth function $f : \mathbb{R} \to \mathbb{R}$, the following (informally stated) second-order Taylor approximation is valid at all states:

$$f(B_{t+h}) - f(B_t) \approx f'(B_t)(B_{t+h} - B_t) + \frac{1}{2}f''(B_t)(B_{t+h} - B_t)^2. \quad (2.7.1)$$

Using this approximation, we can now compute the limiting distribution of B (as $N \to \infty$ with T fixed). We assume familiarity with the normal (or Gaussian) distribution and its characteristic function (or Fourier transform). Consider approximation (2.7.1) with

$$f(x) \equiv \exp(i\theta x) \equiv \cos(\theta x) + i\sin(\theta x), \quad i \equiv \sqrt{-1}.$$

The fact that f is complex valued presents no problem, since the approximation can be applied on the real and imaginary parts separately. Fixing

---

[12] A probabilistically small risk we are excluding is one where $B_{t+h} - B_t$ takes the value $1 - h$ with probability h and the value $-h$ with probability $1 - h$. In this case, the limiting process is $B_t = C_t - t$, where C_t is a Poisson process with unit arrival rate.

any time s, we compute $m_s(t) \equiv \mathbb{E}_s f(B_t)$ for $t \in [s, T]$, which defines the characteristic function of the distribution of B_t conditionally on time-s information. By the law of iterated expectations, $\mathbb{E}_s \left[B_{t+h} - B_t \right] = 0$ and $\mathbb{E}_s \left[(B_{t+h} - B_t)^2 \right] = h$. Similarly, applying $\mathbb{E}_s$ on both sides of (2.7.1) and using the fact that $f'' = -\theta^2 f$, we find

$$\frac{m_s(t + h) - m_s(t)}{h} \approx -\frac{\theta^2}{2} m_s(t) .$$

Letting h approach zero, we conclude that the derivative of $\log(m_s(t))$ with respect to t equals $-\theta^2/2$. Since $m_s(s) = f(B_s)$, we can integrate from s to t to find that in the limit[13] as $N \to \infty$,

$$\mathbb{E}_s \exp(i\theta(B_t - B_s)) = \exp\left(-\frac{\theta^2}{2} (t - s) \right) .$$

This is the characteristic function of a normal distribution with mean zero and variance $t - s$. We are therefore led to the usual[14] definition of a (one-dimensional[15]) **standard Brownian motion (SBM)** as *an independent-increments stochastic process with continuous paths that start at zero, whose increment over any time interval of length Δ is normally distributed with mean zero and variance Δ.*

To get a sense of what Brownian paths look like, consider first the finite filtration case with the probability P assigning equal mass to every path of B, which as we noted, implies that over every period the increment of B is $\pm\sqrt{h}$, where $h = T/N$. The sum $N\sqrt{h}$ of the absolute value of all these increments along a path of B is known as the path's **total variation**, which in this case blows up as $N \to \infty$. On the other hand, the sum Nh of the square of the increments of B along a path is known as the path's **quadratic variation**, which in this case is equal to T no matter how big N is. These observations are reflected in properties of the

---

[13] The heuristic argument just given can be converted to a rigorous version, but omits an important sense in which the probability distribution of the paths of what we call martingale bases converge to SBM. A rigorous statement and proof of a more-general convergence result for continuous martingales can be found in the exposition of Whitt (2007), while more elaborate extensions to semimartingale limits that allow for Poisson-type jumps is the subject of Jacod and Shiryaev (2003).

[14] Revuz and Yor (1999) and Mörters and Peres (2010) are some excellent accounts of Brownian motion (or Wiener process) and its fascinating properties, some of which are informally discussed in this section.

[15] If the number of immediate successor spots to each nonterminal spot were any positive integer $1 + d$, the limiting version of a dynamically orthonormal basis would be a d-dimensional vector of mutually stochastically independent SBMs.

paths of an SBM B. Every path of B is continuous but has infinite total variation (and hence infinite length) but quadratic variation equal to T. Omitting rigorous definitions, we write the quadratic variation property as $\int_0^t (dB_s)^2 = t$ for all $t \in [0, T]$, or $(dB_t)^2 = dt$. Since $t = \text{var}(B_t)$, the variance of B_t is perfectly revealed by its time-series estimate along any path of B up to time t, no matter how small t is. On the other hand, paths of B do not reveal the mean of B_t, which can only be estimated in an infinite-horizon version of the model through the law of large numbers: $\lim_{t \to \infty} B_t/t = 0$ with probability one. Brownian paths can also be shown to be nowhere differentiable and have further intricate structure not covered here. These properties may cause some suspicion as to whether Brownian motion exists as a rigorous mathematical object. Its existence, however, has been established in a variety of ways within the conventional set-theoretic foundations of real analysis.

A key assumption in our heuristic derivation of the Brownian motion distribution has been that B is a zero-mean martingale satisfying $\mathbb{E}_t\big[(B_{t+h} - B_t)^2\big] = h$ or, equivalently, $B_t^2 - t = \mathbb{E}_t\big[B_{t+h}^2 - (t+h)\big]$. This insight is reflected in **Lévy's characterization of SBM**: *A process B with continuous paths*[16] *starting at zero is an SBM if and only if both B_t and $B_t^2 - t$ are martingales.* In fact, for the "if" part, it is sufficient to assume that B_t and $B_t^2 - t$ are *local* martingales. The distinction between a martingale and a local martingale is not present in the finite information setting, but is essential in the continuous-time limit. A **martingale** over the continuous-time interval $[0, T]$ is an adapted stochastic process M such that $\mathbb{E}|M_t| < \infty$ for every time t, and $s < t$ implies $\mathbb{E}_s M_t = M_s$. A **local martingale** is an adapted process M for which there exists an increasing sequence of stopping times $\tau_1 \leq \tau_2 \leq \cdots$ converging to T with probability one and such that for every n, the process M is a martingale up to time τ_n (meaning that the process that equals M_t on $\{t \leq \tau_n\}$ and M_{τ_n} on $\{t > \tau_n\}$ is a martingale). On a three-date infinite state-space setting, an example[17] of a local martingale that is not a martingale is any adapted process M whose increment $\Delta \equiv M_2 - M_1$ satisfies $\mathbb{E}_1|\Delta| < \infty$ and $\mathbb{E}_1\Delta = 0$, but $\mathbb{E}|\Delta| = \infty$. This type of example fully encapsulates the distinction

---

[16] The continuity of paths is essential here. Suppose instead we applied an analogous argument for a process $B_t = C_t - t$, where the paths of C are right-continuous and constant except for jumps of size $+1$. Think of C_t as counting the number of arrivals over $[0, t]$. Then the martingale property of B characterizes C as a Poisson process with unit arrival rate. This is Watanabe's characterization of Poisson processes.

[17] See example 1.49 of chapter 1 in Jacod and Shiryaev (2003).

between a martingale and a local martingale in discrete time,[18] but the issue is more subtle in the continuous-time limit.

The fact that the paths of an SBM B have infinite length implies that an integral $\sigma \bullet B$ cannot be defined path by path in a conventional way, even if σ is assumed to be bounded. A key insight of Ito's calculus is that $\sigma \bullet B$ must be defined as a genuinely stochastic integral, taking into account the entire filtration and the requirement that σ is predictable. In the Brownian continuous-time limit, predictability of σ can be thought of as the requirement that σ is adapted, since the distinction between information at times t and $t - dt$ is negligible. For example, the SBM B is predictable. Where in discrete time we would recursively express the relationship $M = \sigma \bullet B$ as $\Delta M_t = \sigma_t \Delta B_t$, in continuous time, we write $dM_t = \sigma_t dB_t$, where the differential notation dM_t can be thought of as standing for the infinitesimal increment $M_t - M_{t-dt}$, and analogously for dB_t. We also write $M_t = \int_0^t \sigma_s dB_s$, $t \in [0, T]$. Given that σ is predictable and B is a martingale, in the finite state-space model $\sigma \bullet B$ is a martingale, but for an SBM B, we can only claim that $\sigma \bullet B$ is a *local* martingale. As the heuristic $(dM_t)^2 = \sigma_t^2 (dB_t)^2 = \sigma_t^2 dt$ suggests, in order for the process $M \equiv \sigma \bullet B$ to be well defined as a process of finite quadratic variation, it is necessary that the process $Q_t \equiv \int_0^t \sigma_s^2 ds$ is well defined and finite, in which case, Q is the quadratic variation process of M.

Ito's calculus corresponds to the limiting case of our earlier approximations in terms of h, which can be stated as the exact relationships

$$(dB_t)^2 = dt \quad \text{and} \quad dB_t dt = (dt)^2 = 0, \tag{2.7.2}$$

while the Taylor approximation (2.7.1), for twice continuously differentiable $f : \mathbb{R} \to \mathbb{R}$, becomes

$$df(B_t) = f'(B_t)\,dB_t + \frac{1}{2}f''(B_t)\,dt. \tag{2.7.3}$$

For example, for $f(x) = x^2$, we find $dB_t^2 = 2B_t dB_t + dt$, or

$$B_t^2 - t = 2\int_0^t B_s dB_s. \tag{2.7.4}$$

Since B is predictable, the above integral defines a local martingale. For an SBM B, we already know that $B_t^2 - t$ is a martingale. The argument leading to equation (2.7.4), however, applies to any local martingale B with continuous paths and quadratic variation $(dB_t)^2 = dt$. By Lévy's characterization, *any such local martingale starting at zero must necessarily be an SBM.*

---

[18] See proposition 1.64 of chapter 1 in Jacod and Shiryaev (2003).

The preceding argument leads to an interesting insight on the relationship between volatility and time. Consider the local martingale $M_t \equiv \int_0^t \sigma_s dB_s$, for predictable σ, defining the volatility of M. The process M is close to being a Brownian motion—it is a local martingale with continuous paths, but it does not necessarily have unit volatility: $(dM_t)^2 = \sigma_t^2 dt$. Imagine now that M appears in a movie. Think of t as real time for the duration $[0, T]$ of the movie. Think of u as time shown on a clock within the movie. At real time t, the clock in the movie shows $u = \int_0^t \sigma_s^2 ds$, which is the quadratic variation of M up to time t. With u and t so related, let $W_u = M_t$. For example, if W represents a price process from the perspective of a character in the movie, M represents the same price process from the perspective of the viewer. The movie character observes that W is a local martingale with continuous paths that satisfies $(dW_u)^2/du = (dM_t)^2/(\sigma_t^2 dt) = 1$, and therefore can correctly claim that W is an SBM.

To get a glimpse of phenomena that arise in the continuous-time limit but are not present in a finite-tree model, suppose $\int_0^T \sigma_s^2 ds = \infty$ (for example, $\sigma_t = (T - t)^{-1/2}$). The movie is the complete biography of an immortal god, who correctly argues that with probability one W eventually hits the value one. One way to see this is through the reflection principle for SBM. For each (continuous) path ω of W that crosses 1 before time t, there is an equally likely path $\tilde{\omega}$ that coincides with ω up the first time t_1 that ω hits 1, and thereafter it is the vertical reflection of ω relative to the horizontal line through 1, that is, for $s > t_1$, $\tilde{\omega}_s \equiv 1 - (\omega_s - 1)$. By construction, $\tilde{\omega}_t < 1$ if and only if $\omega_t > 1$. Both ω and $\tilde{\omega}$ have the property that they cross 1. Conversely, every path that crosses 1 before time t belongs to such a pair $\{\omega, \tilde{\omega}\}$. Therefore, the probability that W crosses 1 by time t is twice the probability of the event $\{W_t > 1\} = \{W_t t^{-1/2} > t^{-1/2}\}$, which converges to one as $t \to \infty$ since $W_t t^{-1/2}$ is normally distributed with zero mean and unit variance. Equivalently, the stopping time $\tilde{\tau}$ defined as the first time that W takes the value one is finite with probability one. But then the viewer of the movie must also conclude that there is a $[0, T]$-valued stopping time τ such that $M_\tau = 1$ with probability one. Doob's optional stopping theorem states that if M is a zero-mean martingale, then $\mathbb{E}M_\tau = 0$ for every stopping time τ that is bounded by some deterministic time. Within the movie, we have an example of a zero-mean martingale W and a stopping time $\tilde{\tau}$ such that $\mathbb{E}W_{\tilde{\tau}} = 1$, which is consistent with Doob's theorem, since $\tilde{\tau}$ is not bounded—it can take an arbitrarily long time for W to hit one. The movie viewer's stopping time τ, however, is bounded and therefore it must be the case that M is a local martingale that is not a martingale.

In a market interpretation, we can think of σ as a trading strategy in a contract with gain process B, while within the movie the god holds one share in a contract whose gain process is W. The cumulative gains of σ at real time t match the cumulative gains of the god's strategy at god time $u = \int_0^t \sigma_s^2 ds$. In a finite model, the expected gains from trading a contract whose gain process is a zero-mean martingale are zero. The god who holds the contract until W hits one is guaranteed a unit expected gain. Likewise, the trading strategy σ can guarantee a unit gain by time T. Both strategies are a form of arbitrage. Within the movie, the arbitrage disappears if we place a bound on how negative the god's wealth can become before liquidation. Alternatively, the arbitrage disappears if we assume the god has a finite expected lifetime after all. In real time, an appropriate lower bound on wealth eliminates the arbitrage, and that's the modeling choice we will adopt in Section 2.8. Limiting the god's expected horizon corresponds to the square integrability condition $\mathbb{E}\int_0^T \sigma_s^2 ds < \infty$, which implies that $M \equiv \sigma \bullet B$ is a zero-mean finite-variance martingale. Conversely, every such martingale M can be represented as $M \equiv \sigma \bullet B$ for some square-integrable predictable σ.

In a Brownian model, we represent prices and budget equations in terms of Ito processes. An **Ito process** is a process of the form

$$x_t = x_0 + \int_0^t \mu_s ds + \int_0^t \sigma_s dB_s,$$

for adapted processes μ and σ, respectively referred to as the **drift** and **volatility** of x, where the integrals $\int_0^T |\mu_t|\, dt$ and $\int_0^T \sigma_t^2 dt$ are well defined and finite with probability one. We alternatively denote Ito processes in differential form:

$$dx_t = \mu_t dt + \sigma_t dB_t \quad \text{or} \quad dx = \mu dt + \sigma dB.$$

The Ito decomposition into a drift and a volatility term is unique (modulo zero-probability event technicalities we ignore). To see why, suppose $\mu dt + \sigma dB = \tilde{\mu} dt + \tilde{\sigma} dB$. Take the square of $(\mu - \tilde{\mu})\, dt + (\sigma - \tilde{\sigma})\, dB = 0$ and use (2.7.2) to find $(\sigma - \tilde{\sigma})^2 dt = 0$ and therefore $\sigma = \tilde{\sigma}$ (essentially), which in turn implies $\mu = \tilde{\mu}$ (essentially). For an adapted process θ, the integral $\theta \bullet x$ can be expressed as

$$\int_0^t \theta_s dx_s = \int_0^t \theta_s \mu_s ds + \int_0^t \theta_s \sigma_s dB_s \quad \text{or} \quad \theta dx = \theta \mu dt + \theta \sigma dB,$$

provided $\int_0^T |\theta_t \mu_t| \, dt$ and $\int_0^T (\theta_t \sigma_t)^2 \, dt$ are well defined and finite with probability one.

The type of second-order Taylor approximation that led to identity (2.7.3) applies to any Ito process x and time-dependent function $f \colon [0, T] \times \mathbb{R} \to \mathbb{R}$ that is continuously differentiable with respect to time and twice continuously differentiable with respect to the second argument:

$$df(t, x_t) = \frac{\partial f(t, x_t)}{\partial t} dt + \frac{\partial f(t, x_t)}{\partial x} dx_t + \frac{1}{2} \frac{\partial^2 f(t, x_t)}{\partial x^2} (dx_t)^2, \quad (2.7.5)$$

where $(dx_t)^2 = \sigma_t^2 dt$, according to the Ito multiplication rules (2.7.2). This is known as **Ito's formula** or **lemma** and extends to multidimensional Ito processes (and beyond that, to general semimartingales). We will not go further in this direction here, except state an important special case, known as the **integration by parts** formula: If x and y are Ito processes, then

$$d(x_t y_t) = y_t dx_t + x_t dy_t + dx_t dy_t.$$

By the Ito multiplication rules (2.7.2), if x and y have volatility σ^x and σ^y, respectively, then $dx_t dy_t = \sigma_t^x \sigma_t^y dt$. For example, for $x = y = B$, we recover[19] identity (2.7.4). For strictly positive x and y, the following version of integration by parts is often more convenient to use:

$$\frac{d(xy)}{xy} = \frac{dx}{x} + \frac{dy}{y} + \frac{dx}{x} \frac{dy}{y}.$$

We close this section with some comments on the interchange of limits and expectations which will arise in our discussion of state pricing in Section 2.8. Consider first the example of the state space $\Omega = (0, 1]$ with the underlying probability P being the uniform distribution: $0 < a < b \le 1$ implies $P((a, b]) = b - a$. The sequence of random variables $X_n(\omega) \equiv n 1_{\{\omega \le 1/n\}}$ converges to zero for every $\omega \in \Omega$, yet $\mathbb{E} X_n = 1$ for all n. Unlike the finite-Ω case, we cannot freely interchange limits and

---

[19] Integration by parts can be used inductively to show that every polynomial in (t, x) satisfies Ito's formula, which can then form the basis for proving the Ito formula as stated above. The idea is to stop the process x before it leaves a compact interval and then uniformly approximate f and the relevant derivatives on that interval by polynomials.

expectation. An essential positive result in this regard, which is not far from the way expectations are defined, is the **monotone convergence theorem**: *For all $[0, \infty)$-valued random variables $X, X_1, X_2, \ldots$, if $X_n \uparrow X$ then* $\mathbb{E}X_n \uparrow \mathbb{E}X$ (where $X_n \uparrow X$ means that, with probability one, $X_{n+1} \geq X_n$ for all n, and $\lim_{n \to \infty} X_n = X$). The conclusion applies even if $\mathbb{E}X = \infty$. The monotone convergence theorem implies **Fatou's lemma**.

LEMMA 2.7.1 (Fatou) *If $X_n \geq 0$ for all n, then*

$$\liminf_n \mathbb{E}X_n \geq \mathbb{E}\liminf_n X_n.$$

PROOF Let $Y_n = \inf_{k \geq n} X_k$ and note that $0 \leq Y_n \uparrow Y \equiv \liminf_n X_n$. For all $m \geq n$, $\mathbb{E}X_m \geq \mathbb{E}Y_n$ and therefore $\inf_{m \geq n} \mathbb{E}X_m \geq \mathbb{E}Y_n$. Taking the limit as $n \to \infty$ and using the monotone convergence theorem, we conclude that $\liminf_n \mathbb{E}X_n \geq \mathbb{E}Y$. □

2.8 Brownian Market Example

As a simple example, in this section we discuss an arbitrage-free market with the underlying filtration generated by an SBM B over the continuous time interval $[0, T]$. In other words, the information available at time t is the realization of the path of B up to time t. The process B is one-dimensional, which means that the filtration can be thought of as a high-frequency limiting version of the binomial filtration of Example 1.1.1. The extension to multidimensional B is mainly a matter of notation and will not be given here. The arguments that follow are for the most part similar to finite-information counterparts we have encountered already, with simplifications resulting due to small-risk approximations which are codified as exact infinitesimal relationships by the Ito calculus. As in Section 2.7, we omit some mathematical details. For example, the mere definition of the information generated by the SBM B requires notions of σ-algebras and measurability, as well as an underlying probability measure that defines the expectation operator $\mathbb{E}$. The hope is that once the main economic arguments are established, and the formalities are tied to finite-tree constructs, the more-advanced literature that includes all these mathematical details will make a lot more sense.

A general continuous-time cash flow on the given filtration can be defined analogously to the way cumulative distribution functions are used to specify probability distributions on the real line. This being a simple introduction, we instead consider a special type of market in which all cash flows consist of just two lump-sum payments at times 0 and T. More precisely, a **cash flow** is any pair $(c_0, c_T) \in \mathbb{R} \times L_2$, where L_2 is the set of every random variable z such that $\mathbb{E}[z^2] < \infty$ (with the usual convention of identifying any two random variables x and y such that $\mathbb{E}[(x - y)^2] = 0$). If x is a traded cash flow, the time-0 payment x_0 is used to purchase an initial portfolio of value $-x_0$, which is then updated through a trading strategy up to time T, when the terminal portfolio value x_T is liquidated. Every trading strategy is assumed to be self-financing, meaning that no cash is generated or injected after the initial purchase and prior to the terminal liquidation.

The market is implemented by two contracts: a money-market account and a stock. The **money-market account (MMA)** has a value process identically equal to one unit of account, let's call it a dollar, and a continuously compounded constant **interest rate** $r \in \mathbb{R}$. A dollar in the MMA at time t generates interest $r\,dt$ at time $t + dt$. Equivalently, a dollar invested in the MMA at time zero, with all interest continuously reinvested, results in a time-t account balance of e^{rt} dollars. The **stock** has a price process S which follows geometric Brownian motion with drift: For given constants $a \in \mathbb{R}$ and $S_0, \sigma \in (0, \infty)$,

$$S_t = S_0 \exp(at + \sigma B_t).$$

The assumption that σ is positive is just a convention, since the implications of the model are invariant to replacing B with the SBM $-B$. The stock's **dividend yield** is a constant $y \in \mathbb{R}$, meaning that a share purchased at time t pays $y\,dt$ stock shares (or $y S_t dt$ dollars) in dividends at $t + dt$. Equivalently, if a share of the stock is purchased at time zero and all dividends are reinvested in the stock, then the time-t stock position is e^{yt} shares, which is worth $e^{yt} S_t$ dollars. (Note that by this definition, we could have called r the dividend yield of the MMA.) Since interest and dividends are paid out continuously, the cum and ex-dividend price processes for each contract are the same. Ito's formula (2.7.5) implies that S is an Ito process satisfying

$$\frac{dS_t}{S_t} + y\,dt = \mu\,dt + \sigma\,dB_t, \quad \mu \equiv a + y + \frac{1}{2}\sigma^2. \tag{2.8.1}$$

A trading strategy takes the form of a pair of predictable processes (θ^0, θ), where θ_t^0 is the time-t dollar balance in the MMA and θ_t is the the number of stock shares held at time t. The gain processes for the two contracts are defined analogously to the finite case by

$$G_t^0 \equiv 1 + rt \quad \text{and} \quad G_t \equiv S_t + \int_0^t y S_u \, du.$$

The trading strategy (θ^0, θ) is **self-financing** if it satisfies the **budget equation** expressing the fact that the portfolio value is equal to the initial portfolio value plus accumulated gains from trading:

$$\theta_t^0 + \theta_t S_t = \theta_0^0 + \theta_0 S_0 + \int_0^t \theta_u^0 \, dG_u^0 + \int_0^t \theta_u \, dG_u. \qquad (2.8.2)$$

Stated more simply, the self-financing condition is

$$d(\theta^0 + \theta S) = \theta^0 r \, dt + \theta(dS + yS \, dt).$$

The trading strategy (θ^0, θ) is **admissible** if it is self-financing (implying the integrals of the budget equation are well defined) and there exist constants $\bar{\theta}^0$ and $\bar{\theta}$ such that, with probability one, the following **solvency constraint** is satisfied:

$$\bar{\theta}^0 e^{rt} + \bar{\theta} e^{yt} S_t + \theta_t^0 + \theta_t S_t \geq 0 \quad \text{for all } t \in [0, T]. \qquad (2.8.3)$$

The constraint can be thought of as a simple form of a collateral requirement. A brokerage carries out your trades, but must ensure that at all times your total account balance can cover your trading losses. At time zero, you are asked to deposit $\bar{\theta}^0$ dollars in the MMA and $\bar{\theta}$ shares of the stock, with interest and dividends reinvested in the respective contracts. The solvency constraint requires that the net value of all your positions with the brokerage remains positive. While redundant in our earlier finite state-space model, it is needed here in some form in order to preclude arbitrage trades based on increasingly larger bets of the type discussed in the last section. (As suggested by the discussion in Section 2.7, a variant of this example replaces the solvency constraint with a square-integrability restriction on admissible trading strategies. Either approach supports the example of Black–Scholes option pricing and hedging presented later in this section.)

The budget equation should be invariant relative to a change of the unit of account implied by the strictly positive Ito process π. At time t a dollar is the same as π_t of the new unit of account. The corresponding gain processes in the new unit are

$$G_t^{0\pi} \equiv \pi_t + \int_0^t r\pi_u \, du \quad \text{and} \quad G_t^\pi \equiv \pi_t S_t + \int_0^t y\pi_u S_u \, du.$$

The integration by parts formula implies that the budget equation (2.8.2) is equivalent to

$$\theta_t^0 \pi_t + \theta_t \pi_t S_t = \theta_0^0 \pi_0 + \theta_0 \pi_0 S_0 + \int_0^t \theta_u^0 \, dG_u^{0\pi} + \int_0^t \theta_u \, dG_u^\pi. \qquad (2.8.4)$$

The solvency constraint under the new units is inequality (2.8.3) multiplied through by π_t, and is therefore also invariant under a change of the unit of account.

We say that the cash flow $x = (x_0, x_T)$ is **traded** if there exists an admissible trading strategy (θ^0, θ) such that $x_0 = -(\theta_0^0 + \theta_0 S_0)$ and $x_T = \theta_T^0 + \theta_T S_T$. A **state-price density (SPD)** process in this context is a strictly positive Ito process π such that $\pi_T \in L_2$ and for every traded cash flow (x_0, x_T),

$$\pi_0 x_0 + \mathbb{E}\left[\pi_T x_T\right] \leq 0. \tag{2.8.5}$$

Note that the assumption $x_T, \pi_T \in L_2$ ensures that $\mathbb{E}\left|\pi_T x_T\right| < \infty$, thanks to the Cauchy–Schwarz inequality.

Based on the analysis for the finite case, we can guess the state-price dynamics in terms of the short-rate process r and the market-price-of-risk process η, both of which are constant in this example:

$$\frac{d\pi_t}{\pi_t} = -r\,dt - \eta\,dB_t, \quad \text{where} \quad \eta \equiv \frac{\mu - r}{\sigma}. \tag{2.8.6}$$

In integrated form, the proposed SPD is

$$\pi_t = \pi_0 e^{-rt} \xi_t, \quad \xi_t = \exp\left(-\frac{\eta^2}{2}t - \eta B_t\right). \tag{2.8.7}$$

To confirm that this form of π indeed follows the claimed dynamics, apply integration by parts and Ito's formula, which implies that ξ solves

$$\frac{d\xi_t}{\xi_t} = -\eta\,dB_t, \quad \xi_0 = 1.$$

This Ito decomposition implies that ξ is a local martingale. Given that η is a constant, the process ξ is in fact a martingale, which can be confirmed using (2.8.7) and the fact that B has independent and normally distributed increments.

As in the finite case, the coefficients of the Ito expansion of π are determined by the requirement that $G^{0\pi}$ and G^{π} are local martingales. To see that, suppose $d\pi/\pi = a\,dt + b\,dB$ and use integration by parts:

$$\frac{dG^{0\pi}}{\pi} = (a + r)\,dt + b\,dB_t,$$

$$\frac{dG_t^{\pi}}{\pi S} = (a + \mu + b\sigma)\,dt + (b + \sigma)\,dB.$$

Setting the drift term in the first equation to zero is equivalent to setting $a = -r$, and given that, setting the drift term in the second equation to zero is equivalent to setting $b = -\eta$. In the following proposition we prove that π is a state-price density by showing that the local martingale property of $G^{0\pi}$ and G^π implies the state-price density property of π.

PROPOSITION 2.8.1 *The process π defined in (2.8.7) is a state-price density, for any choice of $\pi_0 \in (0, \infty)$.*

PROOF Suppose the cash flow (x_0, x_T) is generated by the admissible trading strategy (θ_t^0, θ_t), satisfying the solvency constraint (2.8.3). Let

$$\bar{W}_t \equiv \bar{\theta}^0 e^{rt} + \bar{\theta} e^{yt} S_t \quad \text{and} \quad W_t \equiv \bar{W}_t + \theta_t^0 + \theta_t S_t.$$

We remarked earlier that $\pi_0 \xi_t = \pi_t e^{rt}$ is a martingale. Similarly, applying integration by parts to $\zeta_t \equiv \pi_t e^{yt} S_t$, we find that

$$\frac{d\zeta_t}{\zeta_t} = (\sigma - \eta)\, dB_t,$$

which implies that ζ is a martingale (for the same reason that ξ is a martingale). The local martingale property of $G^{0\pi}$ and G^π, the budget equation (2.8.4), and the fact that $\pi \bar{W}$ is a martingale together imply that πW is a local martingale. There exists, therefore, an increasing sequence of stopping times τ_n that converges to T with probability one such that πW is a martingale up to time τ_n and therefore

$$\pi_0 W_0 = \mathbb{E}\left[\pi_{\tau_n} W_{\tau_n}\right], \quad n = 1, 2, \ldots$$

By the solvency constraint, $\pi W \geq 0$. By Fatou's Lemma 2.7.1,

$$\pi_0 W_0 = \liminf_{n \to \infty} \mathbb{E}\left[\pi_{\tau_n} W_{\tau_n}\right] \geq \mathbb{E}\left[\liminf_{n \to \infty} \pi_{\tau_n} W_{\tau_n}\right] = \mathbb{E}\left[\pi_T W_T\right].$$

Subtracting the martingale identity $\pi_0 \bar{W}_0 = \mathbb{E}\left[\pi_T \bar{W}_T\right]$ on both sides and using the fact that $W_0 = \bar{W}_0 - x_0$ and $W_T = \bar{W}_T + x_T$, we obtain the state pricing inequality (2.8.5). $\qquad\square$

The relationship between an EMM Q and SPD π extends to the current context as expected, with the added bonus of Girsanov's theorem. Let the probability Q be defined by $Q(F) = \mathbb{E}[1_F \xi_T]$, $F \in \mathcal{F}_T$, which is possible since ξ is a strictly positive unit-mean martingale. Equivalently, Q is characterized by the condition $\xi_T = dQ/dP$, the density of Q with respect to P. **Girsanov's theorem** states that

$$B_t^Q \equiv B_t + \eta t \text{ is a standard Brownian motion under } Q.$$

To get a sense of why this is true, start with the counterpart of (2.5.8), which is that B^Q is a local martingale under Q. By the Ito multiplication rules (2.7.2), $(dB^Q)^2 = dt$ and therefore B^Q must be an SBM under Q, as we saw in Section 2.7.

REMARK 2.8.2 (Normal density) Girsanov's theorem is closely related to the functional form of the standard normal density function ϕ. Girsanov's theorem implies $\int_{-\infty}^{x} \phi(y)\, dy = Q(B_1^Q \leq x)$. The change of measure formula $\mathbb{E}^Q z = \mathbb{E}[z\xi_1]$, for any $\mathcal{F}_1$-measurable bounded random variable z, and (2.8.7) imply

$$Q(B_1^Q \leq x) = \mathbb{E}\left[1_{\left\{ B_1^Q \leq x \right\}} \xi_1 \right] = \int_{-\infty}^{x} \phi(y - \eta)\, e^{-\frac{\eta^2}{2} - \eta(y-\eta)} dy.$$

Therefore, $\phi(y) = \phi(y - \eta) \exp(\eta^2/2 - \eta y)$. For $\eta = y$, this gives the standard normal density as $\phi(y) = \phi(0) \exp(-y^2/2)$, where $\phi(0)$ is set to $(2\pi)^{-1/2}$ so that $\int_{-\infty}^{\infty} \phi(y)\, dy = 1$. Conversely, the functional form of ϕ can be used to show that under Q, for all $s > t$, $B_s^Q - B_t^Q$ has a normal distribution of zero mean and variance $s - t$, independently of $\mathcal{F}_t$, and therefore B^Q is an SBM under Q. $\diamond$

The stock price dynamics (2.8.1) can be equivalently written as

$$\frac{dS_t}{S_t} + y\,dt = r\,dt + \sigma\,dB_t^Q. \tag{2.8.8}$$

Repeating the previous analysis with Q as the underlying probability, B^Q in place of B, and r in place of μ results in the EMM-pricing condition $x_0 + \mathbb{E}^Q\left[e^{-rT} x_T \right] \leq 0$ for every traded cash flow x. Alternatively, we can reach this conclusion by applying Proposition 2.8.1 and the identity $\mathbb{E}\left[(\pi_T/\pi_0)\, x_T \right] = \mathbb{E}^Q\left[e^{-rT} x_T \right]$, which follows from the identities $\mathbb{E}^Q x_T = \mathbb{E}\left[(dQ/dP)\, x_T \right]$ and $\pi_T = \pi_0 e^{-rT} dQ/dP$.

REMARK 2.8.3 The reader should have no difficulty formally extending this section's results to allow for stochastic and time-varying market parameters r, μ, σ and $\eta \equiv (\mu - r)/\sigma$. In such an extension, the integrated version of the state-price density dynamics (2.8.6) is $\pi_t = \pi_0 \exp\left(- \int_0^t r_s ds \right) \xi_t$, where

$$\xi_t \equiv \exp\left(-\frac{1}{2} \int_0^t \eta_s^2 ds - \int_0^t \eta_s dB_s \right), \tag{2.8.9}$$

which is the continuous-time counterpart[20] of (2.5.10). One should be cautioned, however, on the need for regularity conditions, beyond what is obviously needed for integrals to make sense. For example, the existence of an EMM Q and corresponding statement of Girsanov's theorem require a regularity condition on η. Even if η is such that ξ is well defined by (2.8.9), it is not generally necessary that ξ is a martingale. Since $d\xi/\xi = -\eta dB$, we know that ξ is a positive local martingale, a property that implies (via Fatou's Lemma 2.7.1) that ξ is a supermartingale: $s > t$ implies $\mathbb{E}_t \xi_s \le \xi_t$. The process ξ is a martingale if and only if $\mathbb{E}\xi_T = 1$, in which case $dQ/dP \equiv \xi_T$ defines a probability Q and $B_t + \int_0^t \eta_s ds$ is an SBM under Q. A sufficient, but not necessary, condition for ξ to be a martingale is the *Novikov condition*: $\mathbb{E}\exp\left(\frac{1}{2}\int_0^T \eta_s^2 ds\right) < \infty$. $\diamond$

EXAMPLE 2.8.4 (Forward pricing) Suppose a forward contract for delivery of one share of the stock at time T is traded at time zero, with corresponding **forward price** $F_{0,T} \in \mathbb{R}$. The cash flow (x_0, x_T) generated by entering a long position in the forward contract is $(0, S_T - F_{0,T})$. The same traded cash flow can be generated by borrowing $S_0 e^{-yT}$ from the MMA to purchase e^{-yT} stock shares at time zero, continuously rolling over the loan and reinvesting all dividends. At time-T the result is a long position of one share of the stock and a short position of $-S_0 e^{(r-y)T}$ in the MMA. Barring arbitrage, it follows that

$$F_{0,T} = S_0 e^{(r-y)T}. \tag{2.8.10}$$

This expression can also be dually derived using the pricing restriction $x_0 + \mathbb{E}^Q\left[e^{-rT}x_T\right] = 0$ with $x_0 = 0$ and $x_T = S_T - F_{0,T}$ to conclude that $F_{0,T} = \mathbb{E}^Q S_T$ (a conclusion which critically depends on the assumption that r is constant). To show directly that the two expressions for $F_{0,T}$ are consistent, suppose $F_{0,T}$ is given by (2.8.10) and note that

$$S_T = F_{0,T} \exp\left(-\frac{\sigma^2}{2}T + \sigma B_T^Q\right). \tag{2.8.11}$$

The exponential factor has unit mean under Q for the same reason that ξ_T as defined in (2.8.7) has unit mean under P. $\diamond$

---

[20] To heuristically see the correspondence, take logarithms on both sides of (2.5.10), approximate each $\log(1 - \eta_s \Delta B_s)$ term on the right-hand side with a second-order Taylor expansion, and use the quadratic variation formula $(dB_s)^2 = ds$.

REMARK 2.8.5 (Stock price drift under EMM) The forward pricing argument of the preceding example sheds light onto the form of the stock price drift term of equation (2.8.8), which can be restated heuristically as $\mathbb{E}_t^Q S_{t+dt} - S_t = (r - y) S_t dt$. The price S_t represents the time-t cash payment for immediate delivery of one share of the stock at time t. The quantity $\mathbb{E}_t^Q S_{t+dt}$ represents the time-t forward price for delivery of a share at time $t + dt$. In delaying the stock delivery and payment from time t to time $t + dt$, the long position forgoes the dividend payment $y S_t dt$ but saves interest $r S_t dt$. The net $(r - y) S_t dt$ is therefore the adjustment to the spot price S_t required to obtain the forward price $\mathbb{E}_t^Q S_{t+dt}$. Note that while the conclusions of Example 2.8.4 require constant r and y, the argument just given applies for stochastic and time-varying parameters r and y (as well as μ and σ). ◇

The replication and arbitrage-pricing argument of Example 2.8.4 follows a line of reasoning that was introduced in Section 2.5 and extends as follows in the current context. For simplicity, we consider a contract we call the *star* contract, whose value process is V^* and whose only dividend is a terminal lump-sum payment $V_T^* \in L_2$. (In Example 2.8.4, $V_T^* \equiv S_T - F_{0,T}$.) Using the EMM Q, let

$$V_t^* \equiv \mathbb{E}_t^Q \big[e^{-r(T-t)} V_T^* \big], \tag{2.8.12}$$

which defines the Q-martingale $V_t^* e^{-rt}$ thanks to the law of iterated expectations. A martingale representation theorem implies the existence of a predictable process σ^* such that

$$V_t^* e^{-rt} = \mathbb{E}_t^Q \big[e^{-rT} V_T^* \big] = V_0^* + \int_0^t e^{-ru} \sigma_u^* dB_u^Q.$$

Integration by parts leads to the Ito decomposition

$$dV_t^* = r V_t^* dt + \sigma_t^* dB_t^Q = \big(r V_t^* + \sigma_t^* \eta \big) dt + \sigma_t^* dB_t. \tag{2.8.13}$$

In our discussion of equation (2.5.12), we saw that on a finite information tree the analogous relationship between the drift and volatility of a gain process is equivalent to a backward recursion. Equation (2.8.13) can be thought of as the continuous-time version of such a recursion, applied dt by dt backward in time, starting with the given terminal value V_T^*. Mathematically, (2.8.13) is an example of a backward stochastic differential equation (BSDE) to be solved jointly in V^* and σ^* given the terminal value V_T^*. The fact that the drift term in (2.8.13) is linear in V^* and σ^* makes this a linear BSDE, which allows a closed-form solution for V^* in (2.8.12).

The star contract can be thought of as a synthetic contract, generated by the trading strategy (θ^0, θ), where

$$\theta^0 + \theta S = V^* \quad \text{and} \quad \theta S \sigma = \sigma^*. \tag{2.8.14}$$

The budget equation is satisfied, since, using (2.8.8) and (2.8.13),

$$\theta^0 r \, dt + \theta(dS + y S \, dt) = r V^* \, dt + \sigma^* dB^Q = dV^*.$$

Unlike the finite information tree case of Section 2.5, we must further verify the solvency constraint. Moreover, in order to arrive to an arbitrage pricing equation, as opposed to an inequality, we must confirm that both $\pm (\theta^0, \theta)$ are admissible. For instance, the condition is obviously satisfied for the forward contract of Example 2.8.4, as well as the case of a European call reviewed below.

EXAMPLE 2.8.6 (European call and the Black–Scholes formula) Suppose the star contract is a European call option on the stock with strike $K \in (0, \infty)$ and maturity T, which means that $V_T^* \equiv (S_T - K)^+$. The solvency constraint can be satisfied for both the long and short position and therefore the time-zero option price, or **premium**, is $V_0^* = e^{-rT} \mathbb{E}^Q [(S_T - K)^+]$. By identity (2.8.11), the latter can be written as the following expression for the forward call premium per unit strike:

$$p \equiv \frac{V_0^* e^{rT}}{K} = \mathbb{E}^Q \left[\left(\exp\left(m - \frac{v^2}{2} + v \frac{B_T^Q}{\sqrt{T}} \right) - 1 \right)^+ \right], \tag{2.8.15}$$

where, with the stock forward price $F_{0,T}$ given in (2.8.10),

$$m \equiv \log\left(\frac{F_{0,T}}{K} \right) \quad \text{and} \quad v \equiv \sigma \sqrt{T}.$$

Let $\Phi(x) \equiv (2\pi)^{-1/2} \int_{-\infty}^{x} \exp(-y^2/2) \, dy$ denote the standard normal cumulative distribution function. Using the fact that the latter is the distribution of $B^Q/\sqrt{T}$ under Q, the expectation in (2.8.15) can be calculated to be

$$p = e^m \Phi\left(\frac{m}{v} + \frac{v}{2} \right) - \Phi\left(\frac{m}{v} - \frac{v}{2} \right).$$

This is the Black and Scholes (1973) formula for pricing the European call. One can think of pK as a marginal forward cost of insurance against a future price rise of the stock that, for example, someone who will buy the stock at T may demand. The quantity m, known as the option's *moneyness*, represents the amount of insurance, while $v \equiv \sigma \sqrt{T}$ quantifies the total risk to maturity. Another way of looking at the Black–Scholes formula is as an adjustment to the present value $S_0 e^{-yT} - K e^{-rT}$ of a forward contract to exchange one share of the stock for K dollars at time T:

$$V_0^* = S_0 e^{-yT} \Phi\left(\frac{m}{v} + \frac{v}{2}\right) - K e^{-rT} \Phi\left(\frac{m}{v} - \frac{v}{2}\right). \qquad (2.8.16)$$

◇

REMARK 2.8.7 Continuing Remark 2.8.3, the reader can formally extend this analysis to allow for nonconstant market parameters and a star contract that at each time t pays out dividends at a rate δ_t^*. Reversing our earlier line of reasoning, if the star contract is the synthetic contract generated by the trading strategy (θ^0, θ), then the replication condition (2.8.14) must hold along with the budget equation $dV^* = (\theta^0 r - \delta^*)\,dt + \theta\,(dS + ySdt)$, which together with the stock price dynamics (2.8.8) implies that (V^*, σ^*) solves the linear BSDE

$$dV_t^* = -\left(\delta_t^* - r_t V_t^* - \sigma_t^* \eta_t\right)dt + \sigma_t^* dB_t, \quad \text{given } V_T^*.$$

The corresponding solution for V^* is

$$V_t^* = \mathbb{E}_t^Q \left[\int_t^T \exp\left(-\int_t^s r_u du\right) \delta_s^* \, ds + \exp\left(-\int_t^T r_u du\right) V_T^* \right],$$

but only under regularity assumptions that we have avoided so far thanks to our constant-parameter assumptions. If (θ^0, θ) is value-destroying, analogous to a reverse doubling strategy that guarantees a loss, the BSDE still holds, but V_t^* must be strictly greater than the present value of the generated dividends. We will return to this discussion in the following chapter, in the context of optimal consumption and portfolio choice. ◇

The Markovian structure introduced in Section 2.6 is also present in the current market model, with the stock price being the Markov state. The more-general analog of the forward recursion (2.6.2) is a (forward) stochastic differential equation

$$dZ_t = a(t, Z_t)\,dt + b(t, Z_t)\,dB_t, \quad \text{given } Z_0 \in \mathbb{R},$$

with the functions $a, b \colon [0, T] \times \mathbb{R} \to \mathbb{R}$ appropriately restricted to ensure a unique solution, specifying a Markov process Z. Here $Z = S$, $a(t, Z) = (\mu - y)\,Z$ and $b(t, Z) = \sigma Z$.

We henceforth assume that $V_T^* = f_T(S_T)$ for given $f_T \colon \mathbb{R}_+ \to \mathbb{R}$. Suppose the (suitably regular) function $f \colon [0, T] \times \mathbb{R}_+ \to \mathbb{R}$ solves the partial differential equation (PDE)

$$\frac{\partial f}{\partial t} + (r - y)\,S\,\frac{\partial f}{\partial S} + \frac{1}{2}\sigma^2 S^2 \frac{\partial^2 f}{\partial S^2} = rf, \quad f(T, \cdot) = f_T. \qquad (2.8.17)$$

By Ito's lemma, it follows that

$$V_t^* = f(t, S_t) \quad \text{and} \quad \sigma_t^* = \frac{\partial f(t, S_t)}{\partial S} S_t \sigma$$

defines a solution to the linear BSDE (2.8.13). Conversely, the linear BSDE solution corresponding to the present-value expression (2.8.12) defines a solution to the above PDE, a solution that is sometimes referred to as a *Feynman–Kac formula*. (The reader can provide the generalization of PDE (2.8.17) and corresponding Feynman–Kac formula in the context of Remark 2.8.7 and more-general Markovian formulations.) This is an illustration of a close connection between BSDEs and corresponding PDEs, which allows one to leverage the rich theoretical and computational insights of the PDE literature. For example, PDE (2.8.17) can be transformed to the classical heat equation in Physics. Conversely, probabilistic arguments in the BSDE literature can help shed light on the corresponding PDEs.

Given f, we can write the trading strategy defined in (2.8.14) as a function of the Markov state: $\theta_t^0 = \theta^0(t, S_t)$ and $\theta_t = \theta(t, S_t)$, where

$$\theta(t, S) = \frac{\partial f(t, S)}{\partial S}, \quad \theta^0(t, S) = f(t, S) - \theta(t, S) S. \tag{2.8.18}$$

EXAMPLE 2.8.8 (European call replication) Suppose the star contract is the European call option of Example 2.8.6 and therefore $f_T(S) \equiv (S - K)^+$. Expressing the Black–Scholes equation (2.8.16) in the form $V_0^* = f(0, S_0)$ results in an expression for $f(0, \cdot)$. By the Markov property, an expression for $f(t, \cdot)$ is obtained from that for $f(0, \cdot)$ by substituting $T - t$ for T. Using the notation

$$m(t, S) \equiv \log\left(\frac{S e^{(r-y)(T-t)}}{K}\right), \quad v(t) \equiv \sigma\sqrt{T - t},$$

the result is $f(t, S) = \theta^0(t, S) + \theta(t, S) S$, where

$$\theta^0(t, S) = -e^{-r(T-t)} K \Phi\left(\frac{m(t, S)}{v(t)} - \frac{v(t)}{2}\right),$$

$$\theta(t, S) = e^{-y(T-t)} \Phi\left(\frac{m(t, S)}{v(t)} + \frac{v(t)}{2}\right).$$

These two functions satisfy (2.8.18) and therefore define the trading strategy that implements the European call as a synthetic contract in the MMA and the stock.[21] ◇

---

[21] The derivation of the Black–Scholes formula through the computation of a replicating trading strategy using Ito's calculus is due to Merton (1973).

Market practitioners use the notation

$$\Theta \equiv \frac{\partial f}{\partial t}, \quad \Delta \equiv \frac{\partial f}{\partial S}, \quad \Gamma \equiv \frac{\partial^2 f}{\partial S^2}.$$

A market maker (or arbitrageur) who sells the star contract can use f to price the contract, and buy Δ stock shares to hedge the position, an activity known as *delta hedging*. Γ quantifies how aggressively the market maker must trade to maintain the hedging position as a consequence of changes in the underlying stock price. By Ito's lemma, (Θ, Δ, Γ) encapsulates a snapshot of the value

$$V_{t+dt}^* = V_t^* + \Theta \, dt + \Delta \, dS_t + \frac{\Gamma}{2} (dS_t)^2$$

a short time interval in the future as a function of the stock price change, and PDE (2.8.17) expresses the fact that $V_t^* = e^{-rdt}\mathbb{E}_t^Q V_{t+dt}^*$. Notice that Δ represents a directional exposure that delta hedging seeks to cancel out, while Γ represents an exposure to volatility that can be a source of error in the delta hedging strategy. In theory, trading sufficiently frequently makes the error negligible, but in practice the market maker faces transaction costs and may wish to control the overall Γ to reduce the need for frequent rebalancing, a practice known as *gamma hedging*. Moreover, what we called the star contract may be a portfolio of contracts with the same Markov structure, for example, various options with different characteristics on the same stock. Since the derivatives defining Θ, Δ and Γ are linear, these parameters can be added up across portfolio positions, allowing the market maker to easily assess the effect of new trades on the current portfolio.

2.9 Exercises

EXERCISE 2.9.1 Consider the setting of Section 2.2 with $T = 1$. Since there is only one period, we drop all time subscripts. The model can be thought of as a single period on the information tree conditional on a given beginning-of-period spot. Recall that the market is arbitrage-free, $\mathcal{R}$ is the set of all traded returns, and $R^0 \in \mathcal{R}$ is the (zero-variance) MMA return. Assume that the market is *not* priced risk neutrally: There exists some $R \in \mathcal{R}$ such that $\mathbb{E}R \neq R^0$. Finally, assume that the positive-variance return $R^* \in \mathcal{R}$ maximizes the Sharpe ratio (relative to R^0) over $\mathcal{R}$.

(a) Give a direct argument showing that R^* must be a minimum variance frontier return.

(b) Fix any $R \in \mathcal{R}$ such that $\text{var}[R] > 0$ and $\text{cov}[R - R^*, R^*] \neq 0$. On the plane, consider the locus

$$H \equiv \left\{ \left(\text{stdev}[(1 - w) R^* + wR], \mathbb{E}[(1 - w) R^* + wR] \right) \mid w \in \mathbb{R} \right\}$$

as well as the line L that connects $h^* \equiv \left(\text{stdev}[R^*], \mathbb{E}R^* \right)$ to $(0, R^0)$. For every point $(s, m) \in H$ with $s > 0$, the slope of the line that connects $(0, R^0)$ to (s, m) is the Sharpe ratio $(m - R^0) / s$ of the corresponding return. Since the latter is maximized by R^* over all of $\mathcal{R}$ and h^* lies on H, it follows that H lies below L on the plane and it touches L at h^*. Since H is smooth, it must then be the case that H is tangent to L at h^*. Compute the slope of the tangent of H at h^* and show that the tangency condition is equivalent to the beta-pricing equation (2.2.2) (with time subscripts omitted, since $T = 1$).

(c) How would the arguments of parts (a) and (b) change if R^* were instead assumed to *minimize* the Sharpe ratio over all traded returns? Finally, explain why the minimum Sharpe ratio case applied to R^* is essentially the same as the maximum Sharpe ratio case applied to a suitably defined symmetric traded return.

EXERCISE 2.9.2 This exercise outlines a version of the theory of beta pricing without a traded money-market account. As in Section 2.2, the argument applies separately over a single period conditionally on each nonterminal spot. To simplify notation, and without loss of generality, we assume there is a single period $(T = 1)$, so every cash flow x is a pair of a scalar x_0 and a random variable x_1. We fix throughout a reference market, which is arbitrage-free and therefore every marketed cash flow has a uniquely defined present value. Call the random variable x a **marketed payoff** if $(0, x)$ is a marketed cash flow, in which case $\Pi(x)$ denotes the present value of the cash flow $(0, x)$. This defines a linear functional Π on the set of marketed payoffs. In terms of the present-value notation of Chapter 1, $\Pi(x)$ can be thought of as shorthand for $\Pi((0, x))$. The market is implemented by $J \geq 2$ contracts with well-defined and linearly independent returns $R^1, \ldots, R^J$. Fixing an underlying full-support probability, let $\Sigma = [\Sigma_{ij}]$ denote the return variance–covariance matrix: $\Sigma_{ij} \equiv \text{cov}(R^i, R^j)$, $i, j \in \{1, \ldots, J\}$. A **portfolio allocation** is any $\psi = (\psi^1, \ldots, \psi^J) \in \mathbb{R}^J$ such that $\sum_j \psi^j = 1$, generating the return $R^\psi \equiv \sum_j \psi^j R^j$. The set of traded returns is the linear manifold $\mathcal{R} \equiv \{ R^\psi \mid \psi \in \mathbb{R}^J \}$. Assume throughout that there is *no traded money-market account:* $\text{var}[R] > 0$ for all $R \in \mathcal{R}$. Let $L^\mathcal{R}$ denote the linear span of $\mathcal{R}$.

(a) Explain why $L^{\mathcal{R}}$ equals the set of marketed payoffs, and $x/\Pi(x) \in \mathcal{R}$ for every $x \in L^{\mathcal{R}}$ such that $\Pi(x) \neq 0$.

(b) Verify that covariance defines an inner product for the vector space $L^{\mathcal{R}}$. Treat $L^{\mathcal{R}}$ as an inner product space with the covariance inner product for the remainder of this exercise.

(c) Let x^{Π} denote the Riesz representation of the present-value functional Π in $L^{\mathcal{R}}$:

$$x^{\Pi} \in L^{\mathcal{R}} \quad \text{and} \quad \Pi(x) = \text{cov}\big[x^{\Pi}, x\big] \text{ for all } x \in L^{\mathcal{R}}.$$

Explain why $\Pi(x^{\Pi}) > 0$ and therefore R^{Π} is well defined by

$$R^{\Pi} \equiv \frac{x^{\Pi}}{\Pi(x^{\Pi})} \in \mathcal{R}.$$

Use Proposition B.2.6 to derive closed-form expressions, in terms of Σ, for the portfolio allocation that generates R^{Π} and the variance of R^{Π}.

(d) Show that $\text{var}\big[R^{\Pi}\big] = \min\{\text{var}[R] \mid R \in \mathcal{R}\}$ by using an orthogonal projection argument to argue that the minimum exists, is unique and is characterized by the orthogonality condition $R^{\Pi} \perp R - R^{\Pi}$ for all $R \in \mathcal{R}$ or, equivalently,

$$\text{for all } R \in \mathcal{R}, \ \text{cov}\big[R^{\Pi}, R\big] = \text{var}\big[R^{\Pi}\big].$$

Verify that R^{Π} satisfies this condition and is therefore the traded return of least variance.

(e) The return $R^* \in \mathcal{R}$ is a **frontier** return if

$$\text{var}\big[R^*\big] = \min\big\{\text{var}[R] \mid \mathbb{E}R = \mathbb{E}R^*, R \in \mathcal{R}\big\}.$$

Give a geometric interpretation of the property as an orthogonal projection and briefly explain why $R^* \in \mathcal{R}$ is a frontier return if and only if $R^* \perp R - R^*$ for every traded return R such that $\mathbb{E}R = \mathbb{E}R^*$. Call the market **degenerate** if all traded returns have the same mean, that is, there exists an m such that $\mathbb{E}R = m$ for all $R \in \mathcal{R}$. What is the set of frontier returns if the market is degenerate? Show that if the market is *not* degenerate, then the set of frontier returns is a line in $L^{\mathcal{R}}$ that contains and is orthogonal to R^{Π}.

Hint The idea is to show that given a frontier return $R^* \neq R^{\Pi}$ and any given mean m, there is a point on the line through R^* and R^{Π} that is the projection of zero to the linear manifold $\{R \in \mathcal{R} \mid \mathbb{E}R = m\}$. The requisite orthogonality condition should follow from the corresponding orthogonality condition for R^* and the fact that R^{Π} is orthogonal to $\mathcal{R}$.

(f) Let $x^{\mathbb{E}}$ be the Riesz representation of $\mathbb{E}$ restricted to $L^{\mathcal{R}}$, defined by the requirements

$$x^{\mathbb{E}} \in L^{\mathcal{R}} \quad \text{and} \quad \mathbb{E}x = \text{cov}\big[x^{\mathbb{E}}, x\big] \text{ for all } x \in L^{\mathcal{R}}.$$

Use Proposition B.2.6 to derive closed-form expressions, in terms of Σ and $(\mathbb{E}R^1, \ldots, \mathbb{E}R^J)$, for the row vector $\beta^{\mathbb{E}}$ such that $x^{\mathbb{E}} = \sum_j \beta_j^{\mathbb{E}} R^j$ as well as $\Pi(x^{\mathbb{E}})$.

(g) Prove that the market is degenerate if and only if there exists a scalar m such that $x^{\mathbb{E}} = m x^{\Pi}$. Assume that the market is *not* degenerate for the remainder of this exercise.

(h) Show that $R^* \in \mathcal{R}$ is a frontier return if and only if it takes the form $\alpha x^{\Pi} + \beta x^{\mathbb{E}}$ for scalars α and β. Use this fact to give another proof that the set of frontier returns is a line in $L^{\mathcal{R}}$.

(i) Fix an arbitrary frontier return R^* other than R^{Π} and show that there exists a unique frontier return R^0 that is uncorrelated with R^*. Give a geometric interpretation of this result. How is R^0 positioned relative to R^{Π} and R^* on the frontier line? How does R^0 behave as R^* approaches R^{Π}?

(j) Suppose that R^* is a frontier return and R^0 is the unique frontier return that is uncorrelated with R^*. Show that

$$\mathbb{E}\big[R - R^0\big] = \frac{\text{cov}\big[R^*, R\big]}{\text{var}\big[R^*\big]} \mathbb{E}\big[R^* - R^0\big], \quad R \in \mathcal{R}.$$

Conversely, show that if this beta pricing equation holds for some $R^*, R^0 \in \mathcal{R}$, then R^* and R^0 are uncorrelated, R^* is necessarily a frontier return, and R^0 can always be selected to be a frontier return.

EXERCISE 2.9.3 Assume the setting of Section 2.2 with a single period $(T = 1)$. The (arbitrage-free) market is implemented by $1 + J$ linearly independent contracts: an MMA with return $R^0 \equiv 1 + r$ and J risky contracts with returns $R^1, \ldots, R^J$. Define

$$1 + \mu^j \equiv \mathbb{E}R^j \quad \text{and} \quad \Sigma_{ij} \equiv \text{cov}\big[R^i, R^j\big],$$

and write $\mu \equiv (\mu^1, \ldots, \mu^J)$, a row vector, and $\Sigma \equiv [\Sigma_{ij}]$, a $J \times J$ matrix. A **portfolio allocation** is any $\psi = (\psi^1, \ldots, \psi^J) \in \mathbb{R}^J$, which generates the return $R^{\psi} \equiv R^0 + \sum_j \psi^j (R^j - R^0)$. The set of traded returns is the linear manifold $\mathcal{R} \equiv \{R^{\psi} \mid \psi \in \mathbb{R}^J\}$. As in Exercise 2.9.2, the set of marketed payoffs is $L^{\mathcal{R}} \equiv \text{span}(\mathcal{R})$ and for every $x \in L^{\mathcal{R}}$, we write $\Pi(x)$, rather than $\Pi((0, x))$, for the present value of x. Moreover, by the argument of Exercise 2.9.2(a), $x/\Pi(x) \in \mathcal{R}$ for every $x \in L^{\mathcal{R}}$ such that $\Pi(x) \neq 0$.

(a) Explain why covariance is *not* an inner product in $L^{\mathcal{R}}$. Instead, for the remainder of this exercise assume $L^{\mathcal{R}}$ is an inner-product space with the inner product $\langle x \mid y \rangle \equiv \mathbb{E}[xy]$.

(b) Let x^{Π} denote the Riesz representation of Π in $L^{\mathcal{R}}$:

$$x^{\Pi} \in L^{\mathcal{R}} \quad \text{and} \quad \Pi(x) = \mathbb{E}\left[x^{\Pi}x\right] \text{ for all } x \in L^{\mathcal{R}}.$$

Define the corresponding return

$$R^{\Pi} \equiv \frac{x^{\Pi}}{\Pi\left(x^{\Pi}\right)} \in \mathcal{R}.$$

Use Proposition B.2.6 to derive closed-form expressions, in terms of r, μ and Σ, for the portfolio allocation that generates R^{Π}.

Hint Note that $x^{\Pi} - \mathbb{E}x^{\Pi}$ is the Riesz representation of the present-value functional in the linear subspace $\{x - \mathbb{E}x \mid x \in L^{\mathcal{R}}\}$.

(c) Use an orthogonal projection argument to show that

$$\mathbb{E}\left(R^{\Pi}\right)^2 = \min\left\{\mathbb{E}R^2 \mid R \in \mathcal{R}\right\}.$$

What is the associated orthogonality condition satisfied by R^{Π}?

(d) Show that the set of frontier returns is

$$\left\{\alpha R^0 + (1 - \alpha)R^{\Pi} \mid \alpha \in \mathbb{R}\right\}.$$

(e) Call the market **degenerate** if all traded returns have the same mean. Show that the set of frontier returns is a point if and only if the market is degenerate, and otherwise it is a line.

(f) Arguing directly from the definition of frontier returns, show that the set of frontier returns other than R^0 is exactly the set of traded returns of maximum absolute Sharpe ratio.

(g) The locus of all pairs $\left(\text{stdev}\left[R^*\right], \mathbb{E}R^*\right)$ as R^* ranges over all frontier returns is known as the **minimum-variance frontier**. Describe and plot the minimum-variance frontier, and specify geometrically the location of the point $\left(\text{stdev}\left[R^{\Pi}\right], \mathbb{E}R^{\Pi}\right)$.

EXERCISE 2.9.4 Consider the setting of Section 2.2. To simplify notation, assume there is a single period ($T = 1$), which entails no loss of generality since the argument of this exercise can be made over a single period conditionally on any nonterminal spot. In Proposition 2.2.3 we encountered the **beta** of a traded return R with respect to the traded return R^*:

$$\beta \equiv \frac{\text{cov}\left[R^*, R\right]}{\text{var}\left[R^*\right]}.$$

Suppose that in an empirical implementation of the beta-pricing equation of Proposition 2.2.3, a proxy $R^* + \varepsilon$ is used instead of a frontier return R^*, where the error ε is judged to be small. The corresponding beta is

$$\beta^\varepsilon \equiv \frac{\text{cov}\left[R^* + \varepsilon, R\right]}{\text{var}\left[R^* + \varepsilon\right]}.$$

Give a simple example illustrating the claim that an arbitrarily small value of $\mathbb{E}\varepsilon^2$ is consistent with an arbitrarily large value of $|\beta^\varepsilon - \beta|$.

EXERCISE 2.9.5 (a) Show that an adapted process M is a martingale if and only if $M_0 = \mathbb{E}M_\tau$ for every stopping time $\tau : \Omega \to \{0, \ldots, T\}$. (Note that τ is not allowed the value ∞.)

(b) Suppose that the market is implemented by the contracts $(\delta, V) \in \mathcal{L}^J \times \mathcal{L}^J$. Show that a strictly positive adapted process π is an SPD if and only if for every finite stopping time $\tau : \Omega \to \{0, \ldots, T\}$,

$$V_0 = \frac{1}{\pi_0} \mathbb{E}\left[\sum_{t=0}^{\tau-1} \pi_t \delta_t + \pi_\tau V_\tau\right].$$

EXERCISE 2.9.6 (Binomial pricing) This is a continuation of Exercise 1.9.9, whose setting and notation is assumed here. The condition of part (b) of Exercise 1.9.9, which is implied by the no-arbitrage assumption, is equivalent to

$$q \equiv \frac{\frac{1+r}{1+y} - D}{U - D} \in (0, 1). \tag{2.9.1}$$

Assume that this condition holds for the remainder of this exercise.

(a) Compute all EMM-discount pairs for the given market. Is the market complete, and why? Show that for the market to be arbitrage-free, it is sufficient that $1 + r \in (0, \infty)$ and condition (2.9.1) holds.

(b) For the remainder of this exercise, assume that (Q, ρ) is an EMM-discount pair. Is Z a Markov process under Q, and why?

(c) Consider a contract (δ^*, V^*) that is specified in terms of the payoff function $f_T : \mathcal{Z}_T \to \mathbb{R}$ by

$$\delta_T^* = V_T^* = f_T(Z_T) \quad \text{and} \quad \delta_-^* = 0.$$

Use an EMM to show a pricing relationship of the form $V_t^* = f_t(Z_t)$, where the functions $f_t : \mathcal{Z}_t \to (0, \infty)$ are computed recursively, backward in time, starting with the known terminal function f_T. Confirm that your result is consistent with that of Exercise 1.9.9.

(d) What is the order of magnitude of the number of operations needed to compute V_0^* in part (c)? How does that compare to the number of spots on the information tree? What key assumptions make this type of computational efficiency possible?

(e) Postulate an underlying probability P that is defined in terms of the constant $p \in (0, 1)$ by

$$P(\{\omega\}) = p^{N(\omega)} (1 - p)^{T - N(\omega)}, \quad \omega \in \Omega,$$

where $N \equiv \sum_{t=1}^{T} b_t$. The process B is defined recursively by

$$B_0 = 0, \quad \Delta B_t = b_t \sqrt{\frac{1 - p}{p}} - (1 - b_t) \sqrt{\frac{p}{1 - p}}, \quad t = 1, \dots, T.$$

Verify that B is a dynamic orthonormal basis under P, and specify the set of all dynamically orthonormal bases in terms of B.

(f) Compute coefficients α and β so that

$$\frac{\Delta Z}{Z_-} = \alpha + \beta \Delta B.$$

Be as specific as you can, given the primitives of the model.

(g) Define the processes μ^R and σ^R by the return representation

$$\frac{\Delta G}{S_-} = \mu^R + \sigma^R \Delta B, \quad \mu^R \in \mathcal{P}_0, \quad \sigma^R \in \mathcal{P}_0^{1 \times d}.$$

Derive formulas for μ^R and σ^R in terms of α, β and y.

(h) Assume that Q is an EMM with corresponding market-price-of-risk process $\eta \in \mathcal{H}$. Compute $\mathbb{E}_{t-1}^Q [\Delta G_t / S_{t-1}]$ and use the fact that $\eta = -\mathbb{E}_{t-1}^Q \Delta B_t$ to conclude that

$$\eta = \frac{\mu^R - r}{\sigma^R}.$$

Give an expression for η in terms of p and q, and use it together with the definition of ΔB to confirm that

$$1 - \eta \Delta B = \frac{q}{p} b + \frac{1 - q}{1 - p} (1 - b),$$

and that the conditional density process $\xi_t = \mathbb{E}_t [dQ/dP]$ is given by

$$\xi_t = \left(\frac{q}{p} \right)^{N_t} \left(\frac{1 - q}{1 - p} \right)^{t - N_t}, \quad N_t \equiv \sum_{u=0}^{t} b_u.$$

How can you use this result to recover the EMM expression you derived in part (a)?

EXERCISE 2.9.7 Assume the stochastic setting of Section 2.8.

(a) Suppose $x_t = x_0 + \alpha t + \beta B_t$, for constant $\alpha, \beta \in \mathbb{R}$. Using Ito's lemma to derive the Ito decomposition of $\exp(x_t)$, and then use this decomposition to compute $\mathbb{E}\exp(x_t)$. You can use without proof the facts that in this context, the expectation $\mathbb{E}$ and integral $\int_0^t$ can be interchanged, and the local martingale part in the Ito expansion of $\exp(x_t)$ is in fact a martingale. Jensen's inequality requires that the ratio $\mathbb{E}\exp(x_t) / \exp(\mathbb{E}x_t)$ is greater than one. Explain how your calculation quantifies this ratio.

(b) Suppose x_t is a strictly positive Ito process and $\alpha \in \mathbb{R}$. Use Ito's calculus to give expressions for $d \log x_t$ and dx_t^α / x_t^α as a function of dx_t/x_t. Show how these expressions simplify when x is geometric Brownian motion with drift: $dx_t/x_t = \mu dt + \sigma dB_t$ for constant μ and σ.

EXERCISE 2.9.8 Use integration by parts to show the equivalence of the budget equation (2.8.2) and the budget equation (2.8.4) after the change of unit of account implied by the strictly positive Ito process π.

EXERCISE 2.9.9 This exercise provides some insight on the relationship between the Brownian model of Section 2.8 and a high-frequency version of the binomial model of Exercises 1.9.9 and 2.9.6.

(a) As preparation, you will prove a special version of the strong law of large numbers. Suppose $x_{n,N}, n = 1, 2, \ldots N; N = 1, 2, \ldots$, are i.i.d. (that is, identically distributed and stochastically independent) random variables. These are defined on a common state space with a given probability measure P and corresponding expectation operator $\mathbb{E}$. Assume that for all n, N, $\mathbb{E}x_{n,N} = 0$ and $\mathbb{E}[x_{n,N}^2] = 1$. Moreover, assume that there exists a constant C such that for all n, N, $|x_{n,N}| \le C$ everywhere. (The argument you are about to give applies if the last assumption is weakened to $\mathbb{E}[x_{n,N}^4] < \infty$, but we do not need this generality here.) Define the averages

$$\bar{x}_N \equiv \frac{1}{N} \sum_{n=1}^N x_{n,N}, \quad N = 1, 2, \ldots$$

Prove that the random variable $\sum_{N=1}^\infty \bar{x}_N^4$ has finite expectation and explain why it must then be true that $\lim_{N \to \infty} \bar{x}_N = 0$ with probability one. You can use the fact that $\mathbb{E}\sum_{N=1}^\infty \bar{x}_N^4 = \sum_{N=1}^\infty \mathbb{E}[\bar{x}_N^4]$, which is a consequence of the monotone convergence theorem.

(b) For each $N = 1, 2, \ldots$, consider a version of the model of Exercises 1.9.9 and 2.9.6, where instead of normalizing the time unit to correspond to one period, we assume that $T = 1$ and there are N periods. The time length of each period is therefore $1/N$. We wish to select the remaining parameters so that the Brownian model of Section 2.8 is approximated as N gets large (without actually proving a convergence result). The transition probability p is the same for all N. Suppose $z_{n,N}, n = 1, 2, \ldots, N; N = 1, 2, \ldots$ are i.i.d. random variables (under some given probability P on some common state space), and

$$p = P\left[z_{n,N} = \sqrt{\frac{1-p}{p}} \right] = 1 - P\left[z_{n,N} = -\sqrt{\frac{p}{1-p}} \right].$$

The martingale basis for the Nth model is B^N, where

$$B_0^N = 0, \quad B_{\frac{n}{N}}^N - B_{\frac{n-1}{N}}^N = \sqrt{\frac{1}{N}} z_{n,N}, \quad n = 1, \ldots, N.$$

The stock return parameters μ^R and σ^R are specified in terms of constants μ and σ by $\mu^R = \mu/N$ and $\sigma^R = \sigma/\sqrt{N}$. The stock cumulative return process R^N is given by

$$R_{\frac{n}{N}}^N - R_{\frac{n-1}{N}}^N \equiv \frac{S_{\frac{n}{N}} e^{\frac{y}{N}}}{S_{\frac{n-1}{N}}} - 1 = \mu \frac{1}{N} + \sigma \sqrt{\frac{1}{N}} z_{n,N}, \quad R_0^N \equiv 1.$$

The interest rate process of the MMA of the Nth model is analogously scaled, so that one dollar invested in the account at time $(n-1)/N$ becomes $1 + r/N$ at time n/N. The constant parameters (r, μ, σ) do not vary with N. Using part (a), show that with probability one,

$$\lim_{N \to \infty} \sum_{n=1}^{N} \left(R_{\frac{n}{N}}^N - R_{\frac{n-1}{N}}^N \right)^2 = \sigma^2.$$

This shows that no matter what the probability $p \in (0,1)$ is, and given any required level of precision, we can take N large enough so that the quadratic variation of R^N equals σ^2 up to the required precision. In the continuous-time limit, where R is an Ito process, this fact corresponds to the quadratic variation calculation $(dR_t)^2 = \sigma^2 \, dt$.

(c) Explain why in the finite binomial model the choice of the parameter $p \in (0, \infty)$ is irrelevant for the pricing of the European call option. Also explain why in the Brownian model, the value of a (or μ) is irrelevant for the pricing of the European call option. (The absence of μ in the Black–Scholes formula reflects this fact. Here you are asked to provide a deeper reason for this absence.)

(d) Given the insight of part (c), set $p = 1/2$. Recall that following each nonterminal spot, the stock price changes by a factor of either $U \equiv U^N$

or $D \equiv D^N$, where the superscript N has been added to emphasize the dependence of these parameters on N. (Technically speaking, the convention $S_T = 0$ requires us to slightly modify this statement, but the last period becomes infinitesimal in the limit and does not matter.) What are suitable expressions for U^N and D^N in terms of the parameters (r, y, σ) so that, as $N \to \infty$, the binomial model approximates the Brownian model of Section 2.8 with $a = 0$. Provide a brief explanation of your claim, without proving anything formal.

3

Optimality and Equilibrium Pricing

We have so far discussed markets that contain no cash flows that are desirable in the sense of arbitrage. In this chapter, we expand the notion of a desirable cash flow by introducing preferences. The absence of traded desirable cash flows defines optimality, which forms the basis for this chapter's treatment of equilibrium pricing and optimal consumption–portfolio choice. As in previous chapters, ≡ stands for "equal by definition" and terms in boldface are being defined.

3.1 Preferences and Optimality

We adopt the setting of Chapter 1, with information unfolding over times $0, \ldots, T$, as represented by a filtration defining $1 + N$ spots. There is no information at time zero, corresponding to spot zero, and the state is revealed at time T. A **consumption plan** is an adapted process. The set of all consumption plans is therefore $\mathcal{L}$, which can be identified with $\mathbb{R}^{1+N}$. A consumption plan represents an agent's total contingent consumption at each spot. A cash flow available in a market can be incrementally added to a given consumption plan to modify it to a preferred consumption plan. In reality, consumption consists of bundles of multiple goods. By assuming that consumption is one-dimensional at every spot, we are implicitly taking relative spot prices of goods as given and we measure consumption in some unit of account.

A **consumption set** is a set C of consumption plans such that for all $c \in C$, if x is an arbitrage, then $c + x \in C$. Given $c \in C$, we wish to specify a set $\mathcal{D}(c)$ with the interpretation that every $x \in \mathcal{D}(c)$ represents a **desirable** cash flow in the sense that $c + x$ is a consumption plan that is strictly preferred to c from the perspective of time zero. We impose some minimal requirements on preferences, listed below. Further structure will be imposed as needed later on.

DEFINITION 3.1.1 A **preference correspondence** is a function $\mathcal{D}$ whose domain, denoted $\text{dom}(\mathcal{D})$, is a consumption set, and such that for all $c \in \text{dom}(\mathcal{D})$, $\mathcal{D}(c)$ is a set of cash flows with the following properties.

- admissibility: $x \in \mathcal{D}(c)$ implies $c + x \in \text{dom}(\mathcal{D})$.
- irreflexivity: $0 \notin \mathcal{D}(c)$.
- continuity: $\mathcal{D}(c)$ is open.
- monotonicity: For every arbitrage cash flow y, if $x = 0$ or $x \in \mathcal{D}(c)$ then $x + y \in \mathcal{D}(c)$.

The first two properties state that $c + x$ must be an admissible consumption plan in order to be strictly preferred to c, and c cannot be strictly preferred to itself. The third condition states that if $c+x$ is strictly preferred to c then there is a sufficiently small tolerance $\epsilon > 0$ such that $\|x - x'\| < \epsilon$ implies that $c+x'$ is also strictly preferred to c. Finally, the fourth condition means that the agent always strictly prefers to increase consumption at some spot, provided consumption is not reduced at any other spot.

We henceforth fix a reference market X, relative to which we formulate notions of individual and allocational optimality.

DEFINITION 3.1.2 A consumption plan c is **optimal** for the preference correspondence $\mathcal{D}$ given X if $c \in \text{dom}(\mathcal{D})$ and $X \cap \mathcal{D}(c) = \emptyset$.

The first part of the optimality condition requires that c is an admissible consumption plan, while the second part states that there is no trade that results in a desirable incremental cash flow relative to c. Since, by monotonicity, $\mathcal{D}(c)$ contains all arbitrage cash flows, optimality of c implies the no-arbitrage condition $X \cap \mathbb{R}_+^{1+N} = \{0\}$.

Individual optimality extends usefully to a notion of allocational optimality. Fixing a positive integer I, we think of $i \in \{1, \ldots, I\}$ as labeling agents representing potential market participants. A **consumption allocation** is a tuple $c = (c^1, \ldots, c^I)$, with the interpretation that c^i is a consumption plan for agent i. A **preference profile** is a preference correspondence tuple $(\mathcal{D}^1, \ldots, \mathcal{D}^I)$, with $\mathcal{D}^i$ representing the admissible consumption plans and preferences of agent i. The sum of the $\mathcal{D}^i$ is the correspondence $\mathcal{D}$ on $\text{dom}(\mathcal{D}) \equiv \prod_{i=1}^{I} \text{dom}(\mathcal{D}^i)$ defined by

$$\mathcal{D}(c) \equiv \sum_{i=1}^{I} \mathcal{D}^i(c^i) \equiv \left\{ \sum_{i=1}^{I} x^i \mid x^i \in \mathcal{D}^i(c^i) \text{ for all } i \right\}. \tag{3.1.1}$$

DEFINITION 3.1.3 An allocation $c = (c^1, \ldots, c^I)$ is **optimal** for the preference profile $(\mathcal{D}^1, \ldots, \mathcal{D}^I)$ given the market X if $c \in \mathrm{dom}(\mathcal{D})$ and $X \cap \mathcal{D}(c) = \emptyset$.

The existence of some $x \in X \cap \mathcal{D}(c)$ can be thought of as a profitable market-making opportunity. Suppose $x = \sum_{i=1}^I x^i$ where $x^i \in \mathcal{D}^i(c^i)$. A market maker can offer x^i to each agent i in exchange for some positive amount, since agent i strictly prefers to add x^i to c^i. The market maker can then simply trade away the aggregate cash flow x, since $x \in X$. The net result is a positive profit for the market maker, a type of arbitrage that is not in X but can be obtained by using X and simultaneously contracting with the I agents in an incentive compatible way. Optimality of the allocation c given X means there are no market-making opportunities of this sort.

The monotonicity and continuity properties of preference correspondences imply that the optimality of an allocation c is equivalent to the apparently stronger condition of optimality of the allocation for every subset of the agents.

PROPOSITION 3.1.4 *The allocation $(c^1, \ldots, c^I)$ is optimal for the preference profile $(\mathcal{D}^1, \ldots, \mathcal{D}^I)$ given X if and only if for all $A \subseteq \{1, \ldots, I\}$, the allocation $(c^i)_{i \in A}$ is optimal for $(\mathcal{D}^i)_{i \in A}$ given X.*

PROOF Suppose $x = \sum_{i \in A} x^i \in X$, where $x^i \in \mathcal{D}^i(c^i)$ for all $i \in A \subseteq \{1, \ldots, I\}$. For $i \notin A$ let $x^i = 0$. The idea is to increase every x^i at the expense of a single agent $\alpha \in A$ by a sufficiently small amount so that agent α still finds the resulting incremental cash flow desirable. More formally, let y be any arbitrage cash flow, say $y = 1$, and fix any agent $\alpha \in A$. Given that $\mathcal{D}^\alpha(c^\alpha)$ is an open set, choose scalar $\epsilon > 0$ small enough so that $\bar{x}^\alpha \equiv x^\alpha - (I - 1)\epsilon y \in \mathcal{D}^\alpha(c^\alpha)$, and for every agent $i \neq \alpha$ let $\bar{x}^i \equiv x^i + \epsilon y$. By preference monotonicity, $\bar{x}^i \in \mathcal{D}(c^i)$ for all i, and by construction $\sum_i \bar{x}^i = x \in X$. Therefore, $(c^1, \ldots, c^I)$ is not optimal given X. The converse is immediate. $\qquad\square$

Since individual optimality is the same as the degenerate case of single-agent allocational optimality, it follows that allocational optimality implies individual optimality.

COROLLARY 3.1.5 *If $(c^1, \ldots, c^I)$ is optimal for $(\mathcal{D}^1, \ldots, \mathcal{D}^I)$ given X, then for all $i \in \{1, \ldots, I\}$, c^i is optimal for $\mathcal{D}^i$ given X.*

We proceed under the assumption that c represents either a consumption plan of a single agent with preference correspondence $\mathcal{D}$, or an allocation

$c = (c^1, \ldots, c^I)$ for I agents with preference profile $(\mathcal{D}^1, \ldots, \mathcal{D}^I)$. In the latter case, $\mathcal{D}$ denotes the sum of the $\mathcal{D}^i$ as specified in (3.1.1) with $\mathrm{dom}(\mathcal{D}) \equiv \prod_{i=1}^{I} \mathrm{dom}(\mathcal{D}^i)$. In the following discussion, any statement of optimality of c is understood to be for $\mathcal{D}$ if c is a consumption plan and for $(\mathcal{D}^1, \ldots, \mathcal{D}^I)$ if c is an allocation. With these conventions, we introduce a dual notion of optimality that will be key in relating optimality given a market to optimality given a budget constraint in terms of present values.

DEFINITION 3.1.6 For any linear functional $\Pi \colon \mathcal{L} \to \mathbb{R}$, the consumption plan or allocation c is Π-**optimal** if $c \in \mathrm{dom}(\mathcal{D})$ and $x \in \mathcal{D}(c)$ implies $\Pi(x) > 0$.

Since $\mathcal{D}(c)$ contains all arbitrage cash flows, if c is Π-optimal, then Π is necessarily positive ($\Pi(x) > 0$ if x is an arbitrage).

LEMMA 3.1.7 *For any positive linear functional* $\Pi \colon \mathcal{L} \to \mathbb{R}$, *a consumption plan or allocation is* Π-*optimal if and only if it is optimal given the market* $X^{\Pi} \equiv \{x \in \mathcal{L} \mid \Pi(x) = 0\}$.

PROOF Suppose c is not Π-optimal and therefore $\Pi(x) \leq 0$ for some $x \in \mathcal{D}(c)$. Then the constant cash flow $\delta \equiv -\Pi(x)/\Pi(1)$ is either zero or an arbitrage and therefore $x + \delta \in \mathcal{D}(c) \cap X^{\Pi}$, which implies c is not optimal given X^{Π}. The converse is immediate. $\square$

Lemma 3.1.7 allows us to apply Proposition 3.1.4 and its corollary to Π optimality. Therefore, if an allocation $c = (c^1, \ldots, c^I)$ is Π-optimal, then c^i is Π-optimal for every agent i. The fact that the Π in this statement is the same for every agent will be key.

The duality between optimality given a market and Π-optimality for a present-value function Π has the geometric structure of the duality between an arbitrage-free market and a present-value function (Theorem 1.4.3). The set $\mathcal{D}(c)$ enlarges the set of arbitrage cash flows and the condition that $\Pi(x)$ is positive for all $x \in \mathcal{D}(c)$ strengthens the condition that Π is positive. A present-value function Π (strictly) separates the market from the set of arbitrage cash flows, while a present-value function Π such that c is Π-optimal separates the market from $\mathcal{D}(c)$. This picture is the basis for the following proposition, where we take the reference market X as given, and we say that the consumption plan or allocation c is **optimal** to mean that c is optimal given the market X. Note that the second part assumes convexity of $\mathcal{D}(c)$, which in the multiagent case is implied by the convexity of all $\mathcal{D}^i(c)$.

PROPOSITION 3.1.8 *The following are true for all $c \in \mathrm{dom}(\mathcal{D})$.*

(1) *If c is Π-optimal for some present-value function Π, then c is optimal.*
(2) *If c is optimal and $\mathcal{D}(c)$ is convex, then there exists a present-value function Π such that c is Π-optimal.*
(3) *Suppose the market is complete and Π is the unique present-value function. Then c is optimal if and only if it is Π-optimal.*

PROOF (1) If $x \in \mathcal{D}(c)$, then $\Pi(x) > 0$ and therefore $x \notin X$.

(2) Suppose c is optimal and $\mathcal{D}(c)$ is convex. Viewing the space of adapted processes as $\mathbb{R}^{1+N}$ with the Euclidean inner product, the separating hyperplane Theorem B.6.2 allows us to select a nonzero $p \in \mathbb{R}^{1+N}$ such that $x \in X$ implies $p \cdot x \leq 0$, and $x \in \mathcal{D}(c)$ implies $p \cdot x \geq 0$. Suppose that $p \cdot x = 0$ for some $x \in \mathcal{D}(c)$. Since $\mathcal{D}(c)$ is open, there exists an $\epsilon > 0$ such that $x - \epsilon p \in \mathcal{D}(c)$, leading to the absurdity $0 > p \cdot (x - \epsilon p) \geq 0$. Therefore, $p \cdot x > 0$ for all $x \in \mathcal{D}(c)$, which implies that p is strictly positive (since $\mathcal{D}(c)$ contains every arbitrage). It follows that $\Pi(x) = p \cdot x / p_0$ defines a present-value function such that c is Π-optimal.

(3) This is Lemma 3.1.7 with $X = X^{\Pi}$.   □

A preference correspondence $\mathcal{D}$ has been defined from the perspective of time zero. A preference correspondence $\mathcal{D}_{F,t}$ can be analogously defined from the perspective of spot (F, t), with $\mathcal{D}_{F,t}(c) \subseteq \mathcal{L}_{F,t}$ for all $c \in C$. Given a corresponding spot-(F, t) market $X_{F,t} \subseteq \mathcal{L}_{F,t}$, a key observation is that if $X_{F,t} \subseteq X$ and $\mathcal{D}_{F,t}(c) \subseteq \mathcal{D}(c)$, then $X \cap \mathcal{D}(c) = \emptyset$ implies $X_{F,t} \cap \mathcal{D}_{F,t}(c) = \emptyset$. This condition states that if c is optimal from the perspective of spot zero, then it is also optimal from the perspective of spot (F, t). The condition guarantees that an agent who selects an optimal consumption plan c at time zero has no incentive to deviate from that choice as uncertainty unfolds. In discussing Proposition 1.2.4, we saw that $X_{F,t} \subseteq X$ is a dynamic consistency assumption on the market: If an incremental cash flow x is available in the market at spot (F, t), then it is also available at time zero, since the agent can make contingent plans to carry out whatever trades generate x if spot (F, t) is realized. Analogously, $\mathcal{D}_{F,t}(c) \subseteq \mathcal{D}(c)$ is a **dynamic consistency** assumption on preferences: If from the perspective of spot (F, t), consumption plan $c + x$ is strictly preferred to c for some incremental cash flow $x \in \mathcal{L}_{F,t}$, then the agent at time zero anticipates this preference contingent on spot (F, t) and therefore decides that $c + x$ is strictly preferred to c from the perspective of time zero as well.

A useful way of thinking about dynamic choice is as if the decision maker is multiple agents, one for every spot. Spot-(F, t) agent has preferences represented by $\mathcal{D}_{F,t}$ and selects trades in $X_{F,t}$. It is not hard to envision situations where dynamic consistency is violated. For example, suppose that the time-zero copy of the agent considers a contingent trade at spot (F, t) optimal, let's say a trade to rebalance into a falling stock market, yet, the spot-(F, t) copy of the same agent, faced with immediate market losses, becomes afraid and stays out of the market. (As former heavyweight world champion Mike Tyson put it, "Everyone has a plan 'till they get punched in the mouth.") Perhaps the spot-zero agent has poor foresight and would have chosen differently given a better understanding of the future self. Alternatively, the spot-zero agent may be well aware of the role of future temptations and may therefore seek to commit to a plan at time zero in a way that cannot be reversed at spot (F, t).[1] The strategic interaction among copies of the same agent can become complex (and interesting). Here we bypass this complexity, by removing any conflict among copies of the same agent. Dynamic consistency ensures that what is best for the spot-zero copy of the agent is also best for the spot-(F, t) agent, and is therefore sufficient to consider only time-zero optimal decisions. Analogous reasoning applies to the notions of allocational optimality and equilibrium, which we therefore only discuss from the perspective of time zero.

3.2 Equilibrium

In order to introduce a simple notion of (competitive) equilibrium, we formally define an **agent** to be a pair of a preference correspondence and a consumption plan, called the agent's **endowment**. We take as primitive the I agents

$$(\mathcal{D}^1, e^1), \ldots, (\mathcal{D}^I, e^I).$$

The initial allocation $(e^1, \ldots, e^I)$ can be modified to a new allocation $c \equiv (c^1, \ldots, c^I)$ through access to a market X.

DEFINITION 3.2.1 An allocation c is X**-feasible** if $c^i - e^i \in X$ for all i, and **clears the market** if $\sum_{i=1}^I c^i = e$, where $e \equiv \sum_{i=1}^I e^i$ is the **aggregate endowment**. An **equilibrium** is a pair (X, c), where X is a market, c is an X-feasible allocation that clears the market, and for all i, c^i is optimal for $\mathcal{D}^i$ given X.

---

[1] The concept is discussed by Strotz (1957), who quotes from Homer's *Odyssey*. Odysseus, being well aware his future self cannot resist the song of the Sirens, instructs his crew to tie him to the mast.

Equilibrium is commonly formulated in the literature in terms of contracts implementing the market, whose dividend processes are given and whose prices are set in equilibrium. The following example outlines the relationship of this approach to the equilibrium notion just introduced.

EXAMPLE 3.2.2 (Contract-market equilibrium) Consider J contracts with given dividend processes $\delta \equiv (\delta^1, \ldots, \delta^J)'$. A **contract-market equilibrium** is a pair of value processes $V \equiv (V^1, \ldots, V^J)'$ and trading strategies $\theta \equiv (\theta^1, \ldots, \theta^I)$ in the contracts (δ, V) such that $\sum_i \theta^i = 0$ and θ^i generates a cash flow x^i such that $c^i \equiv e^i + x^i \in \text{dom}(\mathcal{D}^i)$ and $X(\delta, V) \cap \mathcal{D}^i(c^i) = \emptyset$. Recall that $X(\delta, V)$ denotes the market implemented by (δ, V). The last optimality condition, therefore, states that there is no trading strategy in (δ, V) that agent i desires to add to θ^i. If (V, θ) is a contract-market equilibrium, then $\sum_i \theta^i = 0$ implies $\sum_i x^i = 0$, and therefore $(X(\delta, V), c)$ is an equilibrium. In the converse direction, suppose that $(X(\delta, V), c)$ is an equilibrium and for each agent i, let θ^i be a trading strategy that **finances** c^i, meaning that θ^i generates a cash flow x^i such that $c^i = e^i + x^i$. While it is clear that c^i is optimal for agent i given the market $X(\delta, V)$ and that $\sum_i x^i = 0$, it is not necessary that $\sum_i \theta^i = 0$, since there are potentially more than one trading strategies generating a given traded cash flow. It is possible, however, to choose $\theta = (\theta^1, \ldots, \theta^I)$ so that $\sum_i \theta^i = 0$, and therefore so that (V, θ) is a contract-market equilibrium. The trick is to pick θ^i to finance c^i for $i > 1$, and then define $\theta^1 = -\sum_{i>1} \theta^i$. The fact that c clears the market implies that θ^1 finances c^1, and we have the desired contract-market equilibrium. $\diamond$

Market equilibrium as defined here is closely related to the classical notion of a competitive exchange equilibrium in the absence of time and uncertainty.[2] For example, given an initial allocation of bread and wine, a competitive exchange equilibrium is a price for each good and a new allocation that is resource feasible (no bread or wine is created) and individually optimal (agents cannot do better by selling some wine to buy more bread or vice versa). A key insight of Arrow and Debreu was that the same notion can be applied in contexts with time and uncertainty by reinterpreting the notion of a good, in the example wine or bread, to be what we have called an Arrow cash flow.[3] Bundles of goods correspond to cash flows and prices of goods correspond to state prices. A state price vector p can always be identified with the corresponding linear functional $\Pi(c) = p \cdot c$, which leads to the following equilibrium notion.

[2] Walras (1874).
[3] Arrow (1953, 1964) and Debreu (1959).

DEFINITION 3.2.3 An **Arrow–Debreu equilibrium** is a pair (Π, c) of a linear functional $\Pi \colon \mathcal{L} \to \mathbb{R}$ and an allocation $c \equiv (c^1, \ldots, c^I)$ such that $\sum_{i=1}^{I} c^i \leq e$ and for all i, $c^i \in \operatorname{dom}(\mathcal{D}^i)$, $\Pi(c^i) \leq \Pi(e^i)$ and

$$x^i \in \mathcal{D}^i(c^i) \quad \text{implies} \quad \Pi(c^i + x^i) > \Pi(e^i).$$

The following equivalent form of this definition will be useful for our purposes.

LEMMA 3.2.4 *The pair (Π, c) is an Arrow–Debreu equilibrium if and only if $\Pi \colon \mathcal{L} \to \mathbb{R}$ is a positive linear functional, c clears the market, and for all i, c^i is Π-optimal for $\mathcal{D}^i$ and $\Pi(c^i) = \Pi(e^i)$.*

PROOF Suppose (Π, c) is an Arrow–Debreu equilibrium and x is an arbitrage. For every scalar $\epsilon > 0$, $\epsilon x \in \mathcal{D}^i(c^i)$ and therefore

$$\Pi(c^i) + \epsilon \Pi(x) = \Pi(c^i + \epsilon x) > \Pi(e^i) \geq \Pi(c^i).$$

This implies that $\Pi(x) > 0$ and, since x can be any arbitrage, that Π is positive. Letting $\epsilon \downarrow 0$, we conclude that $\Pi(c^i) = \Pi(e^i)$ for all i. The remaining claims are now immediate. $\qquad\square$

A positive linear functional Π on $\mathcal{L}$ is, after positive scaling, the present-value function for the complete market $X^\Pi \equiv \{x \in \mathcal{L} \mid \Pi(x) = 0\}$. By Proposition 3.1.8 (or Lemma 3.1.7) and Lemma 3.2.4, it follows that (Π, c) is an Arrow–Debreu equilibrium if and only if (X^Π, c) is an equilibrium. How does an equilibrium (X, c) relate to an Arrow–Debreu equilibrium if X is not complete? By Proposition 3.1.8, assuming convex preferences, individual optimality implies Π-optimality for some present-value function Π, which need not be common among all agents, unless the entire allocation c is optimal given the market.

DEFINITION 3.2.5 An **effectively complete market equilibrium** is an equilibrium (X, c) with the property that the allocation c is optimal for $(\mathcal{D}^1, \ldots, \mathcal{D}^I)$ given the market X.

As in Section 3.1, we define $\mathcal{D} \equiv \sum_{i=1}^{I} \mathcal{D}^i$ and note that $\mathcal{D}(c)$ is convex if every $\mathcal{D}^i(c^i)$ is convex. The following claims are a consequence of Proposition 3.1.8 and our earlier discussion.

PROPOSITION 3.2.6 *The following are true for any market X.*

(1) *If the allocation c is X-feasible and there exists a present-value function Π for X such that (Π, c) is an Arrow–Debreu equilibrium, then (X, c) is an effectively complete market equilibrium.*

(2) *Suppose $\mathcal{D}(c)$ is convex. If (X, c) is an effectively complete market equilibrium, then there exists a present-value function Π for X such that (Π, c) is an Arrow–Debreu equilibrium.*

(3) *Suppose the market X is complete with (necessarily unique) present-value function Π. Then (X, c) is an equilibrium if and only if (Π, c) is an Arrow–Debreu equilibrium.*

COROLLARY 3.2.7 *Suppose (X, c) is an equilibrium. If there exists a complete market $\bar{X} \supseteq X$ such that $(\bar{X}, c)$ is an equilibrium, then (X, c) is an effectively complete market equilibrium. The converse claim is also true if $\mathcal{D}(c)$ is convex.*

We conclude this section with two highly stylized examples of (effectively complete) equilibrium pricing that have played an important role in the early development of asset pricing theory. The first one is a version of the CAPM (Capital Asset Pricing Model),[4] which is a myopic, single-period model in which covariance with the market portfolio explains expected excess returns relative to a traded (credit-risk-free) money-market account. The second example, which works just as well in the multiperiod case, is one in which individual optimality can be characterized as the optimality of the aggregate endowment for a suitably constructed "representative" agent. Representative-agent arguments of this type have been used extensively in the literature to justify the formulation of simple single-agent equilibrium models that relate aggregate consumption to market risk premia and interest rates. Both examples rely on special preference and endowment structure and a high degree of similarity among agents.

EXAMPLE 3.2.8 (CAPM) There is a single period ($T = 1$) and an underlying full-support probability, which can be thought of as representing common beliefs among the I agents. Consider an equilibrium (X, c) with a traded money-market account (MMA) defining the short rate r and return $R^0 \equiv 1 + r$. The set of **traded returns** is

$$\mathcal{R} \equiv \{-x_1/x_0 \mid x \in X, \ x_0 \neq 0\}.$$

---

[4] The CAPM, which first appeared in Sharpe (1964), Lintner (1965) and Mossin (1966), has had a transformative impact on how the business and investing world perceives market risk and risk premia. Early ambitions to use the CAPM as an empirical tool were met with the realization that the CAPM is fundamentally untestable, since the market portfolio is not observable. Nevertheless, proxies of the overall market do explain a significant part of risk premia, the CAPM is still part of business school curricula, and estimates of betas relative to market proxies are widely reported.

Each endowment e^i is assumed to be marketed, with present value w^i (which is uniquely defined since X is arbitrage-free). The present value of the aggregate endowment e is therefore $w \equiv \sum_i w^i$. The **market return** R^m is the return of the traded contract whose payoff is the time-one aggregate endowment e_1. Since e_1 has present value $w - e_0$, $R^m = e_1/(w - e_0) \in \mathcal{R}$. To ensure that the market return is well defined and has positive variance, we assume that

$$\mathrm{var}[e_1] > 0 \quad \text{and} \quad e_0 \neq w. \tag{3.2.1}$$

The CAPM states that R^m is a beta-pricing return:

$$\mathbb{E}R - R^0 = \frac{\mathrm{cov}[R, R^m]}{\mathrm{var}[R^m]}(\mathbb{E}R^m - R^0), \quad R \in \mathcal{R}. \tag{3.2.2}$$

We will show the necessity of the CAPM with $\mathbb{E}R^m \neq R^0$ under the assumption that all agents are **variance averse**: For all $c \in \mathrm{dom}(\mathcal{D}^i)$,

$$\mathcal{D}^i(c) \supseteq \{x \in \mathcal{L} \mid (x_0, \mathbb{E}x_1) = (0, 0) \text{ and } \mathrm{var}[c_1 + x_1] < \mathrm{var}[c_1]\}.$$

Moreover, assuming preference transitivity, we will show that the equilibrium is necessarily effectively complete.

The essential argument is simple. Because of the assumptions of common beliefs, marketed endowments and variance aversion, in equilibrium, agents sell their endowments and buy plans on the minimum-variance frontier. By two-fund separation, all agents can finance their equilibrium consumption by holding the MMA and the same frontier risky investment. By market clearing, the market return takes the same form and must therefore also be a frontier return. The CAPM equation (3.2.2) then follows from Proposition 2.2.3.

We proceed with the details, starting with a review of minimum-variance analysis in the current context. On the space of random variables, we use the inner product $\langle x \mid y \rangle = \mathbb{E}[xy]$ with induced norm $\|\cdot\|$. Let x_1^Π denote the Riesz representation of the present-value functional on $X_1 \equiv \{x_1 \mid x \in X\}$, which is the unique $x_1^\Pi \in X_1$ such that $-x_0 = \langle x_1^\Pi \mid x_1 \rangle$ for all $x \in X$. Define $x^\Pi \in X$ by letting $-x_0^\Pi = \langle x_1^\Pi \mid x_1^\Pi \rangle$, a positive number, since $x_1^\Pi \neq 0$ by the MMA pricing equation $\langle x_1^\Pi \mid R^0 \rangle = 1$. The return $R^\Pi \equiv -x_1^\Pi / x_0^\Pi$ is therefore well defined. We call $x \in X$ a **frontier cash flow** if $\mathrm{var}[x_1]$ minimizes $\mathrm{var}[y_1]$ over all $y \in X$ such that $y_0 = x_0$ and $\mathbb{E}y_1 = \mathbb{E}x_1$, a condition that is equivalent to x_1 being the projection of zero onto $M \equiv \{z \in X_1 \mid \langle x_1^\Pi \mid z \rangle = -x_0, \langle 1 \mid z \rangle = \mathbb{E}x_1\}$, which is in turn equivalent to $x_1 \in M^{\perp} = \mathrm{span}(x_1^{\Pi}, 1)$. We have shown that $x \in X$ is a frontier cash flow if and only if $x_1 \in \mathrm{span}(R^{\Pi}, R^0)$.

The main CAPM argument follows. By the marketed-endowment assumption, $c^i = w^i 1_{\Omega \times \{0\}} + x^i$ for some $x^i \in X$. By the optimality of c^i for agent i and variance aversion, x^i is a frontier cash flow and therefore $x_1^i = a^i R^{\Pi} + b^i R^0$ for some $a^i, b^i \in \mathbb{R}$. Let $a \equiv \sum_i a^i$ and $b \equiv \sum_i b^i$. Adding up over all agents and using market clearing and the regularity assumption (3.2.1), we find

$$e = w 1_{\Omega \times \{0\}} + x, \text{ where } x_1 = aR^{\Pi} + bR^0 \text{ and } a \neq 0.$$

By the same regularity assumption, $x_0 \neq 0$ and the market return R^m has positive variance. This shows that $(-1, R^m)$ is a frontier cash flow and therefore R^m is a **frontier return**, in the sense that for all $R \in \mathcal{R}$, $\mathbb{E}R = \mathbb{E}R^m$ implies $\mathrm{var}[R] \geq \mathrm{var}[R^m]$. As in Lemma 2.2.2, this means that R^m is the projection of zero onto the linear manifold $\mathcal{R}_m \equiv \{R \in \mathcal{R} \mid \mathbb{E}R = \mathbb{E}R^m\}$, a condition that is equivalent to the orthogonality of R^m to $\mathcal{R}_m$, which is in turn equivalent to $\mathrm{cov}[R, R^m] = \mathrm{var}[R^m]$ for all $R \in \mathcal{R}_m$. The CAPM equation (3.2.2) with $\mathbb{E}R^m \neq R^0$ now follows exactly as in the $(1 \implies 2)$ part of Proposition 2.2.3.

Finally, we show that (X, c) is an effectively complete market equilibrium, assuming preference **transitivity**:

$$x \in \mathcal{D}^i(c^i) \text{ and } y \in \mathcal{D}^i(c^i + x) \text{ implies } x + y \in \mathcal{D}^i(c^i). \qquad (3.2.3)$$

Suppose $y^i \in \mathcal{D}^i(c^i)$ for every agent i and $\sum_i y^i \in X$. We will show that this violates individual optimality and hence cannot happen in equilibrium. Let $\bar{y}^i \in X$ be defined by the requirement that $\bar{y}_1^i$ is the projection of y_1^i onto X_1. Write $y^i = \bar{y}^i + (\delta^i, \varepsilon^i)$, where $\delta^i \equiv y_0^i - \bar{y}_0^i \in \mathbb{R}$ and $\varepsilon^i \equiv y_1^i - \bar{y}_1^i$ is orthogonal to X_1. Since $1 \in X_1$, $\mathbb{E}\varepsilon^i = 0$. By variance aversion and preference transitivity,

$$\bar{y}^i + (\delta^i, 0) \in \mathcal{D}^i(c^i). \qquad (3.2.4)$$

Let $y \equiv \sum_i y^i$, $\bar{y} \equiv \sum_i \bar{y}^i$ and $\varepsilon \equiv \sum_i \varepsilon^i$. Since $\varepsilon = y_1 - \bar{y}_1 \in X_1$ and ε is also orthogonal to X_1, $\varepsilon = 0$ and therefore $\sum_i \delta^i 1_{\Omega \times \{0\}} = y - \bar{y} \in X$. Since X is arbitrage-free, $\sum_i \delta^i = 0$. If for some i, $\delta^i < 0$, preference monotonicity and (3.2.4) implies $\bar{y}^i \in \mathcal{D}^i(c^i)$, which violates the optimality of c^i. Therefore, $\sum_i \delta^i = 0$ and $\delta^i \geq 0$ for all i, which implies $\delta^i = 0$ for all i. Equation (3.2.4) reduces to $\bar{y}^i \in \mathcal{D}^i(c^i)$, which again violates the optimality of c^i. $\diamond$

EXAMPLE 3.2.9 (Representative-agent pricing) The following is an example of representative-agent pricing based on the assumed scale invariance (also known as homotheticity) of preferences. Other variations are explored in the Exercises (Section 3.10).

We will characterize an equilibrium (X, c) for agents who are specified in terms of a fixed reference agent $(\mathcal{D}^0, e^0)$ with $\mathrm{dom}(\mathcal{D}^0) \equiv \mathcal{L}_{++}$. A key assumption is that $\mathcal{D}^0$ is **scale invariant (SI)**, meaning that the operators of positive scaling and specification of desirable cash flows commute:

$$\mathcal{D}^0(sc) = s\mathcal{D}^0(c) \text{ for all } s \in (0, \infty) \text{ and } c \in \mathrm{dom}(\mathcal{D}^0),$$

where $s\mathcal{D}^0(c) \equiv \{sx \mid x \in \mathcal{D}^0(c)\}$. For $i = 1, \ldots, I$, agent $(\mathcal{D}^i, e^i)$ is defined in terms of the parameters $(b^i, w^i, x^i) \in \mathcal{L} \times (0, \infty) \times X$ by

$$\mathcal{D}^i(c) \equiv \mathcal{D}^0(c - b^i) \text{ for all } c \in \mathrm{dom}(\mathcal{D}^i) \equiv b^i + \mathcal{L}_{++},$$
$$e^i \equiv b^i + w^i e^0 + x^i.$$

The consumption plan b^i can be thought of as a subsistence plan, the scalar w^i as the agent's wealth above subsistence measured in multiples of the reference plan e^0, and x^i as an endowed traded cash flow. For example, if $e^0 = 1_{\Omega \times \{0\}}$ and X is arbitrage-free, then the present value of the endowment in excess of subsistence, $e^i - b^i$, is equal to w^i. We do not assume, however, that e^0 is marketed. Note that the aggregate endowment can be written as $e \equiv \sum_{i=1}^I e^i = b + we^0 + x$, where $b \equiv \sum_i b^i$, $w \equiv \sum_i w^i$ and $x \equiv \sum_i x^i$. Let us now define the allocation c resulting from first allocating to all agents their subsistence plans b^i and then allocating the remaining aggregate endowment $e - b$ in proportion to each agent's wealth, w^i, above subsistence:

$$c^i \equiv b^i + \frac{w^i}{w}(e - b), \quad i = 1, \ldots, I.$$

The **representative agent** $(\mathcal{D}, e)$, whose endowment is the aggregate endowment, is defined by the preference correspondence

$$\mathcal{D}(c) \equiv \mathcal{D}^0(c - b) \text{ for all } c \in \mathrm{dom}(\mathcal{D}) \equiv b + \mathcal{L}_{++}.$$

The claim is that (X, c) is an equilibrium if and only if the aggregate endowment e is optimal for the representative agent given the market X, that is, $X \cap \mathcal{D}(e) = \emptyset$. To verify this claim, one can easily check, using the definition of the endowments and allocation, that the allocation

c is X-feasible and clears the market. Moreover, using the preference and allocation definitions and the scale-invariance of $\mathcal{D}^0$,

$$\frac{1}{w^i}\mathcal{D}^i(c^i) = \mathcal{D}^0\left(\frac{c^i - b^i}{w^i}\right) = \mathcal{D}^0\left(\frac{e - b}{w}\right) = \frac{1}{w}\mathcal{D}(e), \qquad (3.2.5)$$

and therefore $X \cap \mathcal{D}^i(c^i) = \emptyset$ if and only if $X \cap \mathcal{D}(e) = \emptyset$.

Finally, assuming (X, c) is an equilibrium and $\mathcal{D}(e)$ is convex, we show that (X, c) is an effectively complete market equilibrium. Suppose that $y^i \in \mathcal{D}^i(c^i)$ for all i and $y \equiv \sum_i y^i$. By (3.2.5), $\bar{y}^i \equiv (w/w_i)\, y^i \in \mathcal{D}(e)$, and therefore, by the convexity of $\mathcal{D}(e)$, $y = \sum_i \left(w^i/w\right)\bar{y}^i \in \mathcal{D}(e)$. Since e is an optimal consumption plan for the representative agent, it follows that $y \notin X$. $\diamond$

3.3 Utility Functions and Optimality

In the remainder of this chapter we introduce more structured formulations based on utility representations of preferences. For simplicity, we assume throughout that every agent's consumption set is $C \equiv \mathcal{L}_{++}$, although the theory clearly applies more generally and some of the Exercises (Section 3.10) assume specifications that require different consumption sets. We define utility functions to be continuous and monotone, which is not standard in the literature, but is convenient for our purposes. The existence of a utility representation is characterized in Section A.1.

DEFINITION 3.3.1 A **utility** (function) is a continuous function of the form $U : C \to \mathbb{R}$ that is **increasing**: for all $c \in C$, if x is an arbitrage, then $U(c + x) > U(c)$. The function $U : C \to \mathbb{R}$ is a **utility representation** of the preference correspondence $\mathcal{D}$ if $\mathrm{dom}(\mathcal{D}) = C$ and $x \in \mathcal{D}(c)$ is equivalent to $U(c + x) > U(c)$, in which case U is said to **represent** $\mathcal{D}$. Two utility functions are **ordinally equivalent** if they represent the same preference correspondence. A utility function U is **normalized** if $U(s) = s$ for all $s \in (0, \infty)$.

Note that the utility functions U and $\tilde{U}$ on C are ordinally equivalent if and only if $\tilde{U} = f \circ U$ for a (strictly) increasing function f from the range of U onto the range of $\tilde{U}$. (Since these ranges are connected intervals, f is necessarily continuous.) If $\mathcal{D}$ admits a utility representation $\tilde{U}$, the function

$$U(c) \equiv \inf\{s \in \mathbb{R} \mid s - c \in \mathcal{D}(c)\}, \quad c \in \mathrm{dom}(\mathcal{D}), \qquad (3.3.1)$$

is the unique normalized utility that is ordinally equivalent to $\tilde{U}$; it can be expressed as $U = \phi^{-1} \circ \tilde{U}$ where $\phi(s) \equiv \tilde{U}(s)$, $s \in (0, \infty)$. While the numerical value $\tilde{U}(c)$ in isolation is meaningless, since $U(U(c)) = U(c)$, $U(c)$ represents a per-period payment of an annuity that is equally desirable as c.

A preference correspondence $\mathcal{D}$ is **convex** if $\mathcal{D}(c)$ is a convex set for all $c \in \text{dom}(\mathcal{D})$, a condition that can be thought of as preference for consumption smoothing across spots. Concavity of a utility function clearly implies the convexity of the preference correspondence it represents. The converse claim is not generally true—a convex preference correspondence admitting a utility representation may admit no concave utility representation.[5] The following example shows that the converse *is* true for scale-invariant preferences.

EXAMPLE 3.3.2 (Scale-invariant preferences) Suppose the preference correspondence $\mathcal{D}$ is **scale invariant (SI)**:

$$\mathcal{D}(sc) = s\mathcal{D}(c) \quad \text{for all } s \in (0, \infty) \text{ and } c \in \mathcal{L}_{++}.$$

The function U defined in (3.3.1) is **homogeneous of degree one**:

$$U(sc) = sU(c) \quad \text{for all } s \in (0, \infty) \text{ and } c \in \mathcal{L}_{++}.$$

It follows that a preference correspondence admitting a utility representation is SI if and only if it admits a utility representation (the normalized version) that is homogeneous of degree one.[6] If $\mathcal{D}$ is both SI and convex, then U is necessarily concave. Since U is homogeneous of degree one, concavity of U is equivalent to

$$U(x + y) \geq U(x) + U(y) \quad \text{for all } x, y \in \mathcal{L}_{++}. \tag{3.3.2}$$

To show (3.3.2) given convexity of $\mathcal{D}$, fix any $x, y \in \mathcal{L}_{++}$ and define $\alpha, \beta \in (0, \infty)$ so that $S \equiv U(x + y) = U(\alpha x) = U(\beta y)$, that is, $\alpha \equiv S/U(x)$ and $\beta \equiv S/U(y)$. Let $p \equiv \alpha/(\alpha + \beta)$. For all $\epsilon > 0$, utility monotonicity implies $S < U(\alpha x + \epsilon), U(\beta y + \epsilon)$ and therefore $S < U((1 - p)\alpha x + p\beta y + \epsilon)$. Letting $\epsilon \downarrow 0$, we have $S \leq U((1 - p)\alpha x + p\beta y) = S\alpha\beta/(\alpha + \beta)$. Therefore, $\alpha^{-1} + \beta^{-1} \leq 1$, which rearranges to (3.3.2). $\diamondsuit$

Let us now fix a reference market X and a preference correspondence $\mathcal{D}$ with utility representation U. We call a consumption plan c **optimal** if it is

---

[5] For example, Reny (2013) shows that $U(x, y) \equiv x + \sqrt{x^2 + y}$ represents a convex preference correspondence on $(0, \infty)^2$, but there is no strictly increasing function f such that $f \circ U$ is concave.

[6] In the literature, preferences that are representable by a homogeneous-of-degree-one utility function are commonly called "homothetic."

optimal for $\mathcal{D}$ given X. On the space of all adapted processes, we use the inner product

$$\langle x \mid y \rangle = \mathbb{E} \left[\sum_{t=0}^{T} x_t y_t \right], \quad x, y \in \mathcal{L}. \tag{3.3.3}$$

In applications, U is commonly assumed to be differentiable, in which case a simple characterization of optimality can be given in terms of the gradient $\nabla U(c) \in \mathcal{L}$, defined by

$$\langle \nabla U(c) \mid x \rangle = \lim_{\epsilon \downarrow 0} \frac{U(c + \epsilon x) - U(c)}{\epsilon}, \quad x \in \mathcal{L}.$$

EXAMPLE 3.3.3 (Gradient of additive utility) Suppose the utility takes the additive form $U(c) = \mathbb{E}^Q \sum_t u_t(c_t)$, where each $u_t \colon (0, \infty) \to \mathbb{R}$ is a differentiable function and Q is a full-support probability that can be thought of as expressing beliefs. Let ξ denote the conditional density process $\xi_t = \mathbb{E}_t dQ/dP$. Lemma 2.4.4 implies $U(c) = \mathbb{E} \sum_t u_t(c_t) \xi_t$ and therefore $\nabla U(c)_t = u_t'(c_t) \xi_t$. For example, Section A.4 shows that if U represents scale-invariant preferences, then u_t must take a power or logarithmic form, which implies that the utility function is necessarily differentiable. $\diamond$

The role of the utility gradient for optimality is illustrated in Figure 3.1 (which extends Figure 1.2) and is formally shown in Proposition 3.3.4. Note that the assumed strict positivity of $\nabla U(c)$ is not implied by the monotonicity of U. (Set $u_t(c) = (c-1)^3$ in Example 3.3.3 and compute $\nabla U(1)$.) The reader can show, however, that if U is concave, then the fact that U is monotone does imply the strict positivity of $\nabla U(c)$.

PROPOSITION 3.3.4 *Suppose $c \in C$ and the gradient vector $\nabla U(c)$ exists and is strictly positive. If c is optimal, then $\nabla U(c)$ is a state-price density. Conversely, if U is concave and $\nabla U(c)$ is a state-price density, then c is optimal.*

PROOF Suppose c is optimal. Given any $x \in X$, we use the fact that C is open to select $\varepsilon > 0$ so that $c + \alpha x \in C$ for all $\alpha \in [0, \varepsilon]$. The function $f(\alpha) \equiv U(c + \alpha x)$, $\alpha \in [0, \varepsilon]$, is maximized at zero and therefore has a nonpositive right derivative at zero: $\langle \nabla U(c) \mid x \rangle \leq 0$. By assumption, $\nabla U(c) \in \mathcal{L}_{++}$ and therefore $\nabla U(c)$ is a state-price density. Conversely, suppose that U is concave and $\pi \equiv \nabla U(c)$ is a state-price density. For all $x \in X$ such that $c + x \in C$, the gradient inequality and the fact that $\langle \pi \mid x \rangle \leq 0$ imply that $U(c + x) \leq U(c) + \langle \pi \mid x \rangle \leq U(c)$. $\qquad\square$

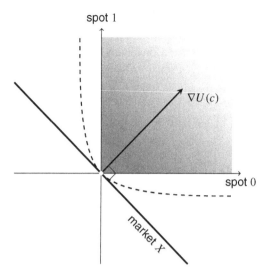

Figure 3.1 The dotted line is the boundary of the set $\mathcal{D}(c) = \{x \mid c + x \in \mathcal{L}_{++}, \; U(c + x) > U(c)\}$, which includes the shaded region of all arbitrage cash flows. The gradient $\nabla U(c)$ is orthogonal to the dotted line and points to the direction of steepest utility gain. Optimality means that X does not intersect $\mathcal{D}(c)$ and is therefore tangent to the dotted line, which in turn implies that $\nabla U(c)$ is orthogonal to the market.

In cases where the utility U is known to be concave but differentiability cannot be guaranteed, optimality of $c \in C$ can be characterized in terms of the superdifferential

$$\partial U(c) \equiv \{ \pi \in \mathcal{L} \mid c + x \in C \implies U(c + x) \leq U(c) + \langle \pi \mid x \rangle \}.$$

As shown in Section B.6, assuming U is concave, $\partial U(c)$ is nonempty, and $d = \nabla U(c)$ if and only if $\partial U(c) = \{d\}$.

PROPOSITION 3.3.5 *Consider any $c \in C$. If there exists a state-price density $\pi \in \partial U(c)$, then c is optimal. Conversely, if c is optimal and U is concave, then there exists a state-price density $\pi \in \partial U(c)$.*

PROOF The result can be viewed as a corollary of Proposition 3.1.8, which relates optimality to Π-optimality, and Proposition 3.3.6, which characterizes Π-optimality. A more-direct argument, using only the separating hyperplane theorem, follows.

Suppose $\pi \in \partial U(c)$ is a state-price density. For all $x \in X$ such that $c + x \in C$, $\langle \pi \,|\, x \rangle \leq 0$ and therefore $U(c + x) \leq U(c) + \langle \pi \,|\, x \rangle \leq U(c)$. This proves the optimality of c. Conversely, suppose c is optimal and U is concave. Consider the convex sets in $\mathcal{L} \times \mathbb{R}$:

$$A \equiv \{ (x, \alpha) \mid x \in X, \, U(c) < \alpha \},$$
$$B \equiv \{ (y, \beta) \mid c + y \in C, \, \beta \leq U(c + y) \}.$$

If $(x, \alpha) \in A \cap B$, then $U(c) < \alpha \leq U(c + x)$ for some $x \in X$, contradicting the optimality of c. Note also that $(0, U(c))$ is in the closure of both sets. Therefore, by the separating hyperplane Theorem B.6.2, there exists some nonzero $(\pi, r) \in \mathcal{L} \times \mathbb{R}$ such that

$$(x, \alpha) \in A \implies \langle \pi \,|\, x \rangle + r\alpha \leq rU(c), \tag{3.3.4}$$
$$(y, \beta) \in B \implies \langle \pi \,|\, y \rangle + r\beta \geq rU(c). \tag{3.3.5}$$

If $r > 0$, condition (3.3.5) is violated by taking β to minus infinity. If $r = 0$, and therefore $\pi \neq 0$, condition (3.3.5) is violated by $y = -\varepsilon\pi$ for $\varepsilon > 0$ small enough so that $c + y \in C$. Therefore, $r < 0$ and, after rescaling, we can set $r = -1$. Given this normalization, condition (3.3.4) implies that $\langle \pi \,|\, x \rangle \leq 0$ for all $x \in X$, and condition (3.3.5) implies that $\pi \in \partial U(c)$. Since U is increasing, π is necessarily strictly positive. $\qquad\square$

We have seen in Proposition 3.1.8 that optimality given the market is closely related to the notion of Π-optimality for a present-value function Π. For example, in a complete market, an agent can sell their endowment for a time-zero financial wealth w and select a consumption plan c to maximize utility subject to the (necessarily binding) wealth constraint $\Pi(c) \leq w$. The Lagrange multiplier λ of this constraint, whose interpretation as a marginal value of wealth is reviewed in the following proposition, will play a prominent role in the sequel.

PROPOSITION 3.3.6 *Suppose Π is a positive linear functional on $\mathcal{L}$ with Riesz representation $\pi \in \mathcal{L}_{++}$ and*

$$\mathcal{V}(w) \equiv \sup \{ U(c) \mid \Pi(c) \leq w, \, c \in \mathcal{L}_{++} \}, \quad w \in (0, \infty).$$

Fix any $c \in \mathcal{L}_{++}$ and $w \in (0, \infty)$ such that $\Pi(c) = w$. If $\lambda\pi \in \partial U(c)$ for some $\lambda \in (0, \infty)$, then c is Π-optimal and $\lambda \in \partial \mathcal{V}(w)$. Assume (for simplicity) that U is concave and therefore $\mathcal{V}$ is concave and $\partial \mathcal{V}(w) \neq \emptyset$. If c is Π-optimal and $\lambda \in \partial \mathcal{V}(w)$, then $\lambda\pi \in \partial U(c)$ and if $\nabla U(c)$ exists, $\mathcal{V}$ is differentiable at w and $\lambda = \mathcal{V}'(w)$.

PROOF The first two claims are an application of Theorem B.7.2 (without U concavity). Condition (2) of Theorem B.7.2 in this context reduces to $\lambda\pi \in \partial U(c)$ for some $\lambda > 0$, where the positivity of λ is a consequence of the monotonicity of U. The claim for the differential case follows from Theorem B.6.4. Suppose U is concave and $\nabla U(c)$ exists. Then $\partial U(c) = \{\nabla U(c)\}$ and therefore $\lambda\pi = \nabla U(c)$ for all $\lambda \in \partial V(w)$, which implies that $\partial V(w)$ is a singleton and therefore $\partial V(w) = \{V'(w)\}$. □

REMARK 3.3.7 Under the proposition's assumptions, suppose U is concave, $V(w) = U(c)$ and $\nabla U(c)$ exists. The fact that $V'(w)\pi = \nabla U(c)$ implies that $V'(w) = \langle \nabla U(c) \mid x \rangle$ for all x such that $\Pi(x) = 1$. Informally speaking, if w is changed by a small amount ϵ, the resulting optimal utility value $V(w + \epsilon)$ is approximately equal to $U(c + \epsilon x)$, where x is any unit-cost cash flow. Once all but the last penny of one's wealth has been optimally allocated, what one does with the last penny is of no quantitative significance. This sort of intuition applies more broadly to a class of so-called envelope theorems and is particularly important in interpreting observed choices involving monetary amounts that are trivial in comparison to one's total wealth. ◇

EXAMPLE 3.3.8 In the context of Proposition 3.3.6, if U is homogeneous of degree one, then $V(w) = V(1)\,w$ and the marginal value of wealth $\lambda = V'(w)$ does not depend on w. ◇

As in Section 3.1, individual optimality characterizations extend to allocational optimality given the market. To see how, let us fix a reference preference correspondence profile $(\mathcal{D}^1, \ldots, \mathcal{D}^I)$ and assume that each $\mathcal{D}^i$ has a utility representation U^i. We call an allocation $c \in C^n$ **optimal** given X if it is optimal for $(\mathcal{D}^1, \ldots, \mathcal{D}^I)$ given X. Thus an allocation c is optimal given X if and only if for every other allocation $c + x \in C^n$,

$$U^i(c^i + x^i) > U^i(c^i) \text{ for all } i \implies \sum_{i=1}^{I} x^i \notin X.$$

Moreover, by Proposition 3.1.4, c is optimal given X if and only if for all $c + x \in C^n$,

$$U^i(c^i + x^i) \geq U^i(c^i) \text{ for all } i \text{ and } U^i(c^i + x^i) > U^i(c^i) \text{ for some } i$$
$$\implies \sum_{i=1}^{I} x^i \notin X.$$

Pareto optimality is allocational optimality given the market $\{0\}$.

PROPOSITION 3.3.9 *Suppose c is an allocation and the gradients* $\nabla U^i(c^i)$
exist and are strictly positive. If c is optimal given X, then there exists a
state-price density π *and some* $\mu \in (0, \infty)^I$ *such that*

$$\pi = \mu_i \nabla U^i(c^i), \quad i = 1, \ldots, I. \tag{3.3.6}$$

Conversely, assuming every U^i *is concave, if* π *is a state-price density*
and (3.3.6) holds for some $\mu \in (0, \infty)^I$, *then c is optimal given X.*

PROOF Suppose $c \in C^n$ is optimal given X. For $i \neq j$, the zero cash
flow $x = 0$ maximizes $U^i(c^i + x)$ subject to $U^j(c^j - x) \geq U^j(c^j)$ as x
ranges over the set of cash flows such that $c^i + x, c^j - x \in C$. Applying
Theorem B.7.4 results in condition (3.3.6) for some $\mu \in (0, \infty)^I$ and π.
Since allocational optimality implies individual optimality, π is a state-
price density by Proposition 3.3.4. Conversely, assuming utility concavity,
consider any $c^i + x^i \in C$ such that $x \equiv \sum_i x^i \in X$. Multiply the gradient
inequality $U^i(c^i + x^i) \leq U^i(c^i) + \langle \nabla U^i(c^i) \mid x^i \rangle$ by μ_i, add up over i, and
use (3.3.6) and $\langle \pi \mid x \rangle = 0$ to conclude that

$$\sum_i \mu_i U^i(c^i + x^i) \leq \sum_i \mu_i U^i(c^i).$$

Since $\mu \in (0, \infty)^I$, it is not the case that $U^i(c^i + x^i) > U^i(c^i)$ for all i. $\square$

The reader is encouraged to attempt a visualization of the preceding
result based on Figure 3.1, but with two agents and a common market X.
The less obvious case is when X is incomplete, which can be visualized by
adding a third orthogonal axis, while preserving X as a line. In this case,
$\nabla U^1(c^1)$ and $\nabla U^2(c^2)$ can both be orthogonal to X without satisfying
the collinearity condition (3.3.6). Individual optimality implies that both
gradients are state-price densities, but not their collinearity. Allocational
optimality given X implies Pareto optimality, which in turn implies the
gradient collinearity. To visualize the last claim, note that if a cash flow x^i
forms an acute angle with $\nabla U^i(c^i)$, then for all sufficiently small $\epsilon > 0, \epsilon x^i$
is strictly desirable for agent i. The noncollinearity of the gradients allows
us to choose such x^i that sum up to zero, implying the violation of Pareto
optimality.

A useful construct for characterizing effectively complete market equilibria is the so-called **central planner**, defined in terms of the parameter
$\mu \in (0, \infty)^I$ as the agent whose endowment is the aggregate endowment e,
and whose preferences are represented by the utility function $U^\mu : \mathcal{L}_{++} \to$
$\mathbb{R}$, where

$$U^{\mu}(c) \equiv \sup \left\{ \sum_{i=1}^{I} \mu_i U^i(c^i) \,\Big|\, \sum_{i=1}^{I} c^i \leq c, \ c^i \in \mathcal{L}_{++} \right\}. \tag{3.3.7}$$

For scalar μ_i, we use the notation

$$\mu_i \partial U^i(c^i) \equiv \left\{ \mu_i d \,|\, d \in \partial U^i(c^i) \right\}.$$

PROPOSITION 3.3.10 *Suppose U^i is concave for every agent i and c is an X-feasible allocation. Then (X, c) is an effectively complete market equilibrium if and only if there exists $\mu \in (0, \infty)^I$ such that $e = \sum_i c^i$ is optimal for the central planner given X and c solves the maximization problem (3.3.7) defining $U^{\mu}(e)$. In this case, there exists a state-price density π for X such that*

$$\pi \in \partial U^{\mu}(e) = \bigcap_{i=1}^{I} \mu_i \partial U^i(c^i),$$

and if the gradient vectors $\nabla U^i(c^i)$ exist, then $\nabla U^{\mu}(e)$ also exists and

$$\pi = \nabla U^{\mu}(e) = \mu_i \nabla U^i(c^i), \quad i = 1, \ldots, I.$$

PROOF Since every U^i is assumed to be concave, U^{μ} is also concave and $\partial U^{\mu}(e) \neq \emptyset$ (by Theorem B.6.3 or Lemma B.7.3). Apply Theorem B.7.2 to (3.3.7) with $c = e$, $F(\delta) = U^{\mu}(e + \delta)$ and $\lambda = \pi$. The theorem's second condition in this application can be stated as

$$U^{\mu}(e) = \sum_{i=1}^{I} \max \left\{ \mu_i U^i(c^i) - \langle \pi \,|\, c^i - e^i \rangle \,|\, c \in \mathcal{L}_{++}^I \right\}, \ \pi \in \mathcal{L}_{++},$$

where strict positivity of π follows from the assumption that every U^i is increasing. The ith term of this sum is maximized by c^i if and only if $\pi \in \mu_i \partial U^i(c^i)$. Therefore, for every $\mu \in (0, \infty)^I$ and allocation c such that $\sum_i c^i = e$, $U^{\mu}(e) = \sum_i \mu_i U^i(c^i)$ and $\pi \in \partial U^{\mu}(e)$ both hold if and only if $\pi \in \bigcap_i \mu_i \partial U^i(c^i)$.

Suppose (X, c) is an effectively complete market equilibrium. By Proposition 3.2.6, there exists a present-value function Π such that (Π, c) is an Arrow–Debreu equilibrium. Given a state-price density π representing Π, the Π-optimality of c^i for agent i implies that there exists $\lambda_i > 0$ such that $\lambda_i \pi \in \partial U^i(c^i)$. The argument of the first paragraph with $\mu_i \equiv 1/\lambda_i$ shows that $U^{\mu}(e) = \sum_i \mu_i U^i(c^i)$ and $\pi \in \partial U^{\mu}(e)$ and therefore, by Proposition 3.3.5, e is optimal for the central planner given X. Reversing these steps yields the converse. The claim involving gradients is a consequence of Theorem B.6.4. $\qquad\square$

EXAMPLE 3.3.11 Suppose that every agent's utility takes the form

$$U^i(c) = \mathbb{E} \sum_{t=0}^{T} u_t^i(c_t),$$

where each $u_t^i \colon (0, \infty) \to \mathbb{R}$ is a concave differentiable function. For all agent weights $\mu \in (0, \infty)^I$, the utility U^μ defined in (3.3.7) is of the same form with $i = \mu$, where

$$u_t^\mu(x) \equiv \sup \left\{ \sum_{i=1}^{I} \mu_i u_t^i(x_i) \mid \sum_{i=1}^{I} x_i \leq x, \ x_i \in (0, \infty) \right\}.$$

Assuming (X, c) is an effectively complete market equilibrium, there exists $\mu \in \mathbb{R}^I_{++}$ such that $\nabla U^\mu(e)$ is a state-price vector and

$$\pi_t = u_t^{\mu\prime}(e_t)$$

defines a corresponding state-price density. In this special context, the state-price density is a deterministic function of the aggregate endowment at every spot. $\diamond$

While Example 3.3.11 suggests that an additive utility structure has some analytical advantages, the following example highlights a limitation of additive utility in capturing aversion to risk associated with the possible persistence of outcomes.

EXAMPLE 3.3.12 Suppose the random variables $\delta_1, \ldots, \delta_T$ are independent and identically distributed. For example, think of each δ_t as taking the value $+1$ or -1, depending on the outcome of a coin toss. (Mathematically, let $\delta_t(\omega) = \omega_t$ in Example 1.1.1 and assign equal probability to every state ω as in Example 2.1.8.) Consider two consumption plans a and b with some common initial value $a_0 = b_0 > 1$ and such that for every time $t > 0$, $a_t = a_0 + \delta_1$ and $b_t = b_0 + \delta_t$. In the coin-toss interpretation, consumption under a is either permanently increased or permanently decreased depending on the outcome of a single coin toss at time one, while consumption under b is increased or decreased in every period, depending on a new coin toss for every period. While both a and b result in the same expected consumption in every period, there is a clear sense in which a is riskier than b. Yet, every utility of the form $U = \mathbb{E} \sum_t u_t$, for any functions $u_t \colon (0, \infty) \to \mathbb{R}$, must necessarily assign the same value to both a and b, since $U = \sum_t \mathbb{E} u_t$ and $\mathbb{E} u_t(a_t) = \mathbb{E} u_t(b_t)$ for all t. $\diamond$

3.4 Dynamic Consistency and Recursive Utility

In this section we show that the dynamic consistency of preferences admitting a utility representation leads to a recursive relationship of utilities on the information tree. The resulting recursive utility class includes additive specifications such as expected discounted utility. Example 3.3.12 gave a first indication of the inadequacy of additive utilities. We will further see in this section that additive utilities require that time preferences in the absence of risk determine attitudes toward risk. This limitation motivates a broader class of recursive utilities, which allow a partial separation of time preferences and attitudes toward risk. Recursive utility is used in the remainder of this chapter to discuss equilibrium pricing and optimal consumption–portfolio choice.

We consider an agent whose preferences at spot (F, t) are represented by a preference correspondence $\mathcal{D}_{F,t}$ with domain $C \equiv \mathcal{L}_{++}$, where $\mathcal{D}_{F,t}(c) \subseteq \mathcal{L}_{F,t}$ for all $c \in C$. Up to Section 3.9, we assume that $\mathcal{D}_{F,t}$ is **forward looking**: For all $a, b \in C$, if $a = b$ on $F \times \{t, \dots, T\}$ then $\mathcal{D}_{F,t}(a) = \mathcal{D}_{F,t}(b)$. In other words, spot-$(F, t)$ preferences do not depend on past or unrealized consumption. While this assumption rules out potentially important aspects of preferences, it is methodologically appropriate to start with the simpler case and layer complexity on top as needed. (For example, Section 3.9 introduces a special type of habit formation or durability of consumption.) Moreover, we assume that every $\mathcal{D}_{F,t}$ is represented by a utility $U_{F,t} \colon C \to \mathbb{R}$, which is necessarily also **forward looking**: For all consumption plans a, b, if $a = b$ on $F \times \{t, \dots, T\}$ then $U_{F,t}(a) = U_{F,t}(b)$. This property allows us to consistently define $U_{F,t}\big(c 1_{F \times \{t, \dots, T\}}\big) \equiv U_{F,t}(c)$ for all $c \in C$, thus extending the domain of $U_{F,t}$ to $C_{F,t} \equiv \big\{c 1_{F \times \{t, \dots, T\}} \mid c \in C\big\}$. We call the function $U_{F,t}$ **increasing on $C_{F,t}$** if for every $c \in C$ and arbitrage $x \in \mathcal{L}_{F,t}$, $U_{F,t}(c + x) > U_{F,t}(c)$.

DEFINITION 3.4.1 A **spot-(F, t) utility** (function) is any forward-looking function $U_{F,t} \colon C \to \mathbb{R}$ that is continuous and increasing on $C_{F,t}$; it is said to **represent** $\mathcal{D}_{F,t}$ if $x \in \mathcal{D}_{F,t}(c)$ is equivalent to $U_{F,t}(c + x) > U_{F,t}(c)$. A **dynamic utility** is a function $U \colon C \to \mathcal{L}$ such that for all spots (F, t), a spot-(F, t) utility $U_{F,t} \colon C \to \mathbb{R}$ is well defined by $U_{F,t}(c) \equiv U(c)(F, t)$. The dynamic utility U is **normalized** if for all $s \in (0, \infty)$, $U(s) = s$.

Let us fix any reference dynamic utility $U \colon C \to \mathcal{L}$. We call $U(c)$ the **utility process** of $c \in C$ and write $U_t(c)$ for its time-t value. Since $U_0(c) = U_{\Omega, 0}(c)$, we write U_0 for the time-zero utility function $U_{\Omega, 0} \colon C \to \mathbb{R}$. (Note the inconsistency with the notation of Section 3.3.

Since we were only concerned with time-zero utilities, in Section 3.3 we used the notation U for what is U_0 in this section.) A dynamic utility $\tilde{U}$ is **ordinally equivalent** to U if for all spots (F, t), $U_{F,t}$ and $\tilde{U}_{F,t}$ represent the same preference correspondence. The discussion of Section 3.3 applied to each $U_{F,t}$ shows that U has a unique ordinally equivalent normalized version. A property of U is **ordinal** if its validity for U is equivalent to its validity for the normalized version of U and is therefore effectively a property of the represented preference correspondences. U is **dynamically consistent** if for all spots (F, t) and $x \in \mathcal{L}_{F,t}$, $U_{F,t}(c + x) > U_{F,t}(c)$ implies $U_0(c + x) > U_0(c)$, provided $c, c + x \in C$. This ordinal property corresponds to the dynamic consistency assumption $\mathcal{D}_{F,t}(c) \subseteq \mathcal{D}_{\Omega,0}(c) \equiv \mathcal{D}(c)$ discussed at the end of Section 3.1. The assumed continuity and monotonicity of utilities implies the following seemingly stronger dynamic consistency condition.

LEMMA 3.4.2 *U is dynamically consistent if and only if for all spots* (F, t), $c \in C$ *and* $x \in \mathcal{L}_{F,t}$ *such that* $c + x \in C$,

$$U_{F,t}(c + x) \geq U_{F,t}(c) \quad \Longleftrightarrow \quad U_0(c + x) \geq U_0(c).$$

PROOF Suppose U is dynamically consistent and $U_{F,t}(c + x) \geq U_{F,t}(c)$. Then $U_{F,t}(c + x + \varepsilon) > U_{F,t}(c)$ for all $\varepsilon > 0$, and therefore $U_0(c + x + \varepsilon) > U_0(c)$ for all $\varepsilon > 0$, and therefore $U_0(c + x) \geq U_0(c)$. The converse is immediate. $\square$

Consider now any nonterminal spot (F, t) and let

$$(F_0, t + 1), \ldots, (F_d, t + 1)$$

denote its immediate successor spots. The number d can vary from spot to spot, although in typical applications it is a constant throughout the filtration. Assuming U is normalized, a consequence of dynamic consistency is that the spot-(F, t) utility remains the same if for each i we replace the restriction of c on the subtree rooted on $(F_i, t + 1)$ with an annuity whose payments are all equal to the (normalized) utility value $U_{F_i,t+1}(c)$. More formally, we have the following claim.

LEMMA 3.4.3 *Suppose the dynamic utility U is a normalized and dynamically consistent. Given any $c \in C_{F,t}$, let $\bar{c} \in C_{F,t}$ be defined by letting $\bar{c}(F, t) = c(F, t)$ and $\bar{c} = U_{F_i,t+1}(c)$ on $F_i \times \{t + 1, \ldots, T\}$. Then $U_{F,t}(c) = U_{F,t}(\bar{c})$.*

PROOF Let $x_i \equiv \left(U_{F_i,t+1}(c) - c \right) 1_{F_i \times \{t+1,\ldots,T\}}$. Adding x_i to c replaces c on the subtree rooted at $(F_i, t + 1)$ with an annuity whose payment is $U_{F_i,t+1}(c)$. We use induction in k to show that

$$U_{F,t}(c) = U_{F,t}\left(c + \sum_{i=0}^{k} x_i\right), \quad k = 0, 1, \ldots, d. \tag{3.4.1}$$

For $k = 0$, the claim is trivially true. For the inductive step, assume $U_{F,t}(c) = U_{F,t}(b)$, where $b = c + \sum_{i=0}^{k-1} x_i$, with $b = c$ for $k = 0$. The utility normalization implies that $U_{F_k,t+1}(c) = U_{F_k,t+1}(c + x_k)$. Since c equals b on the subtree rooted at $(F_k, t+1)$, $U_{F_k,t+1}(b) = U_{F_k,t+1}(b + x_k)$. By Lemma 3.4.2, $U_0(b) = U_0(b + x_k)$ and therefore $U_{F,t}(b) = U_{F,t}(b + x_k)$. The last equation combined with the inductive hypothesis gives $U_{F,t}(c) = U_{F,t}(b + x_k)$, proving (3.4.1). $\qquad\square$

The conclusion of Lemma 3.4.3 is equivalent to the existence of a function $\Phi_{F,t} \colon (0, \infty)^{2+d} \to \mathbb{R}$ such that

$$U_{F,t}(c) = \Phi_{F,t}\big(c(F,t), U_{F_0,t+1}(c), \ldots, U_{F_d,t+1}(c)\big), \quad c \in C. \tag{3.4.2}$$

The functions $\Phi_{F,t}$, as (F,t) ranges over all nonterminal spots, specify a backward recursion on the information tree, which starts with the terminal values $U_T(c)$ (equal to c_T if U is normalized) and computes $U_{F,t}(c)$ in terms of spot-(F,t) consumption and the already computed end-of-period utility values. Let us call a dynamic utility that admits such a recursive specification a **generalized recursive utility**. One can easily check that every generalized recursive utility is dynamically consistent. We have therefore shown the following equivalence.

LEMMA 3.4.4 *A dynamic utility is a generalized recursive utility if and only if it is dynamically consistent.*

We call a (nonnormalized) dynamic utility $\tilde{U}$ **additive** if it takes the form $\tilde{U}_t(c) = \mathbb{E}_t \sum_{s=t}^{T} u_s(c_s)$, for functions $u_t \colon (0, \infty) \to \mathbb{R}$ that are increasing and continuous and hence invertible, and an expectation operator $\mathbb{E}$ under some full-support probability. Every additive utility is dynamically consistent and is therefore a generalized recursive utility. As already noted, utility additivity has some analytical advantages but also some significant limitations. Besides the issue illustrated in Example 3.3.12, risk attitudes represented by additive utilities are entirely determined by preferences in the absence of uncertainty. For every consumption plan c that is **deterministic**, meaning each c_t is constant across states, an additive time-zero utility reduces to $\tilde{U}_0(c) = \sum_t u_t(c_t)$. As a corollary of Theorem A.2.4, we have the following limitation of additive utilities.

PROPOSITION 3.4.5 *Any two additive dynamic utilities that are ordinally equivalent when restricted to deterministic plans are necessarily ordinally equivalent over all consumption plans.*

In order to achieve a partial separation of time preferences and attitudes toward risk, we consider a normalized generalized recursive utility (3.4.2) such that for all $x, x', y_0, \ldots, y_d, y \in (0, \infty)$,

$$\Phi_{F,t}(x, y_0, \ldots, y_d) = \Phi_{F,t}(x, y, \ldots, y) \iff$$
$$\Phi_{F,t}(x', y_0, \ldots, y_d) = \Phi_{F,t}(x', y, \ldots, y). \quad (3.4.3)$$

This is an ordinal property of $U_{F,t}$ (or U_0 given dynamic consistency) since, by Lemma 3.4.3,

$$\Phi_{F,t}(x, y_0, \ldots, y_d) = U_{F,t}\left(x 1_{F \times \{t\}} + \sum_{i=0}^{d} y_i 1_{F_i \times \{t+1, \ldots, T\}}\right). \quad (3.4.4)$$

The value $\Phi_{F,t}(x, y_0, \ldots, y_d)$ can be thought of as a utility value of a single-period uncertain payoff $(y_0, \ldots, y_d) \in (0, \infty)^{1+d}$, with corresponding certainty equivalent value $y \equiv v_{F,t}(y_0, \ldots, y_d)$ defined implicitly by any of the equivalent equations of condition (3.4.3), independently of the value of x. From the perspective of spot (F, t), an agent whose preferences are represented by $U_{F,t}$ is indifferent between the uncertain annuity $\sum_i y_i 1_{F_i \times \{t+1, \ldots, T\}}$ and the certain annuity $v_{F,t}(y_0, \ldots, y_d) 1_{F \times \{t+1, \ldots, T\}}$, independently of spot-$(F, t)$ consumption. The lower the certain annuity payment $v_{F,t}(y_0, \ldots, y_d)$ is, the more risk averse the agent. A dynamic utility U satisfies the **irrelevance of current consumption for risk aversion** if condition (3.4.3) holds at every nonterminal spot (F, t), where $\Phi_{F,t}$ is defined by (3.4.4). If we also define

$$f_{F,t}(x, y) \equiv \Phi_{F,t}(x, y, \ldots, y), \quad x, y \in (0, \infty), \quad (3.4.5)$$

then $\Phi_{F,t}(x, y_0, \ldots, y_d) = f_{F,t}\left(x, v_{F,t}(y_0, \ldots, y_d)\right)$ and (3.4.2) becomes

$$U_{F,t}(c) = f_{F,t}\left(c(F, t), v_{F,t}\left(U_{F_0, t+1}(c), \ldots, U_{F_d, t+1}(c)\right)\right). \quad (3.4.6)$$

By **recursive utility** we mean a dynamic utility that satisfies a recursion of this form. Combining Lemma 3.4.4 with the preceding argument, we have the following ordinal characterization of recursive utility.

PROPOSITION 3.4.6 *A dynamic utility is a recursive utility if and only if it is dynamically consistent and satisfies the irrelevance of current consumption for risk aversion.*

A dynamic utility is recursive if and only if its normalized version is recursive. With a normalized utility in mind, we define the terms **conditional aggregator** and **conditional certainty equivalent (CE)** to mean, respectively, any continuous increasing functions $f_{F,t} : (0, \infty)^2 \to (0, \infty)$

and $v_{F,t} \colon (0,\infty)^{1+d} \to (0,\infty)$ that are **normalized**: For all $s \in (0,\infty)$, $f_{F,t}(s,s) = s$ and $v_{F,t}(s,\dots,s) = s$. A recursive utility U satisfying (3.4.6) is normalized if and only if $U_T(c) = c_T$ and the functions $f_{F,t}$ and $v_{F,t}$ are conditional aggregators and CEs, respectively. The main theory of the remainder of this chapter is presented in terms of normalized recursive utilities.

Our earlier construction of recursion (3.4.6) makes precise the sense in which the conditional aggregator $f_{F,t}$ and conditional CE $v_{F,t}$ represent single-period time preferences and risk aversion from the perspective of spot (F,t). Suppose U' is another normalized recursive utility with spot-(F,t) conditional aggregator $f'_{F,t}$ and conditional CE $v'_{F,t}$. By identity (3.4.5), $f_{F,t} = f'_{F,t}$ if and only if $U_{F,t}$ and $U'_{F,t}$ are equal when restricted to consumption plans of the form $x 1_{F\times\{t\}} + y 1_{F\times\{t+1,\dots,T\}}$. In this sense, $f_{F,t}$ represents time preferences over a single period. A complete separation between time preferences and risk attitudes is not generally possible. We can make the more-modest claim, however, that given single-period time preferences, the conditional CE represents single-period risk attitudes. The conditional CE $v'_{F,t}$ is **more risk averse** than $v_{F,t}$ if $v'_{F,t} \leq v_{F,t}$, meaning that $v'_{F,t}(z) \leq v_{F,t}(z)$ for all $z \in (0,\infty)^{1+d}$. Assuming that $f_{F,t} = f'_{F,t}$, we have seen that making $v'_{F,t}$ more risk averse than $v_{F,t}$ means that from the perspective of spot (F,t), the constant annuity that is of equal utility as the contingent annuity $\sum_i y_i 1_{F_i \times \{t+1,\dots,T\}}$ makes a lower payment $v'_{F,t}(y_0,\dots,y_d)$ under $U'_{F,t}$ compared to the payment $v_{F,t}(y_0,\dots,y_d)$ under $U_{F,t}$. In this sense, risk aversion toward single-period payoffs is higher under $U'_{F,t}$ than $U_{F,t}$.

EXAMPLE 3.4.7 The functional form of $v_{F,t}$ can be founded on any static theory of choice under uncertainty, the most common being expected-utility theory. Suppose $v_{F,t}(z) = u^{-1}\big(\sum_i u(z_i) P[F_i \mid F]\big)$ and $v'_{F,t}(z) = \tilde{u}^{-1}\big(\sum_i \tilde{u}(z_i) P[F_i \mid F]\big)$ for increasing continuous functions $u, \tilde{u} \colon (0,\infty) \to \mathbb{R}$ and a common full-support probability P. Theorem A.6.1 shows that $v'_{F,t} \leq v_{F,t}$ if and only if $\tilde{u} = \phi \circ u$ for a concave function $\phi \colon u(0,\infty) \to \mathbb{R}$. Section A.6 discusses various conditions that express the idea that $v_{F,t}$ is risk averse in an absolute sense, all of which are equivalent to the concavity of u. $\diamond$

Recall that L_t denotes the set of all $\mathcal{F}_t$-measurable random variables. Let L_t^{++} denote the set of strictly positive elements of L_t. Just as it is convenient to represent the conditional expectations $\mathbb{E}[\cdot \mid F]$ for every spot (F,t) by an operator $\mathbb{E}_t$, it will be convenient to represent conditional CEs as operators from L_{t+1}^{++} to L_t^{++} by letting

$$v_t(z)(\omega) \equiv v_{F,t}(z) \equiv v_{F,t}(z(F_0),\ldots,z(F_d)), \ \omega \in F, \ z \in L_{t+1}^{++}, \quad (3.4.7)$$

where $z(F_i)$ denotes the value of the random variable z on the event F_i. For instance, for the expected-utility conditional CE specification of Example 3.4.7, $v_t = u^{-1}\mathbb{E}_t u$, meaning $v_t(z) = u^{-1}\big(\mathbb{E}_t u(z)\big)$ for all $z \in L_{t+1}^{++}$. For conditional aggregators, we write

$$f_t(\omega, x, y) \equiv f(\omega, t, x, y) \equiv f_{F,t}(x, y), \ \omega \in F, \ x, y \in (0, \infty). \quad (3.4.8)$$

As with random variables, the state variable ω is typically elided. These conventions allow us to write recursion (3.4.6) more succinctly as

$$U_t(c) = f_t(c_t, v_t(U_{t+1}(c))), \ U_T(c) = c_T. \quad (3.4.9)$$

EXAMPLE 3.4.8 The normalized version of the additive utility $\tilde{U}_t \equiv \mathbb{E}_t \sum_{s=t}^{T} u_s$ is $U_t \equiv \phi_t^{-1} \circ \tilde{U}_t$, where $\phi_t(\alpha) \equiv \sum_{s=t}^{T} u_s(\alpha), \alpha \in (0, \infty)$. It is not hard to see that U is recursive with an aggregator and CE that are both specified in terms of the same functions u_t, thus tying time preferences to risk aversion as discussed earlier. Rather than go over the general case, the point is made more clearly in the special case where, for some $\beta \in (0, 1)$ and $u: (0, \infty) \to \mathbb{R}$,

$$\tilde{U}_t(c) = \mathbb{E}_t \left[\sum_{s=t}^{T-1} \beta^{s-t} u(c_s) + \frac{\beta^{T-t}}{1-\beta} u(c_T) \right]. \quad (3.4.10)$$

To motivate the utility weight for terminal consumption, define the hypothetical infinite-horizon plan $\bar{c}$ by letting $\bar{c}_t = c_t$ for $t < T$ and $\bar{c}_t = c_T$ for all $t \geq T$. Then $\tilde{U}_t(c) = \mathbb{E}_t \sum_{s=t}^{\infty} \beta^{s-t} u(\bar{c}_s)$. The normalized version U satisfies recursion (3.4.9) with

$$f_t(\omega, x, y) \equiv u^{-1}((1 - \beta) u(x) + \beta u(y)), \quad v_t \equiv u^{-1}\mathbb{E}_t u. \quad (3.4.11)$$

Since preferences over deterministic consumption plans determine f, they also determine u and hence v, which captures attitude toward risk. This illustrates the more-general conclusion of Proposition 3.4.5. By relaxing the additivity assumption, we can use a different function u in specifying f and v, thus freeing the conditional CE specification from any assumptions on preferences in the absence of uncertainty. $\diamond$

Motivated by the streamlined notation of recursion (3.4.9), we henceforth adopt the following terminology.

An **aggregator** f is a mapping that assigns to every state ω and nonterminal time t a normalized, increasing and continuous function

$f(\omega, t, \cdot) : (0, \infty)^2 \to (0, \infty)$ such that the process $(\omega, t) \mapsto f(\omega, t, x, y)$ is adapted, for all (x, y). The aggregator f is **state** (resp. **time**) **independent** if $f(\omega, t, \cdot)$ does not vary with the state ω (resp. time t). For instance, the aggregator in (3.4.11) is state and time independent. For every spot (F, t), the aggregator f **implies** a conditional aggregator $f_{F,t}$ defined by (3.4.8).

A **certainty equivalent (CE)** υ is a mapping that assigns to each nonterminal time t a continuous function $\upsilon_t : L_{t+1}^{++} \to L_t^{++}$ that is normalized ($\upsilon_t(s) = s$ for all $s \in (0, \infty)$) and for every $z \in L_{t+1}^{++}$ and spot (F, t), the value of $\upsilon_t(z)$ on F, which we denote by $\upsilon_{F,t}(z)$, is an increasing function of the restriction[7] of z on F. Given this condition, we consistently extend the domain of $\upsilon_{F,t}$ by letting $\upsilon_{F,t}(z) \equiv \upsilon_{F,t}(z1_F)$ for all $z \in L_{t+1}^{++}$. For every spot (F, t), the CE υ **implies** a conditional CE $\upsilon_{F,t}$ defined by (3.4.7).

Using this terminology, we restate the definition of recursive utility as it applies to a normalized dynamic utility.

DEFINITION 3.4.9 A normalized dynamic utility U is a **recursive utility** if there exist an aggregator f and a CE υ such that for every $c \in C$, the process $U(c)$ solves the backward recursion (3.4.9).

Consider two normalized recursive utilities U and U' with respective CEs υ and υ' and a common state-independent aggregator f. For deterministic c, the utility process $U(c)$ is also deterministic and therefore $\upsilon_t(U_{t+1}(c)) = U_{t+1}(c)$. The analogous claim is true of U'. Since f is common, it follows that $U = U'$ on the set of deterministic plans, and $U_0' \leq U_0$ if and only if υ' is **more risk averse** than υ in the sense that $\upsilon_t' \leq \upsilon_t$ for all t.

We conclude this section with a recursive-utility gradient calculation, which is essential in formulating optimality conditions. As in Section 3.3, the gradient is defined relative to the inner product (3.3.3), where $\mathbb{E}$ is the expectation operator relative to some underlying full-support probability that is fixed throughout.

We will derive a gradient expression in terms of the CE derivative, which we define using the conditional norm notation $\|h\|_t \equiv \left(\mathbb{E}_t[h^2]\right)^{1/2}$.

---

[7] More formally, this means that for all $x, y \in L_{t+1}^{++}$, if $x = y$ on F then $\upsilon_t(x) = \upsilon_t(y)$ on F, and if $x \geq y \neq x$ on F then $\upsilon_t(x) > \upsilon_t(y)$ on F.

DEFINITION 3.4.10 The **derivative** of the CE υ is a mapping κ that assigns to each time $t < T$ and $z \in L_{t+1}^{++}$ a random variable $\kappa_{t+1}(z) \in L_{t+1}$ such that for all $z + h \in L_{t+1}^{++}$,

$$\upsilon_t(z + h) = \upsilon_t(z) + \mathbb{E}_t\big[\kappa_{t+1}(z)\,h\big] + \varepsilon_t(h)\,\|h\|_t,$$

where $\varepsilon_t(h)$ is small for small h in the sense that for all $\epsilon > 0$, there exists $\delta > 0$ such that $\|h\|_t < \delta$ implies $|\varepsilon_t(h)| < \epsilon$. A CE is **differentiable** if it has a derivative.

For a more-concrete expression of the CE derivative, consider the spot-(F, t) conditional CE $\upsilon_{F,t}$ implied by υ, and let z_i denote the value of z on F_i for each immediate successor spot $(F_i, t + 1)$ of (F, t). Writing $\kappa_{F_i, t+1}(z)$ for the value of $\kappa_{t+1}(z)$ on F_i, we have

$$\kappa_{F_i, t+1}(z)\,P[F_i \mid F] = \frac{\partial \upsilon_{F,t}(z_0, \ldots, z_d)}{\partial z_i}. \tag{3.4.12}$$

In the converse direction, by a standard calculus result, if the partial derivatives of $\upsilon_{F,t}$ exist and are continuous for all nonterminal spots (F, t), then κ as specified in (3.4.12) is the derivative of υ.

EXAMPLE 3.4.11 Suppose

$$\upsilon_t = u^{-1}\mathbb{E}_t u$$

for some continuously differentiable increasing function $u \colon (0, \infty) \to \mathbb{R}$. Then the derivative κ of υ exists and is given by

$$\kappa_t(z) = \frac{u'(z)}{u'(\upsilon_{t-1}(z))}, \quad t = 1, \ldots, T,$$

where u' denotes the derivative of u. $\diamond$

The gradient of a recursive utility is computed in Proposition 3.4.12, where $\partial f_t / \partial c$ and $\partial f_t / \partial \upsilon$ denote the partial derivatives of the time-t aggregator f_t with respect to its consumption and CE arguments, respectively.

PROPOSITION 3.4.12 *Suppose U is a normalized recursive utility with aggregator f and CE υ. Suppose also that $f_t(\omega, \cdot)$ is differentiable for every state ω and time $t < T$, and υ has derivative κ. Given a reference consumption plan c, let the processes λ and $\mathcal{E}$ be defined by*

$$\lambda_t \equiv \frac{\partial f_t}{\partial c}(c_t, \upsilon_t(U_{t+1}(c))), \quad t < T, \quad \lambda_T = 1, \tag{3.4.13}$$

$$\mathcal{E}_0 \equiv 1, \quad \frac{\mathcal{E}_t}{\mathcal{E}_{t-1}} \equiv \frac{\partial f_{t-1}}{\partial \upsilon}(c_{t-1}, \upsilon_{t-1}(U_t(c)))\,\kappa_t(U_t(c)). \tag{3.4.14}$$

Then for every adapted process x and time t,

$$\lim_{\epsilon \downarrow 0} \frac{U_t(c + \epsilon x) - U_t(c)}{\epsilon} = \mathbb{E}_t \left[\sum_{s=t}^{T} \frac{\mathcal{E}_s}{\mathcal{E}_t} \lambda_s x_s \right], \tag{3.4.15}$$

and therefore

$$\nabla U_0(c) = \mathcal{E}\lambda.$$

PROOF Fixing $x \in \mathcal{L}$, define, for all sufficiently small $\epsilon \in \mathbb{R}$,

$$\phi_t(\epsilon) \equiv U_t(c + \epsilon x).$$

The left-hand side in (3.4.15) defines the derivative $\phi_t'(0)$. The utility recursion implies

$$\phi_t(\epsilon) = f_t(c_t + \epsilon x_t, \upsilon_t(\phi_{t+1}(\epsilon))).$$

Differentiating at $\epsilon = 0$ using the differentiation chain rule, we have

$$\phi_t'(0) = \lambda_t x_t + \frac{\partial f_t}{\partial \upsilon}(c, \upsilon_t(U_{t+1}(c))) \mathbb{E}_t \left[\kappa_{t+1}(\phi_{t+1}(0)) \phi_{t+1}'(0) \right].$$

Letting $V_t \equiv \phi_t'(0)$ and $\delta_t \equiv \lambda_t x_t$, the recursion can be restated as

$$V_t = \delta_t + \frac{1}{\mathcal{E}_t} \mathbb{E}_t \left[\mathcal{E}_{t+1} V_{t+1} \right], \quad V_T = x_T.$$

As discussed in Section 2.3, this recursion expresses the fact that $\mathcal{E}$ (viewed as a state-price density) prices the contract (δ, V), a condition that can be restated as equation (3.4.15). □

Proposition 3.3.6 gives a context in which λ_0 of Proposition 3.4.12 is the marginal value of wealth $\mathcal{V}'(w)$. The same argument can be applied from the perspective of any other spot. For this reason, we will refer to λ as a **marginal-value-of-wealth** process.

3.5 Scale-Invariant Recursive Utility

The optimality and equilibrium theory to follow assumes a recursive utility that represents scale-invariant (SI) preferences in the sense of Examples 3.2.9 and 3.3.2. In preparation, we analyze a scale-invariant normalized recursive utility U with aggregator f and CE υ (Definition 3.4.9). U is **SI** if $U_{F,t}$ represents SI preferences for every spot (F, t), a condition that, by Lemma 3.4.2, is equivalent to U_0 being SI. For every spot (F, t), $U_{F,t}$ is assumed to be normalized and is therefore SI if and only if it is homogeneous of degree one: $U_{F,t}(sc) = sU_{F,t}(c)$ for all $s \in (0, \infty)$ and

$c \in C$. A conditional aggregator or CE can also be viewed as a normalized utility function and is therefore defined to be **SI** if it is homogeneous of degree one. We call an aggregator f or CE v **SI** if for every nonterminal spot (F, t), the implied conditional aggregator $f_{F,t}$ or CE $v_{F,t}$ is SI. Given the construction a recursive-utility representation in Section 3.4, it is straightforward to verify that *a recursive utility is SI if and only if the corresponding aggregator and conditional CE are both SI.*

EXAMPLE 3.5.1 (Epstein–Zin–Weil utility) By Theorem A.4.3, the expected-utility CE of Example 3.4.7 is SI if and only if

$$v_t = u_\gamma^{-1} \mathbb{E}_t u_\gamma, \text{ where } u_\gamma(x) \equiv \frac{x^{1-\gamma} - 1}{1 - \gamma} \quad \text{(with } u_1 \equiv \log\,), \quad (3.5.1)$$

for a **coefficient of relative risk aversion (CRRA)** γ. Analogously, assuming an additive utility over deterministic plans (characterized by Theorem A.2.3), the aggregator is SI and state and time independent if and only if it takes the form

$$f_t(\omega, c, v) = u_\delta^{-1}((1 - \beta)\, u_\delta(c) + \beta u_\delta(v)), \quad (3.5.2)$$

for parameters $\beta \in (0, 1)$ and $\delta \in \mathbb{R}$, with u_δ defined in (3.5.1). The normalized recursive utility specified in terms of the constant parameters β, δ and γ by the CE (3.5.1) and aggregator (3.5.2) is known as an **Epstein–Zin–Weil (EZW)** utility.[8] (All our later results with EZW utility apply with minor changes to parameters that are adapted processes, which can be thought of as the result of applying Theorem A.4.3 separately to each conditional aggregator $f_{F,t}$ and CE $v_{F,t}$.) The constant parameters (β, δ) of EZW utility determine and are determined by the utility of deterministic consumption plans. Given (β, δ), increasing γ increases risk aversion.

Epstein–Zin–Weil utility reduces to expected discounted utility if and only if $\gamma = \delta$, in which case

$$\tilde{U} \equiv \frac{u_\delta \circ U}{1 - \beta}$$

takes the additive form (3.4.10) of Example 3.4.8 with $u = u_\delta$. Note that on deterministic plans, $\tilde{U}$ takes the same additive form, with $\mathbb{E}_t$ omitted, no matter what the value of γ is. So EZW utility without uncertainty is ordinally equivalent to time-additive discounted utility.

---

[8] Named after the contributions of Epstein and Zin (1989) and Weil (1989, 1990).

Normalizing the more-general expected discounted utility

$$\tilde{U}_t(c) = \mathbb{E}_t\left[\sum_{s=t}^{T-1} b^{s-t}u_\delta(c_s) + \frac{b^{T-t}}{1-\beta}u_\delta(c_T)\right], \qquad (3.5.3)$$

for some positive constant b, not necessarily equal to β, yields the EZW form, but with a time-dependent parameter β. Assuming $\delta \neq 1$, the additive utility (3.5.3) becomes EZW utility (with constant parameters) after a change of the unit of account. Let us call the original unit a "bushel" and the new unit a "dollar," and let $\mathbf{u}$ represent the bushel-to-dollars unit conversion process. For every consumption plan c in bushels, let $c^{\mathbf{u}} \equiv c\mathbf{u}$ denote the same consumption plan in dollars, and let $\tilde{U}^{\mathbf{u}}(c^{\mathbf{u}}) \equiv \tilde{U}(c)$. Clearly, $\tilde{U}^{\mathbf{u}}$ represents the same preferences over consumption plans in dollars as $\tilde{U}$ does over consumption plans in bushels. For the specific unit conversion choice $\mathbf{u}_t \equiv (1+\alpha)^t$, where α solves $(1+\alpha)^{1-\delta}\beta = b$, $\tilde{U}_t^{\mathbf{u}}(c^{\mathbf{u}})$ is ordinally equivalent to

$$\mathbb{E}_t\left[\sum_{s=t}^{T-1} \beta^{s-t}\frac{(c_s^{\mathbf{u}})^{1-\delta}}{1-\delta} + \frac{\beta^{T-t}}{1-\beta}\frac{(c_T^{\mathbf{u}})^{1-\delta}}{1-\delta}\right], \qquad (\delta \neq 1),$$

and therefore the normalized version $U^{\mathbf{u}}$ of $\tilde{U}^{\mathbf{u}}$ is EZW utility with constant parameters β and $\gamma = \delta$. For example, if one is interested in maximizing $\tilde{U}_0(c)$ subject to $\langle \pi \mid c \rangle \leq w$ for some SPD π with $\pi_0 = 1$, one can change the unit of account and equivalently maximize the EZW utility $U_0^{\mathbf{u}}(c^{\mathbf{u}})$ subject to $\langle \pi^{\mathbf{u}} \mid c^{\mathbf{u}} \rangle \leq w\mathbf{u}_0 = w$, where $\pi^{\mathbf{u}} \equiv \mathbf{u}^{-1}\pi$. Note that $\pi^{\mathbf{u}}$ and π differ only in their respective implied short-rate processes $r^{\mathbf{u}}$ and r, which are related by $(1+r^{\mathbf{u}}) = (1+\alpha)(1+r)$. ◇

EXAMPLE 3.5.2 (Multiple-prior forms of recursive utility) Convex duality can be used to express recursive utility in a variety of equivalent ways. This example illustrates the basic idea,[9] by applying Example A.6.2 to the EZW utility U of Example 3.5.1 with $\gamma > \delta = 1$. The claim is that for every consumption plan c,

$$\frac{\log U_0(c)}{1-\beta} = \min_{Q \in \mathcal{Q}}\left\{\tilde{U}_0^Q(c) + \frac{1}{\gamma-1}\tilde{U}_0^Q(\xi^Q)\right\}, \qquad (3.5.4)$$

---

[9] The example is extended in Skiadas (2015) to allow for a nonunit EIS and risk-source dependent CRRA. A continuous-time version can be found in Skiadas (2003). The basic building tool is that of conjugate (or Fenchel) duality, exposited in Rockafellar (1970), which can be applied to any concave or convex function defining the utility recursion to express the corresponding nonlinearity as minimization or maximization over linear forms.

where $\mathcal{Q}$ denotes the set of all full-support probabilities on 2^{Ω}, ξ^Q is the conditional density process of Q with respect to P, and

$$\tilde{U}_0^Q(c) \equiv \mathbb{E}^Q\left[\sum_{t=0}^{T-1} \beta^t \log c_t + \frac{\beta^T}{1-\beta} \log c_T\right].$$

As usual, $\mathbb{E}^Q$ denotes expectation under Q. To show (3.5.4), we apply Example A.6.2 to the conditional CE at each nonterminal spot. By Remark 2.4.5 and associated expression (2.4.7), the identity of Example A.6.2 can be expressed as

$$\log\left(u_\gamma^{-1}\mathbb{E}_t u_\gamma(U_{t+1})\right) = \min_{Q \in \mathcal{Q}}\left\{\mathbb{E}_t^Q \log U_{t+1} + \theta\mathbb{E}_t^Q \log \frac{\xi_{t+1}^Q}{\xi_t^Q}\right\},$$

with the usual abbreviation $U \equiv U(c)$ and

$$\theta \equiv \frac{1}{\gamma - 1}.$$

Substitute into the utility recursion

$$U_t = \exp\left((1-\beta)\log(c_t) + \beta\log\left(u_\gamma^{-1}\mathbb{E}_t u_\gamma(U_{t+1})\right)\right)$$

to find that $V \equiv (1-\beta)^{-1}\log U$ satisfies the recursion

$$V_t = \min_{Q \in \mathcal{Q}}\left\{\log c_t + \frac{\beta\theta}{1-\beta}\mathbb{E}_t^Q \log \frac{\xi_{t+1}^Q}{\xi_t^Q} + \beta\mathbb{E}_t^Q V_{t+1}\right\}.$$

Starting with the terminal value $V_T = (1-\beta)^{-1}\log c_T$ and iterating backward in time, it follows that

$$\frac{\log U_0(c)}{1-\beta} = \min_{Q \in \mathcal{Q}}\left\{\tilde{U}_0^Q(c) + \frac{\theta}{1-\beta}\mathbb{E}^Q\left[\sum_{t=0}^{T-1} \beta^{t+1}\log \frac{\xi_{t+1}^Q}{\xi_t^Q}\right]\right\}.$$

Identity (3.5.4) follows after noting that, since $\xi_0^Q = 1$, the expression in the last expectation can be expanded as

$$\sum_{t=0}^{T-1} \beta^{t+1}\log \frac{\xi_{t+1}^Q}{\xi_t^Q} = \sum_{t=1}^{T} \beta^t \log \xi_t^Q - \beta\sum_{t=1}^{T-1} \beta^t \log \xi_t^Q$$

$$= (1-\beta)\sum_{t=1}^{T-1} \beta^t \log \xi_t^Q + \beta^T \log \xi_T^Q.$$

$\diamond$

At the center of the equilibrium and optimality theory to follow is the utility gradient expression $\mathcal{E}\lambda$ of Proposition 3.4.12, with added regularity implied by scale invariance that we now review. It is a general property of the gradient of a homogeneous-of-degree-one functional that it is homogeneous of degree zero and satisfies the so-called Euler equation: The functional's rate of change at a vector x in the direction of x equals the functional's value at x. The proof is a matter of observing simple properties of the relevant difference quotient. In the following lemma, we state and prove this claim in the language of SI CEs.

LEMMA 3.5.3 *Suppose υ is an SI CE with derivative κ. Then for all processes $s, U \in \mathcal{L}_{++}$,*

$$\kappa_t(s_{t-1}U_t) = \kappa_t(U_t) \quad and \quad \upsilon_{t-1}(U_t) = \mathbb{E}_{t-1}\big[\kappa_t(U_t)\,U_t\big].$$

PROOF Since the derivative κ of υ is in particular a directional derivative, we have the identity

$$\mathbb{E}_{t-1}\big[\kappa_t(s_{t-1}U_t)\,x_t\big] = \lim_{\epsilon \downarrow 0} \frac{\upsilon_{t-1}(s_{t-1}U_t + \epsilon s_{t-1}x_t) - \upsilon_{t-1}(s_{t-1}U_t)}{\epsilon s_{t-1}}$$
$$= \mathbb{E}_{t-1}\big[\kappa_t(U_t)\,x_t\big],$$

where the second equality follows from the fact that υ is SI and therefore the term s_{t-1} can be factored out in the numerator. Since x_t can be any $\mathcal{F}_t$-measurable random variable, $\kappa_t(s_{t-1}U_t) = \kappa_t(U_t)$. For $x_t = U_t$ and $s_{t-1} = 1$, scale invariance allows us to factor out $(1 + \epsilon)$ in the numerator of the above difference quotient, for all $\epsilon > 0$, and conclude that it must equal $\upsilon_{t-1}(U_t)$. $\qquad\square$

The Euler equation for homogeneous functions applied to SI recursive utility gives the following result.

LEMMA 3.5.4 *Suppose U is an SI recursive utility that satisfies the smoothness assumptions of Proposition 3.4.12 and therefore $\pi \equiv \mathcal{E}\lambda$ is the utility gradient of U_0 at c, where λ and $\mathcal{E}$ are defined by (3.4.13) and (3.4.14), respectively. Then*

$$U(c) = \lambda W,$$

where

$$W_t \equiv \mathbb{E}_t\left[\sum_{s=t}^{T} \frac{\pi_s}{\pi_t} c_s\right], \quad t = 0, \ldots, T. \qquad (3.5.5)$$

PROOF Set $x = c$ in (3.4.15). The left-hand side equals $U_t(c)$. $\qquad\square$

For readability, we abuse notation, writing c to denote either a consumption plan or a dummy variable representing a consumption value, and υ to denote either a CE or a dummy variable representing a conditional CE value. An SI aggregator can be written as

$$f_t(\omega, c, \upsilon) = \upsilon g_t\left(\omega, \frac{c}{\upsilon}\right), \quad c, \upsilon \in (0, \infty), \tag{3.5.6}$$

where $g_t(\omega, x) \equiv f_t(\omega, x, 1)$, which leads us to the following terminology and SI aggregator characterization.

DEFINITION 3.5.5 A **proportional aggregator** is a mapping that assigns to each time $t < T$ a function $g_t : \Omega \times (0, \infty) \to (0, \infty)$, where $g_t(\cdot, x)$ is $\mathcal{F}_t$-measurable for all $x \in (0, \infty)$, and for all $\omega \in \Omega$, the function $g_t(\omega, \cdot)$ is continuous and increasing, $g_t(\omega, 1) = 1$, and the mapping $x \mapsto g_t(\omega, x) / x$ is decreasing. A proportional aggregator g **concave** or **differentiable** if $g_t(\omega, \cdot)$ has the respective property for every (ω, t).

LEMMA 3.5.6 *An aggregator f is SI if and only if it takes the form (3.5.6) for some proportional aggregator g, in which case f is concave if and only if g is concave.*

PROOF The first part of the proposition is a straightforward consequence of the definitions. If f is concave, then clearly so is g. Conversely, suppose f is given by (3.5.6) for a concave proportional aggregator g. We fix the reference pair (ω, t) and abuse notation by writing f and g for the functions $f_t(\omega, \cdot)$ and $g_t(\omega, \cdot)$. Consider any pair of distinct points (c^1, υ^1) and (c^2, υ^2) in $(0, \infty)^2$ and let (c, υ) denote their sum. Concavity of g implies

$$g\left(\frac{c}{\upsilon}\right) \geq \frac{\upsilon^1}{\upsilon^1 + \upsilon^2} g\left(\frac{c^1}{\upsilon^1}\right) + \frac{\upsilon^2}{\upsilon^1 + \upsilon^2} g\left(\frac{c^2}{\upsilon^2}\right).$$

Multiplying through by υ shows that $f(c, \upsilon) > f(c^1, \upsilon^1) + f(c^2, \upsilon^2)$, which implies concavity of f, since f is homogeneous of degree one. $\square$

The terminology and notation for proportional aggregators is analogous to that for aggregators. Thus we say that g is **state independent** if $g_t(\omega, \cdot)$ does not vary with ω, and **time independent** if $g_t(\omega, \cdot)$ does not vary with t. We also omit the state variable in expressions like the utility recursion

$$U_t(c) = \upsilon_t(U_{t+1}(c)) \, g_t\left(\frac{c_t}{\upsilon_t(U_{t+1}(c))}\right), \quad t < T, \quad U_T = c_T. \tag{3.5.7}$$

We next introduce some useful transformations of a proportional aggregator g, which is from now on assumed to be differentiable, with $g'_t(\omega, \cdot)$ denoting the derivative of $g_t(\omega, \cdot)$.

The **elasticity** h of g is defined for every time $t < T$ by

$$h_t(x) \equiv \frac{d \log g_t(x)}{d \log x} = \frac{x g'_t(x)}{g_t(x)}, \quad x \in (0, \infty). \tag{3.5.8}$$

Since x, $g_t(x)$ and $g'_t(x)$ are all positive, so is $h_t(x)$. The fact that $g_t(x)/x$ is decreasing is equivalent to $h_t(x) \in (0, 1)$ for all $x \in (0, \infty)$. Note also that since $g_t(1) = 1$, we have $g'_t(1) = h_t(1) \in (0, 1)$. The functions g'_t and h_t arise naturally in the following calculations.

LEMMA 3.5.7 *Under the assumptions of Lemma 3.5.4,*

$$\lambda_t = g'_t(x_t) \quad and \quad \frac{c_t}{W_t} = h_t(x_t), \quad for \ all \ t < T, \tag{3.5.9}$$

where W_t is defined in (3.5.5) and

$$x_t \equiv \frac{c_t}{v_t(U_{t+1}(c))}. \tag{3.5.10}$$

PROOF The expression for λ_t follows from the definition (3.4.13) of λ and identity (3.5.6). Identity $U(c) = \lambda W$ from Lemma 3.5.4 together with the definitions of x and h, and the utility recursion (3.5.7) imply

$$\frac{c_t}{W_t} = \frac{c_t g'_t(x_t)}{U_t(c)} = \frac{c_t g'_t(x_t)}{v_t(U_{t+1}(c)) g_t(x_t)} = h_t(x_t).$$

$\square$

Another transformation of g_t that is going to be useful in the theory of SI pricing to follow is defined as

$$q_t(x) \equiv \frac{g_t(x) - x g'_t(x)}{g'_t(x)}, \quad x \in (0, \infty). \tag{3.5.11}$$

The following lemma shows that the function q_t arises naturally in computing the gradient of an SI recursive utility.

LEMMA 3.5.8 *Given the assumptions of Proposition 3.4.12, suppose further that U is SI recursive utility with proportional aggregator g, and let q_t be defined by (3.5.11). Then the gradient $\pi \equiv \nabla U_0(c)$ solves the recursion*

$$\pi_0 = \lambda_0, \quad \frac{\pi_t}{\pi_{t-1}} = q_{t-1}(x_{t-1}) \kappa_t(U_t(c)) \lambda_t, \quad t = 1, \ldots, T, \tag{3.5.12}$$

where x_t is defined in (3.5.10), and λ is given in (3.5.9) with $\lambda_T = 1$.

PROOF The claim is a corollary of Proposition 3.4.12 and the observation that

$$q_t(x) = \frac{\partial f_t(c, v)/\partial v}{\partial f_t(c, v)/\partial c}, \quad x \equiv \frac{c}{v} \in (0, \infty).$$

□

The elasticity of g should not be confused with the **elasticity of intertemporal substitution (EIS)**, which in the current context is the inverse of the elasticity of q:

$$\frac{1}{\text{EIS}_t(x)} \equiv \frac{d \log q_t(x)}{d \log x}, \quad x \in (0, \infty). \tag{3.5.13}$$

EXAMPLE 3.5.9 (Constant-EIS aggregator) A constant EIS corresponds to the SI aggregator form (3.5.2) of Example 3.5.1. Suppose the proportional aggregator g implies a constant EIS and let $\delta \equiv 1/\text{EIS}$. Define the parameter β by $1 - \beta \equiv g_t'(1) = h_t(1) \in (0, 1)$. Integrating (3.5.13) and using the identity $h_t(x) = 1/(1 + q_t(x)/x)$, we have

$$q_t(x) = \frac{\beta x^\delta}{1 - \beta} \quad \text{and} \quad h_t(x) = \frac{(1 - \beta) x^{1-\delta}}{(1 - \beta) x^{1-\delta} + \beta}. \tag{3.5.14}$$

Finally, integrating (3.5.8) and using the fact that $g_t(1) = 1$ results in the proportional aggregator $g_t(x) \equiv f_t(x, 1)$, where f is the aggregator (3.5.2) of the EZW specification of Example 3.5.1. ◇

The unit-EIS proportional aggregator is an example of what we will call a regular proportional aggregator, which is defined next and used in simplifying some secondary aspects of the optimality theory to follow.

DEFINITION 3.5.10 The proportional aggregator g is **regular** if it is differentiable and for all $\omega \in \Omega$, the derivative $g_t'(\omega, \cdot) : (0, \infty) \to (0, \infty)$ is decreasing and satisfies

$$\lim_{x \downarrow 0} g_t'(\omega, x) = \infty \quad \text{and} \quad \lim_{x \to \infty} g_t'(\omega, x) = 0.$$

In the remainder of this section we assume that g is a regular proportional aggregator and we follow the usual notational convention of omitting the implied state variable ω. In this case, the derivative g_t' is invertible and therefore $\mathcal{I}_t : \Omega \times (0, \infty) \to (0, \infty)$ is well defined by

$$g_t'(\mathcal{I}_t(\lambda)) = \lambda, \quad \lambda \in (0, \infty). \tag{3.5.15}$$

Analogously to c and v, we abuse notation by using the same symbol for a process, λ in this case, and a dummy variable representing possible values of the process. The definition of $\mathcal{I}_t$ is motivated by the relationship $\lambda_t = g_t'(x_t)$ of Lemma 3.5.7, which is equivalent to $x_t = \mathcal{I}_t(\lambda_t)$.

The regularity assumption on g implies that g_t is a (strictly) concave function. The **convex dual**[10] of g_t is the function $g_t^*: \Omega \times (0, \infty) \to (0, \infty)$ defined for every $\lambda \in (0, \infty)$ (and implied state) by

$$g_t^*(\lambda) \equiv \max_{x \in (0,\infty)} \{g_t(x) - \lambda x\} = g_t(\mathcal{I}_t(\lambda)) - \lambda \mathcal{I}_t(\lambda). \qquad (3.5.16)$$

To gain some geometric understanding of this quantity (also encountered in Example A.6.2), draw the graph of g_t on the real plane and a line of slope λ that intersects the graph. Raise the line as high as possible while intersecting the graph and maintaining the slope λ. At the highest such point, the line is tangent to the graph and intersects the vertical axis at $g_t^*(\lambda)$. The fact that this line is above the graph corresponds to the inequality $g_t(x) \leq g_t^*(\lambda) + \lambda x$, with the inequality becoming an equality at the tangency point where $\lambda = g_t'(x)$.

Note that after the change of variables

$$\lambda \equiv g_t'\left(\frac{c}{v}\right) \quad \text{and} \quad x \equiv \frac{c}{v},$$

we have

$$\frac{\partial f_t(c, v)}{\partial v} = g_t^*(\lambda) \quad \text{and} \quad q_t(x) = \frac{g_t^*(\lambda)}{\lambda}. \qquad (3.5.17)$$

An observation that will be useful later on is that equation (3.5.17) uniquely determines λ given $q_t(x)$.

LEMMA 3.5.11 *Assuming the proportional aggregator g is regular, for every time $t < T$ and $q \in (0, \infty)$, there exists a unique $\lambda \in (0, \infty)$ such that $q = g_t^*(\lambda)/\lambda$.*

PROOF Plot the graph of the concave function g_t and note that as the slope $\lambda = g_t'(x)$ of the tangent line at x decreases, the intercept $g_t^*(\lambda)$ with the vertical axis increases. Therefore, g_t^* is a (strictly) decreasing function. Letting the tangent line's slope go to zero, observe that $g_t^*(0+) > g_t(1) = 1$. Consider now the graph of g_t^* on the plane and, for any given $q \in (0, \infty)$, the line $L = \{(\lambda, \lambda q) \mid \lambda \in (0, \infty)\}$. The graph of g_t^* is downward sloping, it is above L near the vertical axis and therefore crosses L at exactly one point, which defines the unique $\lambda \in (0, \infty)$ such that $q = g_t^*(\lambda)/\lambda$. □

---

[10] In more classical terms, if we define $f_t(x) \equiv -g_t(x)$ and $f_t^*(x^*) \equiv g_t^*(-x^*)$, then f_t^* is the Legendre transform of f_t, also known as the convex conjugate of f_t.

3.6 Equilibrium with Scale-Invariant Recursive Utility

We continue taking as given an underlying full-support probability P, defining the expectation operator $\mathbb{E}$, relative to which state-price densities and utility gradients are defined. Proposition 3.4.12 gives the formula $\mathcal{E}\lambda$ for the gradient of recursive utility at a given reference consumption plan c. If c is optimal for some agent given a market, then $\pi \equiv \mathcal{E}\lambda$ is also an equilibrium state-price density (SPD). Since an SPD prices all traded assets, we have a link between consumption and market prices, which is the basis for so-called consumption-based asset pricing models. A useful way of thinking about equilibrium price restrictions is in terms of the period-t **intertemporal marginal rate of substitution (IMRS)** π_t/π_{t-1}, where π is the utility gradient at a given reference consumption plan c. If c is optimal, the IMRS places recursive restrictions on every traded (or synthetic) contract. For example, as we saw in Section 2.3, assuming the contract (δ, V) has well-defined returns $R_t \equiv V_t/(V_{t-1} - \delta_{t-1})$, its pricing by π can be stated as

$$\mathbb{E}_{t-1}\left[\frac{\pi_t}{\pi_{t-1}}R_t\right] = 1.$$

Applied to a traded money-market account with rate process $r \in \mathcal{P}_0$, we have the IMRS mean restriction $\mathbb{E}_{t-1}\pi_t/\pi_{t-1} = (1+r_t)^{-1}$, while the Hansen–Jagannathan bound (2.3.8) implies a lower bound on the IMRS conditional variance.

REMARK 3.6.1 In interpreting the IMRS, it is important to keep in mind that it is defined relative to P. Suppose π^Q is an SPD relative to another full-support probability Q and let $\xi_t \equiv \mathbb{E}\left[dQ/dP\right]$. (Recall that Q is an EMM if π^Q is predictable, but we do not assume that here.) By Lemma 2.4.4, if the pairs (P, π) and (Q, π^Q) represent the same present-value function, then $\pi_t/\pi_0 = \pi_t^Q\xi_t/\pi_0^Q$. ◇

In the remainder of this section we elaborate on equilibrium pricing restrictions under the assumption of SI recursive utility, which is consistent with the representative-agent argument of Example 3.2.9. The theory presented is agnostic toward this interpretation—the consumption plan c can be either an individual investor's plan or a representative agent's plan.

A formula for the equilibrium IMRS for SI recursive utility was established in Lemma 3.5.8. We now show how the IMRS can be determined by taking as input consumption growth only. As a preliminary example, consider an agent maximizing the additive utility of Example 3.4.8 under the assumption of scale invariance, which implies the EZW specification of

Example 3.5.1 with $\gamma = \delta$. The utility gradient at a reference consumption plan c can be computed spot by spot, resulting in the IMRS:[11]

$$\frac{\pi_t}{\pi_{t-1}} = \beta \left(\frac{c_t}{c_{t-1}} \right)^{-\delta} \qquad \text{(assuming } \gamma = \delta\text{)}. \qquad (3.6.1)$$

The IMRS in this case is determined by consumption growth at every spot on the information tree. The following proposition shows that for a more-general SI recursive utility the mapping from consumption growth to IMRS is less direct, requiring the solution of a backward recursion on the information tree. Although not explicitly stated, the proof shows that the process x corresponds to the consumption-to-CE ratio of equation (3.5.10).

PROPOSITION 3.6.2 (IMRS and consumption growth) *Suppose U is a normalized SI recursive utility with differentiable proportional aggregator g and CE υ with derivative κ. Given a consumption plan c, let the process x be defined by the backward recursion, where $g_T(x) \equiv x$,*

$$x_{t-1} = \upsilon_{t-1} \left(\frac{g_t(x_t)}{x_t} \frac{c_t}{c_{t-1}} \right)^{-1}, \quad t = 1, \dots, T; \quad x_T = 1. \qquad (3.6.2)$$

Then $\pi \in \mathcal{L}_{++}$ is the gradient of U_0 at c if and only if $\pi_0 = g_0'(x_0)$ and

$$\frac{\pi_t}{\pi_{t-1}} = q_{t-1}(x_{t-1}) \kappa_t \left(\frac{g_t(x_t)}{x_t} \frac{c_t}{c_{t-1}} \right) g_t'(x_t), \quad t = 1, \dots, T, \qquad (3.6.3)$$

where the functions q_t are defined by (3.5.11).

PROOF With some abuse of notation, let

$$x_t \equiv \frac{c_t}{\upsilon_t(U_{t+1})}, \quad U \equiv U(c). \qquad (3.6.4)$$

By the recursion defining the utility function,

$$U_t = \upsilon_t(U_{t+1}) g_t(x_t) = c_{t-1} \frac{g_t(x_t)}{x_t} \frac{c_t}{c_{t-1}}.$$

Since υ_{t-1} is homogeneous of degree one,

$$\upsilon_{t-1}(U_t) = c_{t-1} \upsilon_{t-1} \left(\frac{g_t(x_t)}{x_t} \frac{c_t}{c_{t-1}} \right).$$

Substituting into equation (3.6.4) with $t - 1$ in place of t results in recursion (3.6.2). Since recursion (3.6.2) uniquely determines x, the recursion's

---

[11] This direct link between the IMRS and consumption growth was established in an influential paper by Lucas (1978).

solution must equal the x defined by equation (3.6.4). The same expression for U_t and the fact that κ_t is homogeneous of degree zero (Lemma 3.5.3) result in the identity

$$\kappa_t(U_t) = \kappa_t\left(\frac{g_t(x_t)}{x_t}\frac{c_t}{c_{t-1}}\right).$$

Lemma 3.5.8 completes the proof.   □

EXAMPLE 3.6.3 For the constant-CRRA CE (3.5.1), Example 3.4.11 and equation (3.6.2) imply that the middle factor of the IMRS expression (3.6.3) can be written as

$$\kappa_t\left(\frac{g_t(x_t)}{x_t}\frac{c_t}{c_{t-1}}\right) = \left(\frac{x_t}{x_{t-1}}\right)^\gamma\left(g_t(x_t)\frac{c_t}{c_{t-1}}\right)^{-\gamma}.$$

◇

Given scale invariance, pricing in terms of consumption growth is closely related to pricing in terms of the market return. We elaborate in the context of an SI recursive utility. Given a consumption plan $c \in \mathcal{L}_{++}$, suppose π is the gradient of U_0 at c. The following proposition relates the corresponding IMRS to the quantity

$$M_t \equiv \frac{W_t}{W_{t-1} - c_{t-1}}, \quad \text{where} \quad W_t \equiv \mathbb{E}_t\left[\sum_{s=t}^T \frac{\pi_s}{\pi_t}c_s\right]. \tag{3.6.5}$$

In the representative-agent pricing context of Example 3.2.9, c is the aggregate endowment and M is the market return process.

PROPOSITION 3.6.4 (IMRS and market returns) *Suppose U is a normalized SI recursive utility whose CE υ has derivative κ and whose proportional aggregator g is regular, with convex dual g^* as defined in (3.5.16). Given any $c, \pi \in \mathcal{L}_{++}$, let the process M be defined by (3.6.5). Then the process $\lambda \in \mathcal{L}_{++}$ is uniquely defined as the solution to the following backward recursion, which also defines the process q along the way:*

$$q_{t-1} = \frac{g_{t-1}^*(\lambda_{t-1})}{\lambda_{t-1}} = \frac{1}{\upsilon_{t-1}(\lambda_t M_t)}, \quad \lambda_T = 1. \tag{3.6.6}$$

Finally, if π is the gradient of U_0 at c, then

$$\frac{\pi_t}{\pi_{t-1}} = q_{t-1}\kappa_t(\lambda_t M_t)\lambda_t, \quad t = 1,\ldots,T, \quad \pi_0 = \lambda_0. \tag{3.6.7}$$

PROOF Suppose π is the gradient of U_0 at c, and let $U \equiv U(c)$. By Lemma 3.5.4 and the definition of M_t,

$$U_t = \lambda_t W_t = (W_{t-1} - c_{t-1}) \lambda_t M_t. \tag{3.6.8}$$

By Lemma 3.5.7, $\lambda_t = g_t'(x_t)$, where x_t denotes the consumption-to-CE ratio (3.5.10). Given this fact, the combination of identities (3.5.11) and (3.5.17) proves the first equality in (3.6.6) with $q_t = q_t(x_t)$. The remainder of (3.6.6) is a consequence of the following string of equalities

$$q_{t-1} = x_{t-1} \frac{1 - h_{t-1}(x_{t-1})}{h_{t-1}(x_{t-1})} = \frac{c_{t-1}}{v_{t-1}(U_t)} \frac{W_{t-1} - c_{t-1}}{c_{t-1}} = \frac{1}{v_{t-1}(\lambda_t M_t)}.$$

The first equality follows from the definition of h_t and q_t in (3.5.8) and (3.5.11), the second equality follows from the identity $h_t(x_t) = c_t/W_t$ of Lemma 3.5.7, and the last equality follows by inserting expression (3.6.8) for U_t and simplifying using the homogeneity of v_{t-1}. That the process λ uniquely solves (3.6.6) is shown in Lemma 3.5.11.

The IMRS expression (3.6.7) follows from Lemma 3.5.8, expression (3.6.8) for U_t, and the fact that κ_t is homogeneous of degree zero (Lemma 3.5.3). □

EXAMPLE 3.6.5 In addition to the assumptions of Proposition 3.6.4, suppose that $v_t = u_\gamma^{-1} \mathbb{E}_t u_\gamma$, where u_γ is the power or logarithmic function (3.5.1) for some CRRA $\gamma > 0$. The CE derivative calculation of Example 3.4.11 and equation (3.6.6) imply that $\kappa_t(\lambda_t M_t) = (q_{t-1}\lambda_t M_t)^{-\gamma}$. The IMRS expression (3.6.7) in this case reduces to

$$\frac{\pi_t}{\pi_{t-1}} = (q_{t-1}\lambda_t)^{1-\gamma} M_t^{-\gamma},$$

with q and λ given by (3.6.6). Note that

$$\gamma = 1 \quad \text{implies} \quad \frac{\pi_t}{\pi_{t-1}} = \frac{1}{M_t}. \qquad \diamond \tag{3.6.9}$$

Consider now the consumption growth to market return ratio

$$\Phi_t \equiv \frac{c_t}{c_{t-1}} \frac{1}{M_t} = \left(\frac{1}{\varrho_{t-1}} - 1 \right) \varrho_t, \quad \varrho_t \equiv \frac{c_t}{W_t}, \tag{3.6.10}$$

where the second equation follows from the definition of M in (3.6.5). For a strictly concave differentiable proportional aggregator g, the processes x and λ uniquely determine each other, since $\lambda_t = g_t'(x_t)$. By Lemma 3.5.7, $\varrho_t = h_t(x_t)$ and therefore the ratio Φ is determined by either x or λ. Given x or λ, consumption growth and market returns carry the same information. In Propositions 3.6.2 and 3.6.4, x or λ is determined by a backward recursion. For EZW utility with nonunit EIS, the fact that we can invert the identity $\varrho_t = h_t(x_t)$ means that the IMRS expression can be computed jointly in terms of market returns and consumption growth, without having to solve a backward recursion. The resulting IMRS expression[12] is a geometric average of expression (3.6.1) for the additive ($\gamma = \delta$) case, and expression (3.6.9) for the unit CRRA ($\gamma = 1$) case.

PROPOSITION 3.6.6 (EZW pricing with nonunit EIS) *Suppose U is the EZW utility of Example 3.5.1 for some CRRA γ and inverse-EIS $\delta \neq 1$. Fix any consumption plan c and let M_t be defined by (3.6.5). If $\pi \in \mathcal{L}_{++}$ is the gradient of U_0 at c, then*

$$\frac{\pi_t}{\pi_{t-1}} = \left(\beta \left(\frac{c_t}{c_{t-1}} \right)^{-\delta} \right)^{\phi} \left(\frac{1}{M_t} \right)^{1-\phi}, \quad \phi \equiv \frac{1-\gamma}{1-\delta}. \tag{3.6.11}$$

PROOF Once again, we apply Lemma 3.5.8 and we show that the IMRS expression (3.5.12) reduces to the claimed expression. By Lemma 3.5.7, $\varrho_t \equiv c_t/W_t = h_t(x_t)$, where x_t is the consumption-to-CE ratio. Here h_t is given by equation (3.5.14), and we can therefore invert the last equation to compute x_t as a function of ϱ_t:

$$x_t = \left(\frac{\beta}{1-\beta} \frac{\varrho_t}{1-\varrho_t} \right)^{1/(1-\delta)}.$$

Using the expression for q_t in (3.5.14), we then compute

$$q_t(x_t) = \left(\frac{\beta}{1-\beta} \right)^{1/(1-\delta)} \left(\frac{\varrho_t}{1-\varrho_t} \right)^{\delta/(1-\delta)}.$$

Similarly, using the fact that $g_t(x) = ((1-\beta)x^{1-\delta} + \beta)^{1/(1-\delta)}$, we find

$$\lambda_t = g_t'(x_t) = (1-\beta)^{1/(1-\delta)} \left(\frac{1}{\varrho_t} \right)^{\delta/(1-\delta)}.$$

The last two centered equations and equation (3.6.10) together imply

$$q_{t-1}(x_{t-1})\lambda_t = \beta^{1/(1-\delta)} \Phi_t^{-\delta/(1-\delta)}. \tag{3.6.12}$$

---

[12] The proposition's proof shows the more-general IMRS expression (3.6.13), which applies for any, not necessarily additive, differentiable SI CE. Another version is given in Skiadas (2009) for a constant CRRA CE and any regular proportional aggregator g, provided h is invertible.

Finally, we claim that

$$\kappa_t(U_t(c)) = \kappa_t(\lambda_t M_t) = \kappa_t\left(\Phi_t^{-\delta/(1-\delta)} M_t\right).$$

Since κ_t is homogeneous of degree zero (Lemma 3.5.3), the first equation follows from equation (3.6.8) (just as in the proof of Proposition 3.6.4), and the second equation follows from equation (3.6.12). The last two displays and IMRS expression (3.5.12) result in

$$\pi_0 = \lambda_0 \quad \text{and} \quad \frac{\pi_t}{\pi_{t-1}} = \beta^{1/(1-\delta)} \Phi_t^{-\delta/(1-\delta)} \kappa_t\left(\Phi_t^{-\delta/(1-\delta)} M_t\right). \quad (3.6.13)$$

We will also use the identity

$$\upsilon_{t-1}\left(\Phi_t^{-\delta/(1-\delta)} M_t\right) = \beta^{-1/(1-\delta)}. \quad (3.6.14)$$

To show it, let $q_t \equiv q_t(x_t)$. As in the proof of Proposition 3.6.4, $q_{t-1} = 1/\upsilon_{t-1}(\lambda_t M_t)$ and therefore $\upsilon_{t-1}(q_{t-1}\lambda_t M_t) = 1$ (since υ is SI). Substituting expression (3.6.12) for $q_{t-1}\lambda_t$ results in (3.6.14).

All results so far apply for any differentiable SI CE. For a constant CRRA CE, $\kappa_t(z) = (z/\upsilon_{t-1}(z))^{-\gamma}$. Applying this expression with $z = \Phi_t^{-\delta/(1-\delta)} M_t$ in the IMRS expression (3.6.13) and using (3.6.14) the claimed IMRS expression (3.6.11) follows. □

EXAMPLE 3.6.7 (Unit EIS) The key to the proof of Proposition 3.6.6 is the invertibility of the identity $\varrho_t = h_t(x_t)$ of Lemma 3.5.7, which does not hold in the unit-EIS case $g_t(x) = x^{1-\beta}$, $\beta \in (0, 1)$, where $h_t(x) = 1 - \beta$. Equation (3.6.10) in this case gives

$$\frac{c_t}{c_{t-1}} = \beta M_t \qquad \text{(assuming unit EIS).} \qquad \diamond$$

This section's results attribute all IMRS variability to stochastic consumption growth and/or market returns. Another source of IMRS variability can be hard-wired into preferences, for example, through stochastic parameters β, γ, δ, or beliefs (Remark 3.6.1). There are many other possible sources of IMRS variability that do not fit this section's formalism, including agent heterogeneity, nontradeability of labor income due to moral hazard concerns, collateral constraints and associated leverage dynamics, other institutional constraints, market panics or runs, limited ability to model the future leading to model revisions violating dynamic consistency, and trading patterns driven by narrative and influence dynamics that are hard to explain in a Bayesian framework.

3.7 Optimal Consumption and Portfolio Choice

In this section we discuss the problem of finding an optimal trading strategy for an agent that maximizes an SI recursive utility U given a positive initial financial wealth w and no subsequent income. (Any endowment is effectively assumed marketed and already sold.) We assume an arbitrage-free market X that is implemented by $1 + J$ contracts, denoted as in Section 1.8, with every V^j strictly positive. Since the market is arbitrage-free, S_{t-1}^j and $R_t^j \equiv V_t^j / S_{t-1}^j$ are also strictly positive, for all $t > 0$. Contract 0 is a money-market account (MMA) with rate process r and return process $R^0 \equiv 1 + r$.

Suppose the agent converts the initial wealth w to a consumption plan $c \in \mathcal{L}_{++}$ by following a trading strategy (θ^0, θ). The latter defines in (1.8.4) a corresponding portfolio allocation policy $\psi = (\psi^1, \ldots, \psi^J)$, and equation (1.8.5) defines the return process R^ψ. The agent enters period t with financial wealth W_{t-1}, consumes c_{t-1} and invests the remainder according to the allocation ψ_t earning a return R_t^ψ for the period:

$$W_0 = w, \quad W_t = (W_{t-1} - c_{t-1})R_t^\psi, \quad W_T = c_T. \tag{3.7.1}$$

The pair (c, W) is related to the synthetic contract $(\delta^\theta, V^\theta)$ by

$$S_{t-1}^\theta = W_{t-1} - c_{t-1}, \quad W_t = V_t^\theta, \quad c_t = \delta_t^\theta, \quad t = 1, \ldots, T.$$

Note that the middle equation in (3.7.1) states that $R_t^\theta = R_t^\psi$. The agent effectively spends $w - c_0$ at time zero to purchase the synthetic contract $(\delta^\theta, V^\theta)$ and subsequently consumes all dividends. Since every synthetic contract is traded, the contract $(\delta^\theta, V^\theta)$ is priced by every SPD π, which is condition (3.5.5) and the reason we have used the same notation W in both instances.

With this background in mind, we state the optimal consumption–portfolio problem more parsimoniously by taking as primitive the contract returns and the agent's initial wealth, and expressing optimal decisions in terms of wealth allocations in every period.

DEFINITION 3.7.1 A **consumption** (allocation) **policy** is a $(0, 1)$-valued adapted process ϱ such that $\varrho_T = 1$. A **portfolio** (allocation) **policy** is a process $\psi = (\psi^1, \ldots, \psi^J) \in \mathcal{P}_0^{1 \times J}$. An **allocation policy** is a pair (ϱ, ψ) of a consumption policy and a portfolio policy. A **wealth process** is a strictly positive adapted process. The allocation policy (ϱ, ψ) **generates** the wealth process W defined recursively by

$$W_0 = w, \quad W_t = W_{t-1}(1 - \varrho_{t-1})R_t^\psi, \tag{3.7.2}$$

in which case the allocation policy (ϱ, ψ) **finances** the consumption plan $c \equiv \varrho W$, and is **optimal** if c is optimal, that is, there is no $x \in X$ such that $U_0(c + x) > U_0(c)$.

The section's main result follows. For simplicity, we assume a concave utility, a differentiable CE v, and that the period-t portfolio allocation can be any member of the set

$$\Psi_t \equiv \left\{ \psi_t \in L_{t-1}^{1 \times J} \mid R_t^{\psi} \text{ is strictly positive} \right\}. \tag{3.7.3}$$

Although not shown here,[13] it is worth noting that the following result remains true if v is only assumed to be concave (not necessarily differentiable) and Ψ_t is an arbitrary spot-dependent nonempty convex subset of the set in (3.7.3).

THEOREM 3.7.2 *Suppose U is a normalized SI recursive utility with a regular proportional aggregator g and a concave differentiable CE v. Let h, $\mathcal{I}$ and g^* be defined in (3.5.8), (3.5.15) and (3.5.16), respectively. An allocation policy (ϱ, ψ) is optimal if and only if it is generated by the following recursive procedure.*

(1) *(Initialization) Set $\lambda_T = \varrho_T = 1$ and $t = T$.*
(2) *(Recursive step) Given λ_t, select $\psi_t \in \Psi_t$ such that*

$$\upsilon_{t-1}\left(\lambda_t R_t^{\psi}\right) = \max \left\{ \upsilon_{t-1}\left(\lambda_t R_t^{p}\right) \mid p \in \Psi_t \right\}, \tag{3.7.4}$$

let $\lambda_{t-1} \in L_{t-1}^{++}$ be the unique solution to

$$\frac{\lambda_{t-1}}{g_{t-1}^*(\lambda_{t-1})} = \upsilon_{t-1}\left(\lambda_t R_t^{\psi}\right), \tag{3.7.5}$$

and set $\varrho_{t-1} = h_{t-1}(\mathcal{I}_{t-1}(\lambda_{t-1}))$.
(3) *(Loop) While $t > 1$, decrease t by one and repeat (2).*

Assuming (ϱ, ψ) is optimal and finances c, the process λ generated by this algorithm is the marginal-value-of-wealth process defined by (3.4.13), and $U(c) = \lambda W$, where W is the wealth process generated by (ϱ, ψ).

REMARK 3.7.3 Since U is assumed to be SI, the CE v is homogeneous of degree one. We saw in Example 3.3.2 that a homogeneous-of-degree-one utility is concave if and only if the corresponding preference correspondence is convex. The argument also applies to the conditional CE. In particular, a risk-averse SI expected-utility CE is necessarily concave (since, by Theorem A.6.4, the conditional CE at every spot is ordinally equivalent to a concave utility function).

[13] See Skiadas (2013a) for a proof.

REMARK 3.7.4 By Lemma 3.5.6, the assumed concavity of the proportional aggregator implies the concavity of the aggregator. Even without the SI assumption, if both the aggregator and CE are concave, the recursive utility is concave. This follows by a straightforward induction, starting with the terminal date and then showing that U_t is concave given that U_{t+1} is concave, using the utility recursion.

PROOF Setting to zero the partial derivatives of the value being maximized in (3.7.4) with respect to each contract's allocation, we have the optimality condition

$$\mathbb{E}_{t-1}\left[\kappa_t\left(\lambda_t R^\psi\right)\lambda_t\left(R_t^j - R_t^0\right)\right] = 0, \quad j = 1,\ldots,J, \tag{3.7.6}$$

where κ is the derivative of v. Since the CE v is assumed to be concave, condition (3.7.6) is necessary and sufficient for (3.7.4).

We henceforth assume that the allocation policy (ϱ, ψ) generates the wealth process W and finances the consumption plan $c \equiv \varrho W$, which by the budget equation in the form (3.7.1) implies that

$$R_t^\psi = \frac{W_t}{W_{t-1} - c_{t-1}}. \tag{3.7.7}$$

Necessity Suppose c is optimal with corresponding marginal-value-of-wealth process λ. Then $\pi \equiv \nabla U_0(c)$ is an SPD, and therefore

$$\mathbb{E}_{t-1}\left[\frac{\pi_t}{\pi_{t-1}} R_t^j\right] = 1, \quad j = 0, 1,\ldots,J, \tag{3.7.8}$$

which in turn implies

$$\mathbb{E}_{t-1}\left[\frac{\pi_t}{\pi_{t-1}} R_t^\psi\right] = 1, \quad t = 1,\ldots,T. \tag{3.7.9}$$

Substituting expression (3.7.7) for R^ψ into (3.7.9) shows that π prices the contract (c, W) (technically, the contract $(c - c_0 1_{\Omega \times \{0\}}, W - c_0 1_{\Omega \times \{0\}})$, since $c_0 \neq 0$). It follows that W satisfies (3.5.5) and we can apply Proposition 3.6.4 with $M = R^\psi$. Therefore λ uniquely solves recursion (3.7.5) and π_t/π_{t-1} is given by equation (3.6.7), which in conjunction with the state-pricing condition (3.7.8) implies the optimality condition (3.7.6), and hence (3.7.4). The expression for ϱ follows from Lemma 3.5.7.

Sufficiency Suppose λ, q and π are defined by (3.6.6) and (3.6.7), but with R^ψ in place of M. By construction and Lemma 3.5.3,

$$1 = q_{t-1} \upsilon_{t-1}\left(\lambda_t R_t^{\psi}\right) = q_{t-1} \mathbb{E}_{t-1}\left[\kappa_t\left(\lambda_t R_t^{\psi}\right) \lambda_t R_t^{\psi}\right] = \mathbb{E}_{t-1}\left[\frac{\pi_t}{\pi_{t-1}} R_t^{\psi}\right].$$

Therefore, equation (3.7.9) is satisfied, which as we saw earlier implies that W satisfies (3.5.5). The interested reader can now show that $U(c) = \lambda W$ by verifying the utility recursion, and that $\pi = \nabla U_0(c)$ by applying Lemma 3.5.8. Assuming ψ solves (3.7.4), the corresponding optimality conditions (3.7.6) are satisfied, and therefore $\mathbb{E}_{t-1}\left[(\pi_t/\pi_{t-1}) R_t^j\right]$ takes the same value for all j, a value necessarily equal to one, since we have already shown (3.7.9). It follows that π prices every contract implementing the market and is therefore an SPD. By Remark 3.7.4, the utility is concave, and optimality of c follows by Proposition 3.3.4. $\qquad\square$

EXAMPLE 3.7.5 (Unit EIS and optimal consumption policy) Suppose that the proportional aggregator takes the unit-EIS form of Example 3.5.9 with $\delta = 1$. In this case, $h_t(x) = 1 - \beta$ for all x, and therefore the optimal consumption-to-wealth ratio is constant: $\varrho_{t-1} = 1 - \beta$. As we saw in Example 3.5.1, if the conditional CE is an expected-utility one with unit CRRA, then the recursive utility is ordinally equivalent to expected discounted logarithmic utility. By moving away from the expected discounted utility framework, we can vary risk aversion while retaining the simplifying assumption of a constant consumption-to-wealth ratio. $\diamond$

EXAMPLE 3.7.6 (Markovian formulation) Suppose that the contracts implementing the market have the Markovian structure of Section 2.6, which is defined in terms of an underlying martingale basis with stochastically independent increments. In this case, the process λ and the optimal allocation policy can be expressed as functions of the Markov state: $\lambda_t = \lambda(t, Z_t)$, $\varrho_t = \varrho(t, Z_t)$, and $\psi_t = \psi(t, Z_{t-1})$, with the abuse of notation defined in Section 2.6. Analogously to the arbitrage-pricing application of Section 2.6, the significance of the Markovian formulation is that the recursive formula determining ψ_t and λ_{t-1} in terms of λ_t need be evaluated only for every possible value of the Markov state Z_{t-1} rather than every time-$(t-1)$ spot, which can dramatically reduce the problem's computational complexity. $\diamond$

EXAMPLE 3.7.7 (Deterministic marginal value of wealth) Suppose that the short-rate process r is deterministic and period-t excess returns $R_t^j - R_t^0$ are stochastically independent of $\mathcal{F}_{t-1}$, for all $t > 0$. In this case, a backward induction shows that the marginal-value-of-wealth process λ at the optimum is deterministic. As a consequence, the optimal allocation policy (ϱ, ψ) is also deterministic, with the optimal time-t portfolio weights determined as the solution to the myopic problem

$$\upsilon_{t-1}\left(R_t^{\psi}\right) = \max\left\{\upsilon_{t-1}\left(R_t^p\right) \mid p \in \Psi_t\right\}.$$

$\diamondsuit$

The algorithm of Theorem 3.7.2 produces a solution provided there is an optimal portfolio solution to problem (3.7.4). We show below that this is indeed the case for an expected-utility SI CE with CRRA $\gamma > 0$. Problem (3.7.4) in this case can be equivalently expressed as

$$\psi_t \in \arg\max\left\{\mathbb{E}_{t-1}^Q\left[\frac{\left(R_t^p\right)^{1-\gamma} - 1}{1 - \gamma}\right] \mid p \in \Psi_t\right\}, \qquad (3.7.10)$$

provided we define the probability Q to have the conditional density process ξ determined by the recursion

$$\frac{\xi_t}{\xi_{t-1}} = \frac{\lambda_t^{1-\gamma}}{\mathbb{E}_{t-1}\left[\lambda_t^{1-\gamma}\right]}, \qquad \xi_0 = 1. \qquad (3.7.11)$$

The reason for this is the change-of-measure formula of Lemma 2.4.10.

EXAMPLE 3.7.8 (Unit CRRA and optimal portfolio choice) The optimal portfolio problem with unit CRRA ($\gamma = 1$) reduces to

$$\psi_t \in \arg\max\left\{\mathbb{E}_{t-1}\log\left(R_t^p\right) \mid p \in \Psi_t\right\}.$$

In this case, the optimal portfolio weights are the same as for a myopic agent who maximizes conditional expected logarithmic utility over a single period. In contrast to Example 3.7.7, the myopic portfolio rule is optimal even if the marginal-value-of-wealth process is stochastic. Referring back to Example 3.7.5, recall that a unit EIS implies a myopic consumption policy. The intersection of that example and the present one is the case of expected discounted logarithm utility, in which case the entire optimal policy (ϱ, ψ) is myopic. $\diamondsuit$

PROPOSITION 3.7.9 *For an expected-utility SI CE with positive CRRA* γ, *the optimal-portfolio problem* (3.7.4) *has a solution and therefore the algorithm of Theorem 3.7.2 produces a solution.*

PROOF The argument applies over a single period at each nonterminal spot. We therefore assume, without loss of generality, that $T = 1$ and the underlying probability P coincides with the probability Q in (3.7.10). We omit time subscripts and let $u(x) \equiv \left(x^{1-\gamma} - 1 \right) / (1 - \gamma)$. Suppose first that $\gamma \geq 1$. Given any SPD $(1, \pi)$, define the compact set

$$A \equiv \left\{ z \in L^{++} \mid \mathbb{E}u(z) \geq u(1+r) , \; \mathbb{E}\left[\pi z \right] = 1 \right\}$$

and the closed set $B \equiv \left\{ R^p \mid p \in \mathbb{R}^{1 \times J} \right\}$. Since $1 + r \in B$ and $\mathbb{E}[\pi z] = 1$ for all $z \in B$, the optimal portfolio problem is equivalent to maximizing the continuous function $\mathbb{E}u$ over the compact set $A \cap B$. Existence of a maximum follows by Proposition B.3.7. If $\gamma \in (0, 1)$, A as defined above is not compact. We instead define A to be the set of every $[0, \infty)$-valued random variable z such that $\mathbb{E}\left[\pi z \right] = 1$. Arguing as above, we conclude there is an optimal portfolio choice, provided we verify that the optimizing value of z is strictly positive. This follows from the fact that the marginal value $z^{-\gamma}$ becomes infinite at zero and therefore there is always a utility-improving small deviation away from zero.   □

3.8 Recursive Utility and Optimality in Continuous Time

This section introduces recursive utility and the associated problem of optimal consumption and portfolio choice in the Brownian setting of Section 2.8. Technicalities aside, the resulting formulation is simpler than its discrete counterpart thanks to the quadratic approximations encoded in the Ito calculus. A rigorous treatment of this material requires sophisticated and involved mathematics. The objective here is to provide a more-accessible informal introduction that is consistent with the rigorous theory, but omits several mathematical details, focusing instead on essential results and intuition based on the discrete theory. As in Section 2.8, the underlying filtration is generated by an SBM B over the time horizon $[0, T]$, under some probability with expectation operator $\mathbb{E}$. The extension to a filtration generated by multiple independent SBMs is straightforward and mainly a matter of introducing appropriate matrix notation.

A **consumption plan** is a strictly positive adapted process c (satisfying omitted technical requirements for utility processes and related quantities to be well defined). As in the discrete case, we differentiate between consumption over a single period, which in the current context is $c_t dt$ over $[t, t + dt]$, and terminal lump-sum consumption c_T. State-price densities and utility gradients are defined in terms of the inner product

$$\langle x \mid y \rangle = \mathbb{E}\left[\int_0^T x_t y_t \, dt + x_T y_T \right].$$

To formulate recursive utility in continuous time, we fix a reference consumption plan c and, abusing notation, we let $U \equiv U(c)$ denote the corresponding normalized utility process with Ito decomposition

$$dU_t = \alpha_t^U \, dt + \beta_t^U \, dB_t, \quad U_T = c_T.$$

As we saw in Sections 2.5 and 2.8, expressing α_t^U as a function of $\left(U_t, \beta_t^U\right)$ can be thought of as a backward-in-time recursion, which is formally expressed as a backward stochastic differential equation (BSDE) to be solved jointly in $\left(U, \beta^U\right)$. This is how we can specify recursive utility directly in continuous time.

The heuristic argument that follows can be applied to a broader class of recursive utilities, but for concreteness we assume the expected utility CE[14] $v_t \equiv u^{-1} \mathbb{E}_t u$, for some continuous increasing function $u \colon (0, \infty) \to \mathbb{R}$ with Arrow–Pratt coefficient of absolute risk aversion $A \equiv -u''/u'$. Ito's lemma implies

$$u(U_{t+dt}) = u(U_t) + u'(U_t)\left(\alpha_t^U - \frac{A(U_t)}{2}\left(\beta_t^U\right)^2\right) dt + u'(U_t)\, dB_t.$$

Apply the conditional expectation $\mathbb{E}_t$ on both sides, which eliminates the Brownian term, and then apply u^{-1} on both sides, followed by a first-order Taylor expansion on the right-hand side. The result is the **Arrow–Pratt CE approximation**:[15]

$$v_t(U_{t+dt}) = U_t + \left(\alpha_t^U - \frac{A(U_t)}{2}\left(\beta_t^U\right)^2\right) dt.$$

We can heuristically rewrite the Arrow–Pratt approximation as

$$v_t(U_{t+dt}) = \mathbb{E}_t U_{t+dt} - \frac{A(U_t)}{2} \mathrm{var}_t\left[U_{t+dt}\right].$$

As we will see, this is the key in concluding that if a myopic optimal portfolio in a scale-invariant formulation is justified, the optimal portfolio has to be mean-variance efficient in the maximum-Sharpe-ratio sense of Section 2.2 over every infinitesimal time interval.

The aggregator of the recursive utility can also be specified more generally, but for concreteness we assume

---

[14] Skiadas (2013c) shows how expected-utility CEs can approximate a wider class of nonexpected-utility smooth CEs over small risks, which arise naturally with recursive utility in high-frequency models and their continuous-time limit.

[15] Named after the contributions of Arrow (1965, 1971) and Pratt (1964).

$$u_\delta(U_t) = \left(1 - e^{-\beta dt}\right)u_\delta(c_t) + e^{-\beta dt}u_\delta\left(v_t(U_{t+dt})\right), \quad U_T = c_T, \quad (3.8.1)$$

where $\beta > 0$ is a constant (not to be confused with β^U). For now, $u_\delta \colon (0, \infty) \to \mathbb{R}$ can be any increasing and continuously differentiable function. Inserting the Arrow–Pratt CE approximation in this recursion and using a first-order Taylor expansion results in an expression for α_t^U in terms of $\left(U_t, \beta_t^U\right)$ corresponding to the BSDE

$$dU_t = -\left(f(c_t, U_t) - \frac{A(U_t)}{2}\left(\beta_t^U\right)^2\right)dt + \beta_t^U\, dB_t, \quad U_T = c_T, \quad (3.8.2)$$

where

$$f(c, U) \equiv \beta\frac{u_\delta(c) - u_\delta(U)}{u_\delta'(U)}.$$

We proceed assuming the BSDE has a unique solution,[16] and the solution is increasing in c. Analogously to the discrete case, this section's entire analysis applies with only cosmetic changes if the parameters f and A are state and time dependent, provided $f(\omega, t, c, U)$ and $A(\omega, t, U)$ as functions of (ω, t) are adapted processes (and technically sufficiently regular for the relevant integrals and BSDEs to make sense).

REMARK 3.8.1 The quadratic term in the drift of BSDE (3.8.2) can be eliminated by passing to the ordinally equivalent utility $V_t \equiv u(U_t)$. Letting $\varphi \equiv u_\delta \circ u^{-1}$, recursion (3.8.1) can be restated as

$$\varphi(V_t) = \left(1 - e^{-\beta dt}\right)\varphi(u(c_t)) + e^{-\beta dt}\varphi\left(\mathbb{E}_t V_{t+dt}\right), \quad V_T = u(c_T).$$

This is of the same form as (3.8.1). The argument leading to (3.8.2) in this case leads to the BSDE:[17]

$$dV_t = -\phi(c_t, V_t)\, dt + \beta_t^V\, dB_t, \quad V_T = u(c_T), \quad (3.8.3)$$

---

[16] The existence/uniqueness theory for nonlinear BSDEs begins with Pardoux and Peng (1990) and Duffie and Epstein (1992a), the latter in the context of developing continuous-time recursive utility. These papers imposed conditions that are violated in this section's main application with EZW utility. For the latter, the relevant BSDE foundations were developed by Schroder and Skiadas (1999) and Xing (2017).

[17] In the continuous-time counterpart of Exercise 3.10.6, given in Skiadas (1998), the concavity or convexity of ϕ corresponds to monotonicity of preferences for information, or what in a different setting Kreps and Porteus (1978) termed preferences for the timing for resolution of uncertainty.

where $\phi(c, v) \equiv \beta(u_\delta(c) - \varphi(v))/\varphi'(v)$. Alternatively, the equivalence of BSDEs (3.8.2) and (3.8.3) can be shown as an application of Ito's lemma. Assuming sufficient integrability conditions, BSDE (3.8.3) can equivalently be expressed in the Duffie–Epstein[18] form

$$V_t = \mathbb{E}_t\left[\int_t^T \phi(c_s, V_s)\, ds + u(c_T) \right],$$

which should be thought of as a fixed-point problem in V. ◇

In the remainder of this section, we specialize to an SI normalized utility, implying that $f(c, U) = U g(c/U)$ for a function g that is the continuous-time counterpart of the proportional aggregator. By Theorem A.4.3, additivity of the SI aggregator implies that for some constants β and δ (the inverse of the EIS), we can set

$$g = \beta u_\delta \quad \text{where} \quad u_\delta(x) \equiv \frac{x^{1-\delta} - 1}{1 - \delta}, \quad \text{with } u_1 \equiv \log. \qquad (3.8.4)$$

Similarly, an SI expected-utility CE implies a constant coefficient of relative risk aversion γ, and we can therefore set $A(U)\, U = \gamma$. Preference monotonicity implies $\beta > 0$. We also assume strict concavity of both u_δ and u_γ, and therefore $\gamma, \delta > 0$. The utility BSDE becomes

$$\frac{dU_t}{U_t} = -\left(g\left(\frac{c_t}{U_t} \right) - \frac{\gamma}{2}(\sigma_t^U)^2 \right) dt + \sigma_t^U\, dB_t, \quad U_T = c_T. \qquad (3.8.5)$$

For simplicity,[19] we will discuss optimal consumption and portfolio choice under the assumption that U is specified by BSDE (3.8.5), where g is defined by (3.8.4), corresponding to the continuous-time version of the EZW utility of Example 3.5.1. As in the discrete case, g determines and is determined by the utility of deterministic consumption plans, and given g, increasing γ increases risk aversion.

EXAMPLE 3.8.2 (Expected discounted SI utility) Given any $b \in \mathbb{R}$ and $\beta, \delta \in (0, \infty)$, consider the dynamic utility

$$\tilde{U}_t(c) \equiv \mathbb{E}_t\left[\int_t^T e^{-b(s-t)} u_\delta(c_s)\, ds + \frac{e^{-b(T-t)}}{\beta} u_\delta(c_T) \right],$$

where u_δ is defined in (3.8.4). Let U denote the normalized version of this utility function. Fixing c, we write $\tilde{U}$ and U for $\tilde{U}(c)$ and $U(c)$, respectively.

---

[18] Formulated in Duffie and Epstein (1992a).

[19] The extension to an increasing and concave g is formally straightforward as is allowing time and state dependence of g and γ, although technical regularity conditions are needed for a mathematically rigorous treatment.

Normalization means that at every time t the consumption plan c should have the same utility value as a constant annuity with payment rate U_t, that is,

$$\tilde{U}_t = \mathbb{E}_t\left[\int_t^T e^{-b(s-t)}u_\delta(U_t)\,ds + \frac{e^{-b(T-t)}}{\beta}u_\delta(U_t)\right] = \frac{u_\delta(U_t)}{q_t},$$

where

$$\frac{1}{q_t} \equiv \int_t^T e^{-b(s-t)}\,ds + \frac{e^{-b(T-t)}}{\beta} = \frac{1}{b} + \left(\frac{1}{\beta} - \frac{1}{b}\right)e^{-b(T-t)}.$$

By Ito's lemma, if $u_\delta(U_t) = q_t\tilde{U}_t$, then U satisfies the BSDE

$$\frac{dU_t}{U_t} = -\left(q_t u_\delta\left(\frac{c_t}{U_t}\right) - \frac{\delta}{2}(\sigma_t^U)^2\right)dt + \sigma_t^U\,dB_t, \quad U_T = c_T. \quad (3.8.6)$$

The BSDE is of the form (3.8.5) with $\gamma = \delta$, but time-dependent proportional aggregator $g = qu_\delta$, unless $b = \beta$, in which case U is an EZW utility with $\gamma = \delta$.

The time dependence of g presents no problem, but excluding the logarithmic case, there is still a way of reducing $\tilde{U}$ to the EZW specification with time-independent parameters, even if $b \neq \beta$. Assuming $\delta \neq 1$, let $\alpha \equiv (\beta - b)/(1 - \delta)$ and define U to satisfy

$$u_\delta(U_t) = \beta\tilde{U}_t + \frac{\alpha}{b}\left(1 - e^{-b(T-t)}\right). \quad (3.8.7)$$

By Ito's lemma, U solves the BSDE

$$\frac{dU_t}{U_t} = -\left(\alpha + \beta u_\delta\left(\frac{c_t}{U_t}\right) - \frac{\delta}{2}(\sigma_t^U)^2\right)dt + \sigma_t^U\,dB_t, \quad U_T = c_T. \quad (3.8.8)$$

Analogously to Example 3.5.1, we can eliminate α by changing the unit of account from a "bushel" to a "dollar," using the bushel-to-dollar conversion process $\mathbf{u}_t \equiv e^{\alpha t}$. A consumption plan c in bushels is consumption plan $c^{\mathbf{u}} \equiv c\mathbf{u}$ in dollars, and $U^{\mathbf{u}}(c^{\mathbf{u}}) \equiv \mathbf{u}U(c)$ defines a dynamic utility $U^{\mathbf{u}}$ representing the same preferences over consumption plans in dollars as U does over consumption plans in bushels. Integration by parts shows that BSDE (3.8.8) is equivalent to BSDE

$$\frac{dU_t^{\mathbf{u}}}{U_t^{\mathbf{u}}} = -\left(\beta u_\delta\left(\frac{c_t^{\mathbf{u}}}{U_t^{\mathbf{u}}}\right) - \frac{\delta}{2}(\sigma_t^U)^2\right)dt + \sigma_t^U\,dB_t, \quad U_T^{\mathbf{u}} = c_T^{\mathbf{u}},$$

which is of the normalized EZW form with $\gamma = \delta$. As we will see in Remark 3.8.5, on the market side, changing units from bushels to dollars corresponds to adjusting the short-rate process from r to $r + \alpha$. $\diamond$

Consider now an agent[20] whose preferences are represented by the continuous-time SI recursive utility just defined and who has some initial financial wealth $w > 0$ and no other income. The agent has access to the financial market of Section 2.8, but with r, a, y, σ, and therefore μ, representing general predictable processes, rather than constants (subject to the usual integrability restrictions required for Ito processes to be well defined). The assumed stock return dynamics and associated market-price-of-risk process are therefore

$$\frac{dS_t}{S_t} + y_t dt = \mu_t dt + \sigma_t dB_t, \quad \eta_t \equiv \frac{\mu_t - r_t}{\sigma_t}. \tag{3.8.9}$$

We also assume that $\sigma_t > 0$ and $\int_0^T \eta_t^2 dt < \infty$ with probability one. The agent can use the initial wealth w to purchase an initial portfolio in the MMA and the stock, and can rebalance the account over time, provided the account balance W_t stays positive at all times t. The agent withdraws cash from the account at a time-t rate c_t for all $t < T$, followed by a terminal lump-sum payment $c_T = W_T$, thus converting the initial wealth $w = W_0$ to a consumption plan c. The agent seeks to do so in a way that maximizes the utility $U_0(c)$. By dynamic consistency, time-zero optimality implies optimality at every other time. It is therefore sufficient to solve the agent's problem from the perspective of time zero. A useful way of thinking of the agent's consumption and trading strategy is as the time-zero purchase of the synthetic contract (c, W), where c is the dividend process generated by the contract and W is the contract's cum-dividend price process.

The budget equation of Section 2.8 can be extended to the current context as in Remark 2.8.7. Instead, we equivalently describe the budget equation in terms of wealth allocations, analogously to (3.7.2). An **allocation policy** is a pair (ϱ, ψ) of adapted processes, where for every time $t < T$, $\varrho_t > 0$ represents the consumption rate as a proportion of wealth and ψ_t is the proportion of wealth that is allocated to the stock, with the remainder allocated to the MMA. We also assume, as part of an allocation policy definition, that $\varrho_T \equiv 1$ (corresponding to the assumption $c_T = W_T$) and

---

[20] The rest of this section is based on Schroder and Skiadas (2003), which includes trading constraints, more-general recursive utilities, as well as multiple assets and sources of risk, all in the context of an SI formulation with possibly incomplete markets but tradeable income. Nontradeable income is inconsistent with scale invariance. Schroder and Skiadas (2005) give a version of the argument that includes nontradeable income at the cost of translation-invariant preferences, which remove wealth effects.

that the following version of the budget equation uniquely determines the strictly positive **wealth process** W **generated** by (ϱ, ψ):

$$W_0 = w, \quad \frac{dW_t}{W_t} = (r_t - \varrho_t)\, dt + \psi_t \left(\frac{dS_t}{S_t} + (y_t - r_t)\, dt \right). \quad (3.8.10)$$

The consumption plan **financed** by (ϱ, ψ) is $c \equiv \varrho W$. We call a consumption plan **feasible** if it is financed by some allocation policy, and **optimal** if it maximizes U_0 among all feasible consumption plans. We wish to determine an allocation policy that is **optimal** in that it finances an optimal consumption plan.

Given the reference consumption plan c, the relevant market $X(c)$ is the set of all x such that $c + x$ is a feasible consumption plan. For arbitrary $\pi_0 > 0$, define the process $\pi_t \equiv \pi_0 \exp\left(-\int_0^t r_s \right) \xi_t$, where ξ is defined by (2.8.9), as in Remark 2.8.3. By Ito's lemma,

$$\frac{d\pi_t}{\pi_t} = -r_t dt - \eta_t dB_t. \quad (3.8.11)$$

In a variant of Proposition 2.8.1, we will show that π is an SPD for $X(c)$ by analyzing the linear BSDE

$$dW_t = -\left(c_t - r_t W_t - \eta_t \sigma^W \right) dt + \sigma_t^W dB_t, \quad W_T = c_T, \quad (3.8.12)$$

which results from entering $c = \varrho W$ into the budget equation (3.8.10) and using the return dynamics and definition of η in (3.8.9). The BSDE expresses the heuristic backward recursion

$$W_t = c_t + \mathbb{E}_t \left[\frac{\pi_{t+dt}}{\pi_t} W_{t+dt} \right], \quad W_T = c_T.$$

On a finite information tree, such a recursion implies that the present-value function represented by π prices the contract (c, W). This conclusion is not guaranteed in continuous time. As we saw in Chapter 2, in continuous time we have to entertain the possibility that the synthetic contract (c, W) is implemented by a sort of generalized doubling strategy, where wealth can be created from nothing, or a generalized reverse doubling strategy, where wealth is guaranteed to be destroyed. The fact that W is required to stay positive precludes generalized doubling strategies, but not value-destroying strategies. As a consequence W_t should be at least as great as the time-t present value of c. A condition that limits in a suitable sense how big W can get rules out value-destroying strategies, allowing the conclusion that W_t equals the time-t present value of c.

LEMMA 3.8.3 *Suppose* (W, σ^W) *solves BSDE* (3.8.12) *and* $W_t \geq 0$ *for all* t.
Then

$$W_t \geq \frac{1}{\pi_t} \mathbb{E}_t \left[\int_t^T \pi_s c_s ds + \pi_T c_T \right], \quad t \in [0, T]. \qquad (3.8.13)$$

Moreover, if $\mathbb{E} \left[\sup_t \pi_t W_t \right] < \infty$, *the inequality holds as an equality*

PROOF Given the dynamics (3.8.11) for π and (3.8.12) for W, integration by parts implies that $d(\pi_t W_t) = -\pi_t c_t dt + dM_t$, where M is a local martingale. Let τ_n be a sequence of stopping times such that $\tau_n \uparrow T$ with probability one and M is a martingale up to time τ_n, and therefore $M_t = \mathbb{E}_t M_{\tau_n}$ on $\{\tau_n \geq t\}$, which in turn implies

$$\pi_t W_t = \mathbb{E}_t \left[\int_t^{\tau_n} \pi_s c_s ds + \pi_{\tau_n} W_{\tau_n} \right] \quad \text{on } \{\tau_n \geq t\}.$$

As $n \to \infty$, the term $\mathbb{E}_t \left[\int_t^{\tau_n} \pi_s c_s ds \right]$ converges to $\mathbb{E}_t \left[\int_t^T \pi_s c_s ds \right]$ by the monotone convergence theorem (for conditional expectations). We also know that $\pi_{\tau_n} W_{\tau_n}$ converges to $\pi_T W_T = \pi_T c_T$. Inequality (3.8.13) follows by Fatou's lemma (for conditional expectations). If we further know that $\mathbb{E} \left[\sup_t \pi_t W_t \right] < \infty$, then $\lim_n \mathbb{E}_t \left[\pi_{\tau_n} W_{\tau_n} \right] = \mathbb{E}_t [\pi_T W_T]$ and (3.8.13) becomes an equality thanks to Lebesgue's dominated convergence theorem (for conditional expectations), another fundamental result in the theory of integration.[21]   □

Preference monotonicity implies that if the consumption plan c is a candidate to be optimal, it should not be financed by a value-destroying strategy. A sufficient condition for this to be true is the preceding lemma's integrability condition, which we would have to verify in a mathematically complete formulation. The integrability condition allows us to confirm the SPD property of π as we set out to do.

LEMMA 3.8.4 *Suppose an allocation policy generates the wealth process W and finances the consumption plan c. If $\mathbb{E} \left[\sup_t \pi_t W_t \right] < \infty$, then π is an SPD relative to $X(c)$, that is, $\langle \pi \mid x \rangle \leq 0$ for all x such that $c + x$ is a feasible consumption plan.*

PROOF Suppose $c + x$ is a feasible consumption plan. Applying Lemma 3.8.3, $\pi_0 w \geq \langle \pi \mid c + x \rangle$ and $\pi_0 w = \langle \pi \mid c \rangle$. Subtract the latter from the former to conclude that $0 \geq \langle \pi \mid x \rangle$.   □

---

[21] For all the referenced conditional expectation limit results, see, for example, Chapter 10 of Dudley (2002).

REMARK 3.8.5 (Change of unit of account) Consider a change of the unit of account where consumption plan c in "bushels" is consumption plan $c^{\mathbf{u}} \equiv c\mathbf{u}$ in "dollars." If π is an SPD in bushels, then $\pi^{\mathbf{u}} \equiv \mathbf{u}^{-1}\pi$ is the same SPD in dollars. Maximizing $U_0(c)$ subject to $\langle \pi \mid c \rangle \leq w\pi_0$ (as we are effectively doing in the solution method that follows) is equivalent to maximizing $U_0^{\mathbf{u}}(c^{\mathbf{u}})$ subject to $\langle \pi^{\mathbf{u}} \mid c^{\mathbf{u}} \rangle \leq w^{\mathbf{u}}\pi_0^{\mathbf{u}}$, where $U^{\mathbf{u}}(c^{\mathbf{u}}) \equiv \mathbf{u}U(c)$ and $w^{\mathbf{u}} \equiv w\mathbf{u}_0$. In Example 3.8.2, $\mathbf{u}_t \equiv e^{\alpha t}$ and the only difference between π and $\pi^{\mathbf{u}}$ is in their respective implied short-rate processes r and $r + \alpha$, as can be seen by applying integration by parts to $\pi^{\mathbf{u}} \equiv \mathbf{u}^{-1}\pi$ and comparing the resulting Ito expansion to (3.8.11). $\diamond$

Similarly to Proposition 3.3.4, we will formulate optimality conditions based on the state-price density property of a utility gradient. Technical aspects aside, the utility gradient calculation is analogous to Proposition 3.4.12 for the discrete case. Given any feasible direction x, the idea is to use the utility BSDE to write the Ito expansion of $U(c + \epsilon x) - U(c)$ and then linearize the Ito coefficients, assuming small ϵ, resulting in a linear BSDE of the form we used earlier to price the contract (c, W). The following remark outlines the result and a derivation of the associated gradient inequality.[22]

REMARK 3.8.6 Suppose that for a continuously differentiable F, the pair $(U, \Sigma) \equiv (U(c), \Sigma(c))$ uniquely solves the BSDE

$$dU_t = -F(c_t, U_t, \Sigma_t)\, dt + \Sigma_t dB_t, \quad U_T = c_T.$$

Fixing c, define the processes λ, F_U and F_Σ by

$$\lambda_t \equiv \frac{\partial F(c_t, U_t, \Sigma_t)}{\partial c},$$

$$F_U(t) \equiv \frac{\partial F(c_t, U_t, \Sigma_t)}{\partial U}, \quad F_\Sigma(t) \equiv \frac{\partial F(c_t, U_t, \Sigma_t)}{\partial \Sigma}.$$

---

[22] The utility gradient calculation and its use in this chapter originated in a chapter of my doctoral thesis. My advisor, Darrell Duffie, had just pioneered continuous-time recursive utility with Larry Epstein in Duffie and Epstein (1992a), and asset pricing applications using dynamic programming Markovian methods in Duffie and Epstein (1992b). My idea was to characterize optimality in a static way in terms of gradients without a Markovian structure, and to derive utility gradient expressions for recursive utility and extensions. These initial results were published in Duffie and Skiadas (1994). I continued this work at Northwestern University with Mark Schroder, who was a doctoral student at the time, leading to the theory on which this section is based.

Then, omitting technical requirements,

$$\lim_{\epsilon \downarrow 0} \frac{U_t(c + \epsilon x) - U_t(c)}{\epsilon} = \mathbb{E}_t \left[\int_t^T \frac{\mathcal{E}_s}{\mathcal{E}_t} \lambda_s x_s ds + \frac{\mathcal{E}_T}{\mathcal{E}_t} \lambda_T x_T \right], \quad (3.8.14)$$

where $\mathcal{E}$ solves

$$\frac{d\mathcal{E}}{\mathcal{E}} = F_U dt + F_\Sigma dB, \quad \mathcal{E}_0 = 1.$$

Assuming concavity of F and omitted regularity conditions, we outline a derivation of the gradient inequality

$$U_0(c + x) \leq U_0(c) + \langle \mathcal{E}\lambda \,|\, x \rangle. \qquad (3.8.15)$$

Let $y \equiv U(c + x) - U(c)$ and $z \equiv \Sigma(c + x) - \Sigma(c)$. The gradient inequality for F gives

$$0 \leq D \equiv -F(c + x, U + y, \Sigma + z) + F(c, U, \Sigma) + \lambda x + F_U y + F_\Sigma z.$$

Subtracting the BSDE for $U(c)$ from the BSDE for $U(c + x)$, we get

$$dy = -(\lambda x - D + F_U y + F_\Sigma z) \, dt + z \, dB, \quad y_T = x_T.$$

This is a linear BSDE of the form (3.8.12), with $(y, \lambda x - D, F_U, F_\Sigma)$ corresponding to $(W, c, -r, -\eta)$. Given sufficient regularity, the argument of Lemma 3.8.3 in this context implies that $y_0 = \langle \mathcal{E} \,|\, \lambda x - D \rangle$. Since both $\mathcal{E}$ and D are positive, $y_0 \leq \langle \mathcal{E} \,|\, \lambda x \rangle = \langle \mathcal{E}\lambda \,|\, x \rangle$, which is the claimed gradient inequality. $\diamond$

We continue taking as given the reference consumption plan c and associated utility process $U \equiv U(c)$ determined by BSDE (3.8.5) with g defined in (3.8.4), for constant positive parameters β, γ, δ. Although we have used x for a variety of roles, in the remainder of this section x represents the consumption to utility ratio, in terms of which we define the process λ:

$$x_t \equiv \frac{c_t}{U_t} \quad \text{and} \quad \lambda_t \equiv g'(x_t) = \beta x_t^{-\delta} \text{ for } t < T; \quad \lambda_T = 1. \quad (3.8.16)$$

The convex dual g^* is defined by (3.5.16), and therefore

$$g^*(\lambda_t) = g(x_t) - g'(x_t) x_t.$$

Remark 3.8.6 applies to BSDE (3.8.5) with $\Sigma = U\sigma^U$ and $F(c, U, \Sigma) \equiv Ug(c/U) - (\gamma/2) \Sigma^2/U$. As a consequence (omitting technical requirements), the gradient of U_0 is $\mathcal{E}\lambda$, where $\mathcal{E}$ solves

$$\frac{d\mathcal{E}}{\mathcal{E}} = \left(g^*(\lambda) + \frac{\gamma}{2}(\sigma^U)^2\right)dt - \gamma\sigma^U dB, \quad \mathcal{E}_0 = 1. \tag{3.8.17}$$

The concavity of g implies the concavity of F, as can be seen by applying Lemma 3.5.6 to each of the additive terms of F. We assume enough regularity for the gradient inequality (3.8.15) to hold.

LEMMA 3.8.7 *Suppose the allocation policy (ϱ, ψ) finances the plan c and $\mathcal{E}\lambda$ is an SPD for the market $X(c)$. Then (ϱ, ψ) is optimal.*

PROOF If $c + x$ is a feasible consumption plan, then $\langle \mathcal{E}\lambda \mid x \rangle \leq 0$ and the gradient inequality (3.8.15) implies $U_0(c + x) \leq U_0(c)$. $\qquad\square$

Suppose the candidate optimal allocation policy (ϱ, ψ) finances c, generating the wealth process W such that $\mathbb{E}\left[\sup_t \pi_t W_t\right] < \infty$ (which would need to be confirmed as part of a rigorous optimality verification argument). By Lemma 3.8.4, π is an SPD for $X(c)$. The scaling factor $\pi_0 > 0$ is arbitrary, so let us set $\pi_0 = \lambda_0$. Assuming $\pi = \mathcal{E}\lambda$, we proceed to derive optimality conditions, obtaining expressions for ϱ and ψ along the way. Conversely, under omitted regularity assumptions, these optimality conditions imply that $\pi = \mathcal{E}\lambda$ and hence the optimality of (ϱ, ψ) by Lemma 3.8.7.

As a first step we relate the utility process U to W using scale invariance. The Euler equation for homogeneous functions gives

$$U_t = \lim_{\epsilon \downarrow 0} \frac{U_t(c + \epsilon c) - U_t(c)}{\epsilon} = \lambda_t \mathbb{E}_t\left[\int_t^T \frac{\pi_s}{\pi_t} c_s ds + \frac{\pi_T}{\pi_t} c_T\right].$$

The first equation is true because U_t is homogeneous of degree one and the second equation follows from identity (3.8.14) with $\mathcal{E}\lambda = \pi$. By Lemma 3.8.3, the right-hand side equals $\lambda_t W_t$. Using integration by parts, we therefore have the key identities

$$U = \lambda W \quad \text{and} \quad \frac{dU}{U} = \frac{d\lambda}{\lambda} + \frac{dW}{W} + \frac{d\lambda}{\lambda}\frac{dW}{W}. \tag{3.8.18}$$

The first equation and (3.8.16) allow us to compute the optimal consumption policy as a function of λ:

$$\varrho_t \equiv \frac{c_t}{W_t} = \lambda_t \frac{c_t}{U_t} = \lambda_t x_t = \beta^{1/\delta}\lambda_t^{1-1/\delta}, \quad t < T. \tag{3.8.19}$$

As in the discrete case, the determination of the optimal portfolio allocation policy can be performed jointly with a backward recursion that determines λ. To see how, we start with the notation

$$\frac{d\lambda}{\lambda} \equiv \mu^\lambda dt + \sigma^\lambda dB.$$

The budget equation (3.8.10) and (3.8.18) imply

$$\sigma^U = \sigma^\lambda + \psi\sigma. \tag{3.8.20}$$

Applying integration by parts to $\pi = \mathcal{E}\lambda$, using (3.8.17), we have

$$\frac{d\pi}{\pi} = \left(g^*(\lambda) + \frac{\gamma}{2}(\sigma^U)^2 + \mu^\lambda - \gamma\sigma^U\sigma^\lambda \right) dt - \left(\gamma\sigma^U - \sigma^\lambda \right) dB.$$

Matching coefficients with the Ito decomposition (3.8.11) for $d\pi/\pi$ and substituting expression (3.8.20) for σ^U, we find

$$-\mu^\lambda = r + g^*(\lambda) + \frac{\gamma}{2}\left((\psi\sigma)^2 - (\sigma^\lambda)^2 \right),$$

$$\psi\sigma = \frac{1}{\gamma}\left(\eta + (1-\gamma)\sigma^\lambda \right).$$

These two equations are the counterpart of the joint recursive determination of λ and ψ in the second step of the algorithm of Theorem 3.7.2. The Arrow–Pratt CE approximation leads to the closed-form solution for $\psi\sigma$, which can now be substituted into the expression for μ^λ. The result is a stand-alone BSDE for λ as summarized in the following solution method. The claimed optimal allocation rule is obtained by substituting $\eta \equiv (\mu - r)/\sigma$ in the above expression for $\psi\sigma$, and EIS $\equiv 1/\delta$ in expression (3.8.19) for ϱ.

Solution method Determine $(\lambda, \sigma^\lambda)$ by solving the BSDE

$$\frac{d\lambda_t}{\lambda_t} = -\left(r_t + g^*(\lambda_t) - \frac{\gamma}{2} Q_t(\sigma_t^\lambda) \right) dt + \sigma_t^\lambda dB_t, \quad \lambda_T = 1, \tag{3.8.21}$$

where $Q_t(z) \equiv z^2 - \left(\gamma^{-1}\eta_t - (1-\gamma^{-1})z \right)^2$. The optimal allocation policy (ϱ, ψ) is given by $\varrho_T = 1$ and, for $t < T$,

$$\varrho_t = \beta^{\text{EIS}}\lambda_t^{1-\text{EIS}} \quad \text{and} \quad \psi_t = \frac{1}{\gamma}\frac{\tilde{\mu}_t - r_t}{\sigma_t^2}, \tag{3.8.22}$$

where $\tilde{\mu}_t \equiv \mu_t + (1-\gamma)\sigma_t^\lambda\sigma_t$.

An optimality verification argument essentially reverses the steps leading to this solution method. A mathematically complete treatment is technically involved, requiring regularity assumptions along the way.

EXAMPLE 3.8.8 (Unit EIS) Assuming $\delta = 1$, the optimal consumption allocation policy in (3.8.22) reduces to $\varrho_t = \beta$. To relate this to Example 3.7.5, let us write $\bar{\beta}$ and $\bar{\varrho}$ for the parameters β and ϱ in the discrete context of Section 3.7. If we heuristically think of the infinitesimal time interval $[t, t + dt]$ as a single discrete period, then $\bar{\beta} = e^{-\beta dt} = 1 - \beta dt$ and $\bar{\varrho}_t = \varrho_t dt$. The conclusion $\bar{\varrho}_t = 1 - \bar{\beta}$ of Example 3.7.5 corresponds to $\varrho_t = \beta$ in the current context. $\diamond$

EXAMPLE 3.8.9 (Unit CRRA) Assuming $\gamma = 1$, $\tilde{\mu}_t = \mu_t$ and the optimal portfolio allocation in (3.8.22) reduces to

$$\psi_t = \frac{\mu_t - r_t}{\sigma_t^2}.$$

This is a mean-variance efficient portfolio in the sense of Section 2.2 (with $\Sigma_t = \sigma_t^2$), whose single-period analysis can be heuristically applied over an infinitesimal time interval $[t, t + dt]$. In the discrete Example 3.7.8, we concluded that the optimal portfolio is myopic. Here we can add the insight of mean-variance efficiency thanks to the Arrow–Pratt approximation of a logarithmic CE. $\diamond$

We encountered another example of a myopic optimal portfolio in Example 3.7.7, which in the current context corresponds to deterministic r, μ and σ. For simplicity, in the following example we take these parameters to be constant and derive a closed-form solution. As in Example 3.8.9, the combination of a myopic portfolio rule and the Arrow–Pratt CE approximation lead to a mean-variance efficient optimal portfolio.

EXAMPLE 3.8.10 (Constant investment opportunity set) Suppose that $r_t = r$, $\mu_t = \mu$, $\sigma_t = \sigma$, and therefore $\eta_t = \eta \equiv (\mu - r)/\sigma$, are all deterministic constants. Then the marginal-value-of-wealth process λ is also deterministic and BSDE (3.8.21) reduces to an ordinary differential equation. Setting $\sigma^\lambda = 0$ in (3.8.22) results in the mean-variance efficient optimal portfolio[23]

$$\psi_t = \frac{\eta}{\gamma \sigma} = \frac{1}{\gamma} \frac{\mu - r}{\sigma^2}.$$

The optimal ϱ in (3.8.22) is given, for all $t < T$, by

$$\varrho_t = p \left(1 + \left(p \beta^{-1/\delta} - 1 \right) e^{-p(T-t)} \right)^{-1},$$

---

[23] This optimal portfolio allocation (with extensions) was first shown for the additive case ($\gamma = \delta$) by Merton (1969, 1971), whose work has been seminal for the theory of dynamic optimal consumption and portfolio choice. Merton used the Hamilton–Jacobi–Bellman approach, which is also applicable in the current setting.

where

$$p \equiv \frac{\beta}{\delta} + \left(1 - \frac{1}{\delta}\right)\left(r + \frac{\eta^2}{2\gamma}\right).$$

To show this claim for $\delta \neq 1$, set $\sigma^\lambda = 0$ in BSDE (3.8.21) to find that $z_t \equiv (\lambda_t/\beta)^{(1-\delta)/\delta}$ solves $dz_t = (pz_t - \beta)\,dt$ or $d(e^{-pt}z_t) = (\beta/p)\,de^{-pt}$. Integrating from t to T results in the λ_t such that $\varrho_t = \beta^{1/\delta}\lambda_t^{1-1/\delta}$ has the claimed form. For $\delta = 1$, the claimed optimal consumption allocation expression reduces to $\varrho_t = \beta$, which is consistent with Example 3.8.8. The current example applies in particular to the expected discounted utility of Example 3.8.2 and more generally to any recursive utility specified by BSDE (3.8.5) with $g = \alpha + \beta u_\delta$, by simply replacing r with $r + \alpha$ while keeping η and $\psi = \eta/\gamma\sigma$ the same (implying that μ changes to $\mu + \alpha$). The claim follows from Remark 3.8.5 and the change of the unit of account described in Example 3.8.2. $\diamond$

The more-general expression for the optimal portfolio ψ_t in (3.8.22) allows for deviations from mean-variance efficiency due to a stochastically varying marginal value of wealth. To relate it to the discrete theory of Section 3.7, recall that for the CE $v_t = u_\gamma^{-1}\mathbb{E}_t u_\gamma$, the optimal portfolio rule can be expressed as (3.7.10), where Q is the probability with the conditional density process ξ defined by recursion (3.7.11). Heuristically, if the discrete time period is the time interval $[t, t + dt]$, recursion (3.7.11) becomes $d\xi_t/\xi_t = (1 - \gamma)\sigma_t^\lambda dB_t$.[24] By Girsanov's theorem (see Remark 2.8.3), the drift of the stock's cumulative return, which is μ_t under P, becomes $\tilde{\mu}_t \equiv \mu_t + (1 - \gamma)\sigma_t^\lambda\sigma_t$ under Q. The optimal portfolio in (3.8.22) is therefore mean-variance efficient under the probability Q. Just as in the last two examples under P, this is a consequence of the Arrow–Pratt CE approximation in the myopic optimal portfolio problem (3.7.10) under Q.

---

[24] To see why, write recursion (3.7.11) as

$$\frac{\xi_{t+dt}}{\xi_t} = \frac{\lambda_{t+dt}^{1-\gamma}}{\lambda_t^{1-\gamma} + \cdots dt} = \left(\frac{\lambda_{t+dt}}{\lambda_t}\right)^{1-\gamma} + \cdots dt.$$

Take logarithms on both sides and use a first-order Taylor expansion on the right-hand side, to find $d\log\xi_t = (1 - \gamma)\,d\log\lambda_t + \cdots dt$. Applying Ito's lemma,

$$\frac{d\xi_t}{\xi_t} = (1 - \gamma)\sigma_t^\lambda dB_t + \cdots dt.$$

But since we know that ξ is a local martingale, the $\cdots dt$ term is zero.

Examples in which σ^λ is stochastic can be formulated in terms of Markov processes, where BSDE solutions can be related to corresponding PDE solutions, in a generalization of the argument that related the linear BSDE (2.8.13) to PDE (2.8.17).

3.9 Habit Formation and Durability of Consumption

We have defined dynamic utilities to be forward looking, meaning that utility at a spot on the information tree is not dependent on consumption outside the subtree rooted at the given spot. In this section, we conclude with a simple extension in which past consumption can influence utility for current consumption, either as a detraction, which can be thought of as habit formation, or an enhancement, which can be thought of as durability of past consumption. We represent this dependence by letting the utility process $U(c)$ take the form $\hat{U}(\hat{c})$, where $\hat{c}_t$ is a function of the history of the consumption plan c up to time t. In a typical application $\hat{U}$ can be expected discounted utility or any forward-looking dynamically consistent utility (which by Lemma 3.4.4 is a generalized recursive utility). Assuming a special linear dependence of $\hat{c}$ on c, we will show that maximizing U in a given (effectively) complete market X is equivalent to maximizing $\hat{U}$ in a suitably defined modified market $\hat{X}$, thus reducing the original optimal consumption problem to an equivalent problem of the type we have analyzed so far.

The mathematical structure that accommodates this analysis makes no assumption on the functional form of $\hat{U}$, putting the original and transformed problems on a symmetric footing. To see the essential idea, we start with a three-date finite state-space example. On the space $\mathcal{L}$ of adapted processes we use the inner product $\langle x \mid y \rangle \equiv x_0 y_0 + \mathbb{E}[x_1 y_1 + x_2 y_2]$, where $\mathbb{E}$ denotes expectation under a given full-support probability. An agent with consumption set C and endowment $e \in C$ trades in an arbitrage-free complete market X to maximize the utility function $U_0 \colon C \to \mathbb{R}$. (Assuming dynamic consistency, it is sufficient to focus on utility maximization from the perspective of time zero.) A corresponding state-price density $\pi \in \mathcal{L}_{++}$ exists and is unique up to positive scaling. Since $X = \{x \in \mathcal{L} \mid \langle \pi \mid x \rangle = 0\}$, the agent's optimal consumption problem can be stated as

$$\max \{U_0(c) \mid \langle \pi \mid c \rangle = \langle \pi \mid e \rangle, \ c \in C\}. \tag{3.9.1}$$

The **wealth process** $W_t(c) \equiv \mathbb{E}_t \sum_{s \geq t} (\pi_s / \pi_t) c_s$ associated with the consumption plan c is determined by the requirement that $(c, W(c))$ is a traded contract in X. (Although $c_0 \neq 0$, we can still regard $(c, W(c))$ as a contract by identifying it with $(c - c_0 1_{\Omega \times \{0\}}, W(c) - c_0 1_{\Omega \times \{0\}})$.) Also associated with c is, for fixed scalar parameter φ, the **consumption stock** process $H(c)$, where

$$H_0(c) = 0, \quad H_1(c) = c_0, \quad H_2(c) = c_1 + \varphi c_0.$$

We refer to the primitives introduced so far together with the fixed scalar parameter v as the **primal** formulation.

We now introduce a **dual** formulation using the dual parameters

$$\hat{v} = -v \quad \text{and} \quad \hat{\varphi} = \varphi - v.$$

The mapping that takes primal quantities to their dual defines an isomorphism, meaning that it is a bijection that preserves the structure of the primal formulation. The dual of the consumption plan c is

$$\hat{c} = c + v H(c) \tag{3.9.2}$$

and the dual consumption set is $\hat{C} = \{\hat{c} \mid c \in C\}$. The dual consumption stock process is

$$\hat{H}_0(\hat{c}) = 0, \quad \hat{H}_1(\hat{c}) = \hat{c}_0, \quad \hat{H}_2(\hat{c}) = \hat{c}_1 + \hat{\varphi} \hat{c}_0.$$

Notice that

$$\hat{H}(\hat{c}) = H(c)$$

and therefore equation (3.9.2) can be inverted to

$$c = \hat{c} + \hat{v} \hat{H}(\hat{c}). \tag{3.9.3}$$

The dual state-price density $\hat{\pi}$ is defined so that for all $c \in C$,

$$\langle \pi \mid c \rangle = \langle \hat{\pi} \mid \hat{c} \rangle. \tag{3.9.4}$$

Substituting expression (3.9.3) for c and using the law of iterated expectations, we find

$$\langle \pi \mid c \rangle = \left(1 + \mathbb{E}\pi_1 \hat{v} + \mathbb{E}\pi_2 \hat{v}\hat{\varphi} \right) \hat{c}_0 + \mathbb{E}\left[\left(\pi_1 + \mathbb{E}_1 \pi_2 \hat{v} \right) \hat{c}_1 + \pi_2 \hat{c}_2 \right].$$

The desired identity (3.9.4) is therefore achieved by setting

$$\hat{\pi}_0 = \pi_0 \left(1 + P_0^1 \hat{v} + P_0^2 \hat{v}\hat{\varphi} \right), \quad \hat{\pi}_1 = \pi_1 \left(1 + P_1^2 \hat{v} \right), \quad \hat{\pi}_2 = \pi_2,$$

where $P_t^s \equiv \mathbb{E}_t \pi_s / \pi_t$ is the time-t primal-market price of a unit discount bond maturing at time s (that is, a default-free bond paying one unit of account at time s). We assume that $\hat{\pi} \in \mathcal{L}_{++}$ (which, for example, is true if $v < 0$ and $\varphi > 0$).[25] The dual market $\hat{X} = \left\{ \hat{x} \in \mathcal{L} \mid \langle \hat{\pi} \mid \hat{x} \rangle = 0 \right\}$ is the complete market with state-price density $\hat{\pi}$. The dual agent has the dual endowment $\hat{e} \in \hat{C}$ and trades in the dual market $\hat{X}$ to maximize the dual utility function $\hat{U}_0 \colon \hat{C} \to \mathbb{R}$, where

$$\hat{U}_0(\hat{c}) = U_0(c) . \tag{3.9.5}$$

Equivalently, the dual agent solves

$$\max \left\{ \hat{U}_0(\hat{c}) \mid \langle \hat{\pi} \mid \hat{c} \rangle = \langle \hat{\pi} \mid \hat{e} \rangle, \ \hat{c} \in \hat{C} \right\}. \tag{3.9.6}$$

Thanks to identities (3.9.4) and (3.9.5), the primal problem (3.9.1) is equivalent to the dual problem (3.9.6) in the sense that $c \in C$ is optimal for the primal problem if and only if $\hat{c} \in \hat{C}$ is optimal for the dual problem. The wealth process $\hat{W}(\hat{c})$ is defined by the requirement that $(\hat{c}, \hat{W}(\hat{c}))$ is a traded contract in $\hat{X}$, or $\hat{W}_t(\hat{c}) \equiv \mathbb{E}_t \sum_{s \geq t} \left(\hat{\pi}_s / \hat{\pi}_t \right) \hat{c}_s$.

The dual formulation is the primal with hats on every quantity. Conversely, with the convention that double hats cancel out ($\hat{\hat{x}} = x$), adding hats on every quantity of the dual takes us back to the primal. Because of this symmetry, the dual of every valid expression is also valid. For example, the dual of the transformation we used to define $\hat{\pi}$ is

$$\pi_0 = \hat{\pi}_0 \left(1 + \hat{P}_0^1 v + \hat{P}_0^2 v\varphi \right), \quad \pi_1 = \hat{\pi}_1 \left(1 + \hat{P}_1^2 v \right), \quad \pi_2 = \hat{\pi}_2,$$

where $\hat{P}_t^s = \mathbb{E}_t \hat{\pi}_s / \hat{\pi}_t$ are unit discount-bond prices in the dual market. To see the validity of the claimed expression for π_0 directly, note that the dual of $u \equiv (1, 0, 0)$ is $\hat{u} = (1, v, v\varphi)$, which at time zero can be sold in the dual market for its present value $p \equiv 1 + \hat{P}_0^1 v + \hat{P}_0^2 v\varphi$. That is, there is an $\hat{x} \in \hat{X}$ such that $\hat{u} + \hat{x} = (p, 0, 0)$. By identity (3.9.4) and the fact that $\langle \hat{\pi} \mid \hat{x} \rangle = 0$, we have $\pi_0 = \langle \pi \mid u \rangle = \langle \hat{\pi} \mid \hat{u} + \hat{x} \rangle = \hat{\pi}_0 p$. To see the expression for π_1 repeat this argument from the perspective of every time-one spot. The dual of this argument sheds light on our earlier expressions giving $\hat{\pi}$ in terms of π.

---

[25] The duality argument goes through if we relax the strict positivity of π and $\hat{\pi}$ and monotonicity of U and $\hat{U}$.

The isomorphism just described is useful, because if, say, the dual agent's problem is solved, we also have a solution to the primal problem with $U_0(c) = \hat{U}_0 (c + \nu H(c))$. Assuming $\varphi \in (0, 1)$, one can think of $H(c)$ as an inventory level that is replenished one-to-one with consumption and depreciates at a constant rate: $H_{t+1}(c) = c_t + \varphi H_t(c)$. For $\nu < 0$, past consumption, summarized by $H_t(c)$, reduces the impact of consumption c_t on the agent's utility, reflecting habit formation. Symmetrically, if $\nu > 0$ past consumption enhances current consumption, reflecting consumption durability. At this point the reader may wish to relax the assumption that there are only three dates and explore how our earlier results for recursive utility transform to incorporate habit formation or consumption durability in this special form. The isomorphism also applies to Arrow–Debreu equilibria, and hence any effectively complete market equilibrium, provided all agents have common parameters (ν, φ). Leaving further analysis of the discrete case as an exercise, we proceed to a continuous-time version of the isomorphism, which offers the usual benefit of simplified expressions made possible by the Ito calculus.

We adopt the Brownian information setting of Section 3.8, although the approach and most of the expressions that follow apply in any reasonable continuous-time formulation. To keep the exposition short, we assume there is no terminal consumption. Cash flows and state-price processes live in the space of every adapted process x such that the integral $\int_0^T x_t^2 dt$ is well defined and has finite expectation. On this space, we use the inner product

$$\langle x \mid y \rangle \equiv \mathbb{E} \int_0^T x_t y_t dt.$$

A consumption plan is any strictly positive cash flow c, with c_t interpreted as a time-t consumption rate. The consumption stock process $H(c)$ can be thought of as an inventory that starts at zero, is replenished one-to-one with consumption, and continuously depreciates at a constant rate κ per unit of time. More precisely, $H(c)$ is the unique solution to the ordinary differential equation

$$dH_t(c) = (c_t - \kappa H_t(c))\, dt, \quad H_0(c) = 0. \tag{3.9.7}$$

(The argument that follows goes through if we allow κ and ν to be suitably integrable adapted processes and even if we add a zero-mean martingale to $H(c)$ representing random shocks to the consumption stock.)

Table 3.1 indicates the relationship of κ and ν to the dual parameters $\hat{\kappa}$ and $\hat{\nu}$, the closed-form expression for $H(c)$ and its dual $\hat{H}(\hat{c})$, the relationship between a primal consumption plan c and its dual $\hat{c} = c + \nu H(c)$, and the claim $H(c) = \hat{H}(\hat{c})$, which in turn implies the

Table 3.1 *Durability/habit duality in continuous time.*

Primal	Dual
$\kappa = \hat{\kappa} + \hat{v}, \ v = -\hat{v}$	$\hat{\kappa} = \kappa + v, \ \hat{v} = -v$
$H_t(c) \equiv \int_0^t e^{-\kappa(t-s)} c_s ds$	$\hat{H}_t(\hat{c}) \equiv \int_0^t e^{-\hat{\kappa}(t-s)} \hat{c}_s ds$
$H_t(c) = \hat{H}_t(\hat{c})$	
$c = \hat{c} + \hat{v} \hat{H}(\hat{c})$	$\hat{c} = c + v H(c)$
$U(c) = \hat{U}(\hat{c})$	
$P_t^s \equiv \mathbb{E}_t \pi_s / \pi_t$	$\hat{P}_t^s \equiv \mathbb{E}_t \hat{\pi}_s / \hat{\pi}_t$
$\pi_t = \hat{\pi}_t \left(1 + \int_t^T \hat{P}_t^s v e^{-\kappa(s-t)} ds\right)$	$\hat{\pi}_t = \pi_t \left(1 + \int_t^T P_t^s \hat{v} e^{-\hat{\kappa}(s-t)} ds\right)$
$W_t(c) \equiv \mathbb{E}_t \int_t^T (\pi_s / \pi_t) c_s ds$	$\hat{W}_t(\hat{c}) \equiv \mathbb{E}_t \int_t^T (\hat{\pi}_s / \hat{\pi}_t) \hat{c}_s ds$
$\pi \big(W(c) + H(c)\big) = \hat{\pi} \big(\hat{W}(\hat{c}) + \hat{H}(\hat{c})\big)$	

symmetric expression $c = \hat{c} + \hat{v} \hat{H}(\hat{c})$. To show this claim, compare (3.9.7) to its dual form

$$d\hat{H}_t(c) = \big(\hat{c}_t - \hat{\kappa} \hat{H}_t(\hat{c})\big) dt, \quad \hat{H}_0(c) = 0. \tag{3.9.8}$$

It is clear from the definitions that differential equations (3.9.7) and (3.9.8) are identical and therefore have a common solution.

The next item in Table 3.1 corresponds to the utility specification $U(c) = \hat{U}(c + v H(c))$. As in the discrete example, for $v < 0$ the utility function U adds habit formation to the utility $\hat{U}$, which in an application could be taken to be any (forward-looking) recursive utility. For $v > 0$, the utility U allows for durability in consumption relative to $\hat{U}$. (Of course, the primal and dual are symmetric and their role is interchangeable.)

With $\hat{\pi}_t$ and π_t related as in Table 3.1, identity (3.9.4) follows by a continuous-time version of our earlier argument:

$$\langle \pi \mid c \rangle = \mathbb{E} \int_0^T \pi_t \left(\hat{c}_t + \hat{v} \int_0^t e^{-\hat{\kappa}(t-s)} \hat{c}_s \, ds\right) dt$$

$$= \mathbb{E} \left[\int_0^T \pi_t \hat{c}_t \, dt + \int_0^T \int_0^T 1_{\{s \le t\}} e^{-\hat{\kappa}(t-s)} \mathbb{E} \mathbb{E}_s \big[\pi_t \hat{v} \hat{c}_s\big] \, ds dt\right]$$

$$= \mathbb{E}\left[\int_0^T \pi_t \hat{c}_t \, dt + \int_0^T \int_s^T \mathbb{E}_s \pi_t \hat{v} e^{-\hat{\kappa}(t-s)} \, dt \, \hat{c}_s \, ds \right]$$

$$= \mathbb{E}\left[\int_0^T \left(\pi_t + \int_t^T \mathbb{E}_t \pi_s \hat{v} e^{-\hat{\kappa}(s-t)} \, ds \right) \hat{c}_t \, dt \right] = \langle \hat{\pi} \mid \hat{c} \rangle.$$

In the first equation, we substituted $\hat{c} + \hat{v}\hat{H}(\hat{c})$ for c. In the second equation, we interchanged the expectation operator and the double time integral, and applied the law of iterated expectations. In the third equation, we took the unconditional expectation back out of the double time integral and reversed the order of integration with respect to s and t. In the same double integral, we relabeled (s,t) to (t,s) and factored out $\hat{c}_t$ across both terms to arrive at the fourth equation. The final equation reflects the relationship between $\hat{\pi}$ and π in Table 3.1. A similar argument applied conditionally on time-t information results in the more-general last identity of Table 3.1. Just as in the discrete case, we can conclude that the primal problem (3.9.1) is equivalent to the dual problem (3.9.6), an equivalence that extends to Arrow–Debreu equilibria with common parameters κ and v.

The relationship between π and $\hat{\pi}$ and the state-price dynamics

$$\frac{d\pi}{\pi} = -r dt - \eta dB \quad \text{and} \quad \frac{d\hat{\pi}}{\hat{\pi}} = -\hat{r} dt - \hat{\eta} dB, \tag{3.9.9}$$

allow us to relate the short-rate and market-price-of-risk processes in the primal market to those of the dual market.

LEMMA 3.9.1 $\pi_t\left(r_t + \kappa\right) = \hat{\pi}_t\left(\hat{r}_t + \hat{\kappa}\right).$

PROOF Let

$$\delta_t \equiv e^{-\hat{\kappa}t}, \quad p_t \equiv \int_t^T P_t^s \hat{v} e^{-\hat{\kappa}(s-t)} ds = \frac{1}{\pi_t \delta_t} \mathbb{E}_t \int_t^T \pi_s \hat{v} \delta_s ds,$$

and therefore $d(p\pi\delta) = -\pi\hat{v}\delta dt + d\,\mathrm{LM}$, where we write LM to mean "some local martingale" (not necessarily the same one in every occurrence). Since $\hat{\pi}\delta = \pi\delta(1 + p)$,

$$d\left(\hat{\pi}\delta\right) = d(\pi\delta) - \pi\hat{v}\delta dt + d\,\mathrm{LM}. \tag{3.9.10}$$

The state-price dynamics (3.9.9) and integration by parts result in

$$\frac{d(\pi\delta)}{\pi\delta} = \frac{d\pi}{\pi} + \frac{d\delta}{\delta} = -\left(r + \hat{\kappa}\right) dt + d\,\mathrm{LM}. \tag{3.9.11}$$

Since $\hat{\kappa} + \hat{v} = \kappa$, equations (3.9.10) and (3.9.11) imply that

$$d\left(\hat{\pi}\delta\right) = -\pi(r + \kappa)\delta dt + d\,\mathrm{LM},$$

and (3.9.11) with $\hat{\pi}$ in place of π gives

$$d(\hat{\pi}\delta) = -\hat{\pi}(\hat{r} + \hat{\kappa})\,\delta\,dt + d\,\mathrm{LM}.$$

Matching drift terms yields the claimed identity.   □

Leaving the general case as an exercise, in the remainder of this section, we assume that r is deterministic. The discount-bond prices in this case are $P_t^s = \exp(-\int_t^s r_u du)$ and therefore

$$p_t \equiv \int_t^T P_t^s \hat{v} e^{-\hat{\kappa}(s-t)}\,ds \tag{3.9.12}$$

is also deterministic. Lemma 3.9.1 and the fact that $\hat{\pi}_t = \pi_t(1 + p_t)$ give the dual short rate and market price of risk as

$$\hat{r}_t = \frac{r_t + \kappa}{1 + p_t} - \hat{\kappa} \quad \text{and} \quad \hat{\eta}_t = \eta_t.$$

The pricing of risk is the same in the primal and dual markets as a consequence of the assumption that r is deterministic.

EXAMPLE 3.9.2 Assume that r is constant and $r + \hat{\kappa} > 0$. Then

$$p_t = \frac{\hat{v}}{r + \hat{\kappa}}\left(1 - e^{-(r+\hat{\kappa})(T-t)}\right) \to \frac{\hat{v}}{r + \hat{\kappa}} \quad \text{as } T \to \infty.$$

For large values of T, the ratio $\hat{\pi}_t/\pi_t$ is approximately a constant and pricing in the primal and dual markets is approximately the same. In particular, $\hat{r}_t$ is approximately equal to r. ◇

Finally, we relate optimal consumption–portfolio policies in the primal and dual formulations. We adopt the setting of Section 3.8, where the filtration is generated by a single Brownian motion. Again, this is for expositional expediency, the underlying arguments apply in much greater generality. The primal market is implemented by trading in the money-market account (MMA) with deterministic short-rate process r, and a stock with time-t instantaneous excess return

$$dR_t = \frac{dS_t}{S_t} + (y_t - r_t)\,dt = \eta_t \sigma_t dt + \sigma_t dB_t,$$

where S_t is the stock price, y_t the dividend yield, σ_t is the stock's volatility, and η_t is the market price of risk. We saw earlier that, since r is deterministic, $\hat{\eta} = \eta$ and $\hat{r}$ is also deterministic. The dual market is therefore implemented by trading in an MMA with short-rate process $\hat{r}$ and a stock with the same instantaneous excess returns $d\hat{R} = dR$.

The agent of the primal problem sells the endowment e and follows a consumption–portfolio policy (ϱ, ψ) in the primal market generating the wealth process W, where

$$\frac{dW}{W} = (r - \varrho)\,dt + \psi\,dR, \quad W_0 = \frac{1}{\pi_0}\langle \pi \mid e \rangle.$$

At time t the agent consumes at a rate $c_t = \varrho_t W_t$ and invests a proportion ψ_t of the wealth W_t in the stock, with the remaining proportion $1 - \psi_t$ in the MMA. Let us call the policy (ϱ, ψ) **regular** if $W = W(c)$. As discussed in Section 3.8, this condition means that the policy does not destroy wealth and should therefore hold at an optimum. The analogous statements apply for a policy $(\hat{\varrho}, \hat{\psi})$ in the dual market for the dual agent. The following proposition shows how to determine a primal optimal policy, given a known dual optimal policy.

PROPOSITION 3.9.3 *Suppose the regular[26] dual-market consumption–portfolio policy $(\hat{\varrho}, \hat{\psi})$ finances the dual agent's consumption plan $\hat{c}$. Given a deterministic short-rate process r, let p_t be defined by (3.9.12), and let $z_t \equiv H_t(c)/W_t(c)$. Then, assuming it is regular, the primal-market consumption–portfolio policy (ϱ, ψ), where*

$$\varrho = \frac{1 - pz}{1 + p}\hat{\varrho} - vz \quad and \quad \psi = (1 - pz)\,\hat{\psi},$$

finances the primal agent's consumption plan c. In particular, (ϱ, ψ) is optimal for the primal agent if $(\hat{\varrho}, \hat{\psi})$ is optimal for the dual agent.

PROOF Let $\hat{W} \equiv \hat{W}(\hat{c})$ and $H \equiv H(c) = \hat{H}(\hat{c})$. The last identity of Table 3.1 and the fact that $\hat{\pi} = \pi(1 + p)$ imply $(1 + p)(\hat{W} + H) = W + H$, which simplifies to $(1 + p)\hat{W} = W - pH = (1 - pz)W$ and hence

---

[26] By Lemma 3.8.3, a sufficient condition for regularity is $\mathbb{E}\sup_t \pi_t W_t < \infty$, which is equivalent to $\mathbb{E}\sup_t \hat{\pi}_t \hat{W}_t < \infty$, assuming a deterministic short rate r and EMM density $\xi_T \equiv dQ/dP$ such that $\mathbb{E}\left[\xi_T^2\right] < \infty$. To show this equivalence, start with $\pi_t = \exp\left(-\int_0^t r_s ds\right)\xi_t$ and $\hat{\pi}_t = \exp\left(-\int_0^t \hat{r}_s ds\right)\xi_t$, where $\xi_t \equiv \mathbb{E}_t dQ/dP$. Doob's maximal inequality (a standard result in martingale theory) implies that $\mathbb{E}\left[\sup_t \xi_t^2\right] \leq 4\mathbb{E}\left[\xi_T^2\right] < \infty$ and therefore $\mathbb{E}\left[\sup_t \pi_t^2\right], \mathbb{E}\left[\sup_t \hat{\pi}_t^2\right] < \infty$. Since $\mathbb{E}\int_0^T c_t^2 dt < \infty$ by assumption, Jensen's inequality implies $\mathbb{E}\left[H_T^2\right] < \infty$, where $H \equiv H(c) = \hat{H}(\hat{c})$. Note that there is a constant K such that $H_t \leq K H_T$ for all t. Using the Cauchy–Schwarz inequality, it follows that $\mathbb{E}\sup_t \pi_t H_t < \infty$ and similarly $\mathbb{E}\sup_t \hat{\pi}_t H_t < \infty$. Since $\pi(W + H) = \hat{\pi}(\hat{W} + H)$, the claim follows.

$$\frac{\hat{W}}{W} = \frac{1 - pz}{1 + p}. \tag{3.9.13}$$

Using this ratio, we can now relate ϱ to $\hat{\varrho}$:

$$\varrho = \frac{c}{W} = \frac{\hat{c} - vH}{W} = \frac{\hat{W}}{W}\hat{\varrho} - vz = \frac{1 - pz}{1 + p}\hat{\varrho} - vz.$$

To relate ψ to $\hat{\psi}$, first recall the state-price dynamics (3.9.9) with $\eta = \hat{\eta}$. Integration by parts applied to the last identity of Table 3.1 gives

$$\pi \, dW + \cdots dt - \pi \, (W + H) \, \eta \, dB$$
$$= \hat{\pi} \, d\hat{W} + \cdots dt - \hat{\pi} \left(\hat{W} + H\right) \eta \, dB,$$

where $\cdots dt$ denotes a drift term whose details are irrelevant. The dB terms cancel out. Divide by πW and use first $\hat{\pi} = \pi(1 + p)$ and then (3.9.13) to find

$$\frac{dW}{W} = \cdots dt + (1 - pz)\frac{d\hat{W}}{\hat{W}}.$$

Finally, substitute the budget equation $dW/W = \cdots dt + \psi dR$ on the left and $d\hat{W}/\hat{W} = \cdots dt + \hat{\psi} dR$ on the right, and match martingale parts to find $\psi = (1 - pz)\,\hat{\psi}$. ☐

For a stochastic short-rate process r, the relationship between ϱ and $\hat{\varrho}$ of Proposition 3.9.3 still applies, with the same proof, and primal and dual portfolios are related by $\psi = (1 - pz)\,\hat{\psi} + (1 + z)\sigma^{-1}\left(\eta - \hat{\eta}\right)$, with a similar proof. An expression for $\eta - \hat{\eta}$ in terms of the volatility structure of discount bonds is derived in Schroder and Skiadas (2002), who also assume a more-general specification of H and provide extensions of the same idea in other settings. The simple general insight is that the redefinition of consumption through a linear transformation and associated redefinition of state prices can transform a new problem to one that is either simpler or one that is just different but already analyzed.

3.10 Exercises

EXERCISE 3.10.1 Assume the context of the CAPM equilibrium of Example 3.2.8, including preference transitivity (3.2.3). Provide a dual proof that the equilibrium is effectively complete by showing that there exists a complete market $\bar{X} \supseteq X$ such that $(\bar{X}, c)$ is an equilibrium and applying Corollary 3.2.7.

.

Hint Define the complete market $\bar{X}$ by setting to zero the present value of all time-one payoffs that are orthogonal to all traded time-one payoffs.

EXERCISE 3.10.2 This exercise outlines a variant of the representative-agent equilibrium of Example 3.2.9 that includes an example of a CAPM equilibrium. For this purpose, modify the definition of a consumption set and Definition 3.1.1 of a preference correspondence by weakening the monotonicity requirement to introduce an upper bound on allowable consumption: For every preference correspondence $\mathcal{D}$, there exists some $b \in \mathcal{L}$ such that $\operatorname{dom}(\mathcal{D}) = \{c \in \mathcal{L} \mid c < b\}$, and for every arbitrage cash flow y, if $x = 0$ or $x \in \mathcal{D}(c)$ then $x + y \in \mathcal{D}(c)$, provided $x + y < b$. (The notation $c < b$ means $c(\omega, t) < b(\omega, t)$ for all (ω, t).) The technical motivation behind this definition is to allow for a quadratic utility representation, as in this exercise's final part, which implies variance aversion and the CAPM pricing equation. Quadratic utility is increasing only up to a maximum. The idea is to consider only equilibria in which the constraint that consumption is below that maximum is nonbinding. We proceed with a more-general specification that emphasizes the aggregation argument.

Agent preferences are specified in terms of a given scale-invariant preference correspondence $\mathcal{D}^0$ with $\operatorname{dom}(\mathcal{D}^0) \equiv \{c \in \mathcal{L} \mid c < 0\}$. For all $i \in \{1, \dots, I\}$, the preference correspondence of agent $(\mathcal{D}^i, e^i)$ is specified in terms of $b^i \in \mathcal{L}$ by

$$\operatorname{dom}(\mathcal{D}^i) \equiv \left\{ c \in \mathcal{L} \mid c < b^i \right\} \quad \text{and} \quad \mathcal{D}^i(c) \equiv \mathcal{D}^0(c - b^i).$$

Further assume that there exist $v^i, w^i \in \mathbb{R}$ such that

$$e^i - w^i 1_{\Omega \times \{0\}} \in X, \quad b^i - v^i 1_{\Omega \times \{0\}} \in X \quad \text{and} \quad w^i < v^i,$$

and let $b \equiv \sum_i b^i$, $e \equiv \sum_i e^i$, $v \equiv \sum_i v^i$, $w \equiv \sum_i w^i$. The allocation $c \equiv (c^1, \dots, c^I)$ is defined by

$$c^i \equiv b^i - \frac{v^i - w^i}{v - w}(b - e).$$

Assume that $e < b$ and therefore $c^i < b^i$ for all i by construction.

(a) Show that the allocation c is market-clearing and X-feasible, assuming X is arbitrage-free.

(b) The **representative agent** $(\mathcal{D}, e)$ is defined by

$$\operatorname{dom}(\mathcal{D}) \equiv \{c \in \mathcal{L} \mid c < b\} \quad \text{and} \quad \mathcal{D}(c) \equiv \mathcal{D}^0(c - b).$$

Show that (X, c) is an equilibrium if and only if the consumption plan e is optimal for the representative agent given the market X, that is, $X \cap \mathcal{D}(e) = \emptyset$.

(c) Show that if (X, c) is an equilibrium and $\mathcal{D}(e)$ is convex, then (X, c) is an effectively complete market equilibrium.

(d) Specialize the setting by assuming that $T = 1$ and $\mathcal{D}^0$ has the quadratic utility representation $U^0(c) = -c_0^2 - \beta \mathbb{E}[c_1^2]$, where $\mathbb{E}$ is expectation under a given full-support probability and β is a positive scalar. Assume further that the parameters $b^1, \ldots, b^I$ are (deterministic) constants. Verify that the agent preferences are variance averse in the sense of Example 3.2.8. Adding the assumptions of a traded money-market account and the regularity condition $\text{var}[e_1] > 0$ and $e_0 \neq w$, all the CAPM assumptions of Example 3.2.8 are satisfied. In this context, give an alternative derivation by the CAPM beta-pricing equation (3.2.2) by using Proposition 3.3.4 and the optimality of the aggregate endowment for the representative agent.

Hint Construct a state-price density whose time-one value is an affine function of the market return.

EXERCISE 3.10.3 This exercise outlines a variant of the representative-agent equilibrium of Example 3.2.9, where preferences are assumed to be translation invariant instead of scale invariant. (Theorem A.4.2 shows that an additive utility representing a translation-invariant preference correspondence necessarily takes an exponential form, which in the expected-utility case corresponds to constant absolute risk aversion.) Every preference correspondence in this exercise is assumed to have the entire set of adapted processes $\mathcal{L}$ as its domain. Agents are specified in terms of a fixed reference agent $(\mathcal{D}^0, e^0)$. The key assumption is that $\mathcal{D}^0$ is **translation invariant**:

$$\mathcal{D}^0(c) = \mathcal{D}^0(c + \theta) \text{ for all } \theta \in \mathbb{R}.$$

For $i = 1, \ldots, I$, agent $(\mathcal{D}^i, e^i)$ is specified in terms of the parameters $(\alpha^i, \theta^i, x^i) \in (0, \infty) \times \mathbb{R} \times X$ by

$$\mathcal{D}^i(c) \equiv \alpha^i \mathcal{D}^0 \left(\frac{c}{\alpha^i} \right), \quad e^i \equiv \alpha^i e^0 + \theta^i + x^i.$$

Let $\alpha \equiv \sum_i \alpha^i$, $\theta \equiv \sum_i \theta^i$, $x \equiv \sum_i x^i$. The **representative agent** $(\mathcal{D}, e)$ is defined by

$$\mathcal{D}(c) = \alpha \mathcal{D}^0 \left(\frac{c}{\alpha} \right), \quad e \equiv \sum_{i=1}^{I} e^i = \alpha e^0 + \theta + x.$$

The allocation $c = (c^1, \ldots, c^I)$ is defined by

$$c^i = \theta^i + \frac{\alpha^i}{\alpha}(e - \theta).$$

(a) Show that (X, c) is an equilibrium if and only if the consumption plan e is optimal for the representative agent given the market X, that is, $X \cap \mathcal{D}(e) = \emptyset$.

(b) Suppose (X, c) is an equilibrium and $\mathcal{D}(e)$ is convex. Show that (X, c) is an effectively complete market equilibrium.

(c) Fix an underlying full-support probability and assume that for some $\beta \in (0, 1)$ and all i, $\mathcal{D}^i$ has the utility representation

$$U^i(c) = \mathbb{E}\left[-\alpha^i \sum_{t=0}^{T} \beta^t \exp\left(-\frac{c_t}{\alpha^i}\right) \right].$$

Verify that all preference assumptions made so far are satisfied and specify a state-price density π that corresponds to the utility gradient of the representative agent at the aggregate endowment.

(d) Consider a sequence of models as in part (c), parameterized by the time horizon $T = 1, 2, \ldots$, and take as given an infinite filtration $\{\mathcal{F}_t \mid t = 0, 1, \ldots\}$ and corresponding adapted process $e = (e_0, e_1, \ldots)$. The information and endowment of the Tth model is the restriction of the respective quantity over the time set $\{0, 1, \ldots, T\}$, that is, the filtration is $\{\mathcal{F}_t \mid t = 0, 1, \ldots, T\}$ and the endowment is $(e_0, e_1, \ldots, e_T)$. All other parameters are common across all models. The endowment process is strictly positive and follows the dynamics $e_t = e_{t-1}(1 + g_t)$, where the random variable g_t is stochastically independent of $\mathcal{F}_{t-1}$ and takes the value $\varepsilon \in (0, 1)$ with probability $p \in (0.5, 1)$ and $-\varepsilon$ with probability $1 - p$. Assume $m \equiv p \log(1 + \varepsilon) + (1 - p) \log(1 - \varepsilon) > 0$. What is the limit of the equilibrium short rate r_t as $t \to \infty$? You can use the law of large numbers from probability theory, which implies that $t^{-1} \log(e_t / e_0) = t^{-1} \sum_{s=1}^{t} \log(1 + g_s)$ converges (with probability one) to the mean of $\log(1 + g_t)$, which is $m > 0$, and therefore $e_t \to \infty$ as $t \to \infty$ (with probability one).

EXERCISE 3.10.4 Assume a single period ($T = 1$) and fix an underlying full-support probability. An agent's time-one consumption is restricted to take values in a nonempty open interval $D \subseteq \mathbb{R}$. Given K states, D^K denotes the set of all D-valued random variables, and $C_{++}^2(D)$ denotes the set of all strictly increasing and twice-continuously differentiable functions of the form $u : D \to \mathbb{R}$, with the first two derivatives denoted u' and

u'', respectively. Utilities over time-one consumption are assumed to be of the expected-utility form $\mathbb{E}u(c)$, $c \in D^K$, where $u \in C^2_{++}(D)$. The corresponding **coefficient of absolute risk aversion** is the function $A^u : D \to \mathbb{R}$, where

$$A^u(w) \equiv -\frac{u''(w)}{u'(w)}, \quad w \in D.$$

(a) Suppose that $u, \tilde{u} \in C^2_{++}(D)$. Show that $A^u = A^{\tilde{u}}$ if and only if there exist $a \in (0, \infty)$ and $b \in \mathbb{R}$ such that $\tilde{u} = au + b$. Apply Theorem A.2.4 to conclude that $A^{\tilde{u}} = A^u$ if and only if

for all $x, y \in D^K$, $\quad \mathbb{E}\tilde{u}(x) > \mathbb{E}\tilde{u}(y) \iff \mathbb{E}u(x) > \mathbb{E}u(y)$.

(b) Call u **HARA with coefficients** $(\alpha, \beta) \in \mathbb{R}^2$ if $u \in C^2_{++}(D)$ with

$$D \equiv \{w \mid \alpha + \beta w > 0\} \quad \text{and} \quad A^u(w) \equiv \frac{1}{\alpha + \beta w}.$$

(HARA is an acronym for hyperbolic absolute risk aversion. For $\beta < 0$, consumption is bounded above, just as in Exercise 3.10.2.) Show that u is HARA with coefficients (α, β) if and only if there exist $a \in (0, \infty)$ and $b \in \mathbb{R}$ such that

$$au(w) + b = \begin{cases} \frac{1}{\beta-1}(\alpha + \beta w)^{(\beta-1)/\beta}, & \text{if } \beta \neq 0 \text{ and } \beta \neq 1; \\ \log(\alpha + w), & \text{if } \beta = 1; \\ -\alpha \exp(-w/\alpha), & \text{if } \beta = 0 \text{ and } \alpha > 0. \end{cases} \quad (3.10.1)$$

(c) Explain why u HARA implies that u is strictly concave.

EXERCISE 3.10.5 This exercise, which has Exercise 3.10.4 as a prerequisite, formulates and analyzes a representative-agent equilibrium for agents who maximize HARA expected utility. The formulation nests Example 3.2.9 and Exercises 3.10.2 and 3.10.3 for the expected utility case.[27]

---

[27] This exercise follows DeMarzo and Skiadas (1998, 1999) who include asymmetric private information. A common approach in the literature of competitive equilibrium with asymmetric information is to assume constant absolute risk aversion (CARA) and joint normality of payoffs and private signals, and to construct an equilibrium in which prices are a linear function of the private signals, while remaining silent on the question of the possible existence of other, nonlinear, equilibria with different qualitative implications for the informational role of prices. The approach extending this exercise allows the removal of distributional assumptions, the replacement of CARA with HARA, and the characterization of all equilibria, linear or not. The conclusions highlight indeterminacy issues and the nonrobustness of claims of full revelation of private information in equilibrium. More generally, our theoretical understanding of how information diffuses through prices in equilibrium is highly incomplete.

Assume $T = 1$ and fix an underlying full-support probability over K states. For simplicity, agents can only trade in a forward market to modify their time-one consumption. Taken as given is a column vector $V \equiv (V_1, \ldots, V_J)'$, where $V_j \in \mathbb{R}^K$ represents the time-one value of some asset that can be traded in a forward market. A **portfolio** is a row vector $\theta \equiv (\theta_1, \ldots, \theta_J)$, where $\theta_j \in \mathbb{R}$ represents a number of forward contracts in asset j. Given a **forward-price vector** $f \equiv (f_1, \ldots, f_J)' \in \mathbb{R}^{J \times 1}$, the portfolio θ generates the cash flow $(0, \theta(V - f))$. Since all time-zero cash flows are zero, we define the **market** (given f) by

$$X_f \equiv \left\{ \theta(V - f) \mid \theta \in \mathbb{R}^{1 \times J} \right\}.$$

Each agent $i \in \{1, \ldots, I\}$ has a (time-one) endowment of the form

$$e^i \equiv a^i + b^i V, \quad a^i \in \mathbb{R}, \quad b^i \in \mathbb{R}^{1 \times J},$$

and maximizes the utility $\mathbb{E}u_i$ over time-one consumption, where u_i is HARA with coefficients (α^i, β). Note that $\mathbb{E}$ and β are the same across agents. The set of admissible (time-one) consumption plans for agent i is D_i^K, where $D_i \equiv \{w \mid \alpha^i + \beta w > 0\}$. The consumption plan $c^i \in D_i^K$ is **optimal** for agent i given f if there is no $x \in X_f$ such that $c^i + x \in D_i^K$ and $\mathbb{E}u_i(c^i + x) > \mathbb{E}u_i(c^i)$. An **allocation** is any element of $D_1^K \times \cdots \times D_I^K$. The allocation $c \equiv (c^1, \ldots, c^I)$ is f-**feasible** if $c^i - e^i \in X_f$ for all i, and **market-clearing** if $\sum_i (c^i - e^i) = 0$. An **equilibrium** is a pair (f, c) of a forward-price vector f and an f-feasible and market-clearing allocation c such that for all $i \in \{1, \ldots, I\}$, c^i is optimal for agent i given f.

Define the aggregate parameters

$$\alpha \equiv \sum_{i=1}^{I} \alpha^i, \quad a \equiv \sum_{i=1}^{I} a^i, \quad b \equiv \sum_{i=1}^{I} b^i, \quad e \equiv \sum_{i=1}^{I} e^i = a + bV.$$

The **representative agent** is endowed with the aggregate endowment e and maximizes the expected utility $\mathbb{E}u$, where u is HARA with coefficients (α, β). The set of admissible consumption plans for the representative agent is therefore D^K, where $D \equiv \{w \mid \alpha + \beta w > 0\}$. Assume that for every agent i, $e^i \in D_i^K$ and therefore $e \in D^K$.

(a) Fix any forward-price vector f such that X_f is arbitrage-free (meaning $X_f \cap \mathbb{R}_+^K = \{0\}$). Show that

$$c^i \equiv a^i + b^i f + \frac{\alpha^i + \beta(a^i + b^i f)}{\alpha + \beta(a + bf)} b(V - f) \qquad (3.10.2)$$

defines an allocation $c \in D_1^K \times \cdots \times D_I^K$ that is f-feasible and market-clearing, and there exist scalars $\lambda_f^i > 0$ such that

$$\lambda_f^i u_i'(c^i) = u'(e), \quad i = 1, \ldots, I.$$

(b) Show that for all $f \in \mathbb{R}^{J \times 1}$, the aggregate endowment e is optimal for the representative agent given f (that is, there is no $x \in X_f$ such that $e + x \in D^K$ and $\mathbb{E}u(e + x) > \mathbb{E}u(e)$) if and only if

$$f = \frac{\mathbb{E}\left[u'(e)\, V\right]}{\mathbb{E}u'(e)}. \tag{3.10.3}$$

(c) Show that the pair (f, c) defined by equations (3.10.3) and (3.10.2) is an equilibrium.

(d) Show that the equilibrium of part (c) is an effectively complete market equilibrium: For all $x^1, \ldots, x^I$ such that $c^i + x^i \in D_i^K$, if $\mathbb{E}u_i(c^i + x^i) \geq \mathbb{E}u_i(c^i)$ for all i and $\mathbb{E}u_i(c^i + x^i) > \mathbb{E}u_i(c^i)$ for some i, then $\sum_i x^i \notin X_f$. You can base your argument on Proposition 3.3.9, but you should give a proof from first principles.

(e) Show that the equilibrium of part (c) is unique.

Hint If the forward-price vector f is part of some equilibrium and the market-clearing allocation c is Pareto optimal and f-feasible, then (f, c) is an equilibrium.

(f) Assume $\beta \equiv -1$. Given the equilibrium of part (c), call R a **traded forward return** if there exists a portfolio θ such that $\theta f \neq 0$ and $R = \theta V / \theta f$. Assume that $a + bf \neq 0$ and $\mathrm{var}[e] > 0$, and define the forward return of the aggregate endowment:

$$R^m \equiv \frac{a + bV}{a + bf} = \frac{e}{\text{forward price of } e}.$$

Assuming that R^0 is a forward traded return that is uncorrelated with R^m, show that every traded forward return R satisfies the CAPM equation

$$\mathbb{E}[R - R^0] = \frac{\mathrm{cov}[R, R^m]}{\mathrm{var}[R^m]} \mathbb{E}[R^m - R^0].$$

EXERCISE 3.10.6 (Preferences for the timing of resolution of uncertainty) In the main text preferences are defined over consumption plans taking the information tree (filtration) as given. In settings in which the information structure is not fixed, it is natural to consider preferences not only over consumption but also over information.[28] In particular, an agent may

---

[28] A version of the ideas of this exercise appears in Kreps and Porteus (1978), whose formalism is based on preferences over timed probability distributions. The

have preferences for earlier or later resolution of uncertainty about future consumption. For instance, in the context of Example 3.3.12, given the consumption plan b, an agent with preferences for late resolution of uncertainty would prefer to to wait to find out the outcome of the tth coin toss at time t, rather than have all T coin tosses announced at time zero.

More formally, let C be a (nonempty) set of consumption plans and let Φ be the set of every filtration $\{\mathcal{F}_t\}$ satisfying $\mathcal{F}_0 = \{\Omega, \emptyset\}$ and $\mathcal{F}_T = 2^\Omega$. We consider a (nonnormalized) utility function $V_0(\cdot)$ over the set of all pairs $(c, \{\mathcal{F}_t\}) \in C \times \Phi$ such that c is adapted to $\{\mathcal{F}_t\}$. Taking as primitive an underlying probability with expectation operator $\mathbb{E}$ and the functions $F_t : \mathbb{R}^2 \to \mathbb{R}, t = 0, \ldots, T - 1$ and $F_T : \mathbb{R} \to \mathbb{R}$, assume that $V_0(c, \{\mathcal{F}_t\})$ is the initial value of the process V that solves the backward recursion

$$V_t = F_t\big(c_t, \mathbb{E}[V_{t+1} \mid \mathcal{F}_t]\big), \quad t = 0, \ldots, T - 1; \quad V_T = F_T(c_T).$$

We say that the utility function $V_0(\cdot)$ expresses preferences for earlier resolution of uncertainty if for all $(c, \{\mathcal{F}_t^1\})$ and $(c, \{\mathcal{F}_t^2\})$ in its domain,

$$\mathcal{F}_t^1 \subseteq \mathcal{F}_t^2 \text{ for all } t \quad \text{implies} \quad V_0(c, \{\mathcal{F}_t^1\}) \leq V_0(c, \{\mathcal{F}_t^2\}).$$

Preferences for late resolution of uncertainty are defined analogously, with the last inequality reversed.

(a) Use Jensen's inequality to show that if the functions F_t are convex (resp. concave) in their second (utility) argument for all $t < T$, then $V_0(\cdot)$ expresses preferences for earlier (resp. later) resolution of uncertainty.

(b) Suppose the choice of F_t corresponds to the functional form of expected discounted utility. What does that imply about preferences over information?

(c) Suppose the utility process associated with $(c, \{\mathcal{F}_t\})$ is of the EZW form of Example 3.5.1, with CRRA γ and inverse EIS δ. Characterize preferences for the timing of resolution of uncertainty in terms of the relative value of the parameters γ and δ. You can use without proof the fact that if u is of the HARA form (3.10.1) and $\beta \in (0, 1)$, then the function $u^{-1}((1 - \beta) u(x) + \beta u(y))$ is concave if u is concave, and convex if u is convex. (This follows by the argument of Example 3.3.2 for the power-or-logarithmic case. A similar argument applies for the exponential case, based on translation invariance.)

probability-free approach of this exercise is from Skiadas (1998), where a continuous-time version of the main characterization can also be found.

EXERCISE 3.10.7 In this problem assume scale-invariant recursive utility with a proportional aggregator that takes the unit-EIS form $g_t(x) = x^{1-\beta}$ for some $\beta \in (0, 1)$.

(a) Specialize the result of Proposition 3.6.2 (IMRS and consumption growth) to this case.

(b) Specialize the result of Proposition 3.6.4 (IMRS and market returns) to this case.

(c) Show that the procedures for computing π in parts (a) and (b) can be easily derived from each other, given the relationship between consumption growth and market returns of Example 3.6.7 (unit EIS).

(d) Further specialize the results of parts (a) and (b) by assuming unit-EIS Epstein–Zin–Weil utility. (Provide as specific expressions as you can.)

EXERCISE 3.10.8 An alternative approach to proving Theorem 3.7.2 on optimal consumption and portfolio choice is based on the ideas of dynamic programming. While it is not difficult to show the complete Theorem 3.7.2, in this problem you are only asked to use a dynamic programming argument to prove the sufficiency part: If (ϱ, ψ) is constructed according to the theorem's algorithm, then it is optimal. Also, for notational brevity, assume the proportional aggregator $g_t = g$ is time independent.

Define a value function to be a mapping $\mathcal{V}: (0, \infty) \rightarrow \mathcal{L}_{++}$ that assigns to each $w \in (0, \infty)$ a strictly positive adapted process $\mathcal{V}(w)$. For each spot (F, t), think of $\mathcal{V}(w)(F, t)$ as the optimal utility level at spot (F, t) given total wealth w at that spot. (In a typical application, the dependence on (F, t) is through the value $Z(F, t)$ of some Markov process Z on that spot, and the value function is therefore defined as a function of wealth and the Markov state. Here, $Z(F, t)$ can be thought of as being the spot itself, which can be identified with the entire history leading to that spot.)

Write $\mathcal{V}_t(w)$ for the random variable that takes the value $\mathcal{V}(w)(F, t)$ on F for every spot (F, t). The corresponding **Bellman equation** is

$$\mathcal{V}_{t-1}(w) = \max_{\varrho_{t-1}, \psi_t} f\left(\varrho_{t-1}w, \upsilon_{t-1}\left(\mathcal{V}_t\left((1 - \varrho_{t-1}) w R_t^{\psi}\right)\right)\right).$$

Proceed in the following steps.

(a) What exactly is the mathematical content of the Bellman equation at each spot? What is the set over which maximization occurs at each spot?

(b) Conjecture the functional form of $\mathcal{V}(w)$, explaining your reasoning.

(c) Show that the algorithm of Theorem 3.7.2 produces a solution to the Bellman equation.

(d) Provide an optimality verification argument, that is, show that given the right terminal condition, a solution to the Bellman equation defines an optimal policy. You should explain how the Bellman equation defines the optimal policy and then prove optimality using only the budget equation, the utility definition and the Bellman equation.

EXERCISE 3.10.9 In this exercise you will verify two claims of Example 3.8.2, whose context is assumed. You can ignore technicalities involving omitted integrability assumptions. (For example, with D denoting the random variable inside the expectation defining $\tilde{U}_0$, you can assume that $\mathbb{E}\left[D^2\right] < \infty$ and therefore the martingale $\mathbb{E}_t D$ has a unique martingale representation.)

(a) Verify that if $u_\delta(U_t) = q_t \tilde{U}_t$ then U satisfies BSDE (3.8.6).

(b) Assuming $\delta \neq 1$, verify that if U is defined by (3.8.7), then it satisfies BSDE (3.8.8).

Appendix A

Additive Utility Representations

This appendix presents some basic results on the representation of preferences by additive utility functions. Simple ordinal assumptions, mainly separability and continuity, are enough for the existence of an additive utility representation. Additional ordinal restrictions, like preference convexity, scale or translation invariance, and state independence are shown to have strong implications for the structure of an additive utility representation. The appendix concludes with a discussion of risk aversion in the context of expected utility.

A.1 Utility Representations of Preferences

In this appendix we discuss preferences over the set $C \equiv (\ell, \infty)^N$ for some integer $N \geq 2$ and constant $\ell \in [-\infty, \infty)$. (Note that we allow the possibility $\ell = -\infty$, corresponding to $C = \mathbb{R}^N$.) We treat a constant $\alpha \in (\ell, \infty)$ as an element of C by identifying it with $(\alpha, \dots, \alpha)$. As in Definition 3.3.1, a **utility** is a continuous and increasing function of the form $U : C \to \mathbb{R}$. A **preference** is a binary relation on C, that is, a subset $\succ$ of $C \times C$, with the notation $a \succ b$ indicating $(a, b) \in \succ$. The utility U **represents** the preference $\succ$ defined, for all $a, b \in C$, by

$$a \succ b \quad \Longleftrightarrow \quad U(a) > U(b).$$

The preference $\succ$ is another way of expressing the preference correspondence $\mathcal{D}$ represented by U (Definitions 3.1.1, 3.3.1): $a + x \succ a$ if and only if $x \in \mathcal{D}(a)$. Following this section, all preferences in this appendix are assumed to have a utility representation, but we use the preference notation $\succ$ where we wish to emphasize that a property is ordinal and therefore not dependent on any particular choice of a utility representation.

We close this section with a characterization of every preference $\succ$ that admits a utility representation. We define the relation $\succeq$ on C by letting $a \succeq b$ if and only if not $b \succ a$. Clearly, $\succ$ admits a utility representation U

if and only if $\succeq$ admits a utility representation U in the sense (A.1.1) of the following utility representation theorem. The argument is simplified by our standing assumption of monotone preferences.[1]

THEOREM A.1.1 *For any binary relation $\succeq$ on C, there exists a utility function $U : C \to \mathbb{R}$ such that*

$$a \succeq b \quad \Longleftrightarrow \quad U(a) \geq U(b) \tag{A.1.1}$$

if and only if $\succeq$ is

- **total:** *for all $a, b \in C$, either $a \succeq b$ or $b \succeq a$;*
- **transitive:** *$a \succeq b$ and $b \succeq c$ implies $a \succeq c$;*
- **increasing:** *for all $a, b \in C$, if $b \neq a \geq b$ then not $b \succeq a$;*
- **continuous:** *for every sequence (a_n, b_n) such that $a_n \succeq b_n$ for all n, if $\lim_{n \to \infty} (a_n, b_n) = (a, b) \in C \times C$, then $a \succeq b$.*

PROOF The "only if" part is immediate from the definitions. Conversely, suppose that $\succeq$ is total, transitive, increasing and continuous. On C, define the function $U(c) \equiv \inf \{\alpha \in \mathbb{R} \mid \alpha \succeq c\}$ and the relation $a \sim b$ if and only if both $a \succeq b$ and $b \succeq a$. We proceed in six steps.

Step 1: $U(c) \sim c$. Choose any sequence α_n in $\mathbb{R}$ that converges to $U(c)$ and satisfies $\alpha_n \succeq c$ for all n. Since $\succeq$ is continuous, $U(c) \succeq c$ and $U(c) = \min \{\alpha \in \mathbb{R} \mid \alpha \succeq c\}$. For all $n = 1, 2, \ldots$, it is not the case that $U(c) - n^{-1} \succeq c$, and since $\succeq$ is total, $c \succeq U(c) - n^{-1}$. Since $\succeq$ is continuous, $c \succeq U(c)$. This proves that $U(c) \sim c$.

Step 2: $a \succeq b \iff U(a) \geq U(b)$. Since $\succeq$ is transitive, if $a \succeq b$, then $\{\alpha \in \mathbb{R} \mid \alpha \succeq a\} \subseteq \{\alpha \in \mathbb{R} \mid \alpha \succeq b\}$, and therefore $U(a) \geq U(b)$ by the definition of U. Conversely, since $\succeq$ is monotone, $U(a) \geq U(b)$ implies that $U(a) \succeq U(b)$. By Step 1, $a \succeq U(a)$ and $U(b) \succeq b$. Since $\succeq$ is transitive, it follows that $a \succeq b$.

Step 3: U is increasing. Suppose $a, b \in C$ and $b \neq a \geq b$. Since $\succeq$ is increasing, not $b \succeq a$. By Step 2, $U(a) > U(b)$.

---

[1] Debreu (1983) (based on a 1954 working paper) shows the existence of a continuous utility representation of a continuous, total and transitive binary relation, without monotonicity. The claim fails if continuity is also dropped, with the classic counterexample being lexicographic preferences: For $a, b \in \mathbb{R}^2$, suppose $a \succ b$ if and only if $a_1 > b_1$ or $(a_1 = b_1$ and $a_2 > b_2)$. Suppose $\succ$ has a utility representation U. Then for all $x \in \mathbb{R}$, the interval $(U(x, 0), U(x, 1))$ is nonempty and therefore contains a rational $r(x)$. If $y > x$, then $U(y, 0) > U(x, 1)$ and therefore $r(y) > r(x)$. We therefore have a bijective mapping $x \mapsto r(x)$ of the real numbers onto a subset of the rationals, a contradiction, since the rationals can be enumerated $(r_1, r_2, \ldots)$ while, by Cantor's diagonal argument, any enumeration of the reals must omit some real number.

Step 4: For all $u \in (\ell, \infty)$, $U(u) = u$. Since $\succeq$ is total and increasing, for all $\alpha, u \in (\ell, \infty)$, $\alpha \succeq u$ if and only if $\alpha \geq u$. The claim now follows by the definition of U.

Step 5: For all $u \in \mathbb{R}$, if $u \sim c$, then $u = U(c)$. This is a corollary of Steps 2 and 4.

Step 6: U is continuous. Consider any sequence c_n in C converging to $c \in C$. Suppose first that $\{U(c_n)\}$ converges to some u. Since $U(c_n) \sim c_n$ for all n and $\succeq$ is continuous, it follows that $u \sim c$ and therefore $u = U(c)$ by Step 5. Consider now the general case, where we do not know a priori that $\{U(c_n)\}$ converges. Since $\{c_n\}$ converges, it is bounded. There exist, therefore, $\alpha, \beta \in (\ell, \infty)$ such that c and all the components of c_n are valued in $[\alpha, \beta]$ and therefore $U(c)$ and $U(c_n)$ are all valued in the compact set $[U(\alpha), U(\beta)]$. It follows that every subsequence of $\{U(c_n)\}$ has a convergent further subsequence, which as shown earlier must converge to $U(c)$. This proves that $U(c_n)$ also converges to $U(c)$. $\qquad\square$

A.2 Additive Utility Representations

The remainder of this appendix discusses preferences on $(\ell, \infty)^N$, where $\ell \in [-\infty, \infty)$, that admit an additive utility representation in the following sense.

DEFINITION A.2.1 A utility $U: (\ell, \infty)^N \to \mathbb{R}$ is **additive** if there exist functions $U_n: (\ell, \infty) \to \mathbb{R}$ such that

$$U(x) = \sum_{n=1}^{N} U_n(x_n), \quad x \in (\ell, \infty)^N. \tag{A.2.1}$$

Given any $x, y \in (\ell, \infty)^N$ and $A \subseteq \{1, \ldots, N\}$, $x_A y$ denotes the element of $(\ell, \infty)^N$ defined by

$$(x_A y)_n = \begin{cases} x_n, & \text{if } n \in A; \\ y_n, & \text{if } n \notin A. \end{cases}$$

DEFINITION A.2.2 A preference $\succ$ on $(\ell, \infty)^N$ is **separable** if

$$x_A z \succ y_A z \quad \Longleftrightarrow \quad x_A \tilde{z} \succ y_A \tilde{z},$$

for all $x, y, z, \tilde{z} \in (\ell, \infty)^N$ and $A \subseteq \{1, \ldots, N\}$.

A preference that admits an additive utility representation is clearly separable. The following remarkable theorem gives a converse. It is a special case of Debreu's additive representation theorem,[2] but it captures the deeper aspects of Debreu's theorem. The proof would take us too far afield and will not be given here.

THEOREM A.2.3 (Additive utility representation) *Suppose $N > 2$ and $\succ$ is a preference on $(\ell, \infty)^N$ that admits a utility representation. Then $\succ$ admits an additive utility representation if and only if it is separable.*

The assumption $N > 2$ is critical; separability is not sufficient for the existence of an additive representation if $N = 2$. The uniqueness part of Debreu's theorem (which also applies for $N = 2$) is stated and proved in Theorem A.2.4. The result is essential for our discussion of the limitations of additive utilities in Chapter 3, and our later parametric characterization of scale- or translation-invariant additive preferences, and the associated foundation of expected utility with constant relative or absolute risk aversion.

THEOREM A.2.4 (Uniqueness of additive representations) *For any $N \geq 2$, suppose the utilities U and $\tilde{U}$ are additive: For all $x \in (\ell, \infty)^N$, $U(x) = \sum_n U_n(x_n)$ and $\tilde{U}(x) = \sum_n \tilde{U}_n(x_n)$. Then the following two conditions are equivalent.*

(1) *U and $\tilde{U}$ are **ordinally equivalent**: For all $x, y \in (\ell, \infty)^N$,*

$$U(x) > U(y) \iff \tilde{U}(x) > \tilde{U}(y).$$

(2) *U and $\tilde{U}$ are related by a **positive affine transformation**: There exist $a \in (0, \infty)$ and $b \in \mathbb{R}^N$ such that*

$$\tilde{U}_n = aU_n + b_n, \quad n = 1, \ldots, N.$$

PROOF That (2) $\implies$ (1) is immediate. We show the converse, assuming that $\ell = -\infty$. This is without loss of generality: If $\ell > -\infty$, then apply the result for $\ell = -\infty$ to the utility functions that map $z \in \mathbb{R}^N$ to $\sum_n U_n(\ell + e^{z_n})$ and $\sum_n \tilde{U}_n(\ell + e^{z_n})$, respectively.

Consider any ordinally equivalent additive utilities U and $\tilde{U}$ on $\mathbb{R}^N$ that satisfy, for all $n \in \{1, \ldots, N\}$,

$$U_n(0) = \tilde{U}_n(0) = 0 \quad \text{and} \quad U_1(1) = \tilde{U}_1(1) = 1. \tag{A.2.2}$$

The claim is that $\tilde{U}_n = U_n$ for all n, which proves (1) $\implies$ (2), since any additive utility on $\mathbb{R}^N$ can be made to satisfy normalization (A.2.2)

---

[2] Debreu (1983) characterizes continuous additive utility representations in a way that includes Theorems A.2.3 and A.2.4. Krantz et al. (1971) present an algebraic theory of additive representations that generalizes Debreu's topological results (see also Wakker (1988, 1989)).

after a positive affine transformation. Fixing an arbitrary $n \in \{2, \ldots, N\}$, $L \in (-\infty, 0)$ and scalar Δ such that $L + \Delta > 1$, define the functions $f, g \colon [0, 1] \to \mathbb{R}$ by

$$f(z) \equiv \frac{U_1(L + z\Delta) - U_1(L)}{U_1(L + \Delta) - U_1(L)}, \quad g(z) \equiv \frac{U_n(L + z\Delta) - U_n(L)}{U_1(L + \Delta) - U_1(L)}.$$

Define also $\tilde{f}$ and $\tilde{g}$ by putting a tilde over f, g and every instance of U in the previous display. Applying Lemma A.2.5 immediately following this proof to these functions, it follows that

$$U_1(x) = \tilde{U}_1(x) \text{ and } U_n(x) = \tilde{U}_n(x) \text{ for all } x \in [L, L + \Delta].$$

This proves that $\tilde{U}_n = U_n$ for all n, since every $x \in \mathbb{R}$ is in an interval of the form $(L, L + \Delta) \supset [0, 1]$. $\qquad\square$

LEMMA A.2.5 *Suppose the functions* $f, g, \tilde{f}, \tilde{g} \colon [0, 1] \to \mathbb{R}$ *are strictly increasing and continuous, and satisfy*

$$f(0) = g(0) = \tilde{f}(0) = \tilde{g}(0) = 0 \quad and \quad f(1) = \tilde{f}(1) = 1. \quad \text{(A.2.3)}$$

Suppose also that for all $x, y, z, w \in [0, 1]$,

$$f(x) + g(y) = f(z) + g(w) \iff \tilde{f}(x) + \tilde{g}(y) = \tilde{f}(z) + \tilde{g}(w). \quad \text{(A.2.4)}$$

Then $f = \tilde{f}$ *and* $g = \tilde{g}$.

PROOF Let N be any positive integer such that $2^{-N} < g(1)$. Given any $n \in \{N, N + 1, \ldots\}$, define $x_k^n \in [0, 1]$ and $y^n \in (0, g(1))$ by

$$f(x_k^n) = k2^{-n}, \quad k = 0, 1, \ldots, 2^n, \quad and \quad g(y^n) = 2^{-n}. \quad \text{(A.2.5)}$$

Note that $x_0^n = 0$ and $x_{2^n}^n = 1$. Since $g(0) = 0$, we have

$$f(x_k^n) + g(0) = f(x_{k-1}^n) + g(y^n), \quad k = 1, \ldots, 2^n.$$

By assumption (A.2.4), it is also true that

$$\tilde{f}(x_k^n) + \tilde{g}(0) = \tilde{f}(x_{k-1}^n) + \tilde{g}(y^n), \quad k = 1, \ldots, 2^n. \quad \text{(A.2.6)}$$

Since $\tilde{g}(0) = \tilde{f}(0) = 0$, it follows that

$$1 = \tilde{f}(1) = \sum_{k=1}^{2^n} \tilde{f}(x_k^n) - \tilde{f}(x_{k-1}^n) = 2^n \tilde{g}(y^n).$$

This proves that $\tilde{g}(y^n) = 2^{-n}$, which together with (A.2.6) shows that $\tilde{f}(x_k^n) = k2^{-n}$ for $k > 0$. Comparing this conclusion to (A.2.5), we have proved that the functions f^{-1} and $\tilde{f}^{-1}$ are equal on the set $D_n = \{k2^{-n} \mid k = 0, \ldots, 2^n\}$, for all $n \geq N$. Since the set $\bigcup_{n \geq N} D_n$ is dense in $[0, 1]$ and the functions f^{-1} and $\tilde{f}^{-1}$ are continuous, it follows that $f^{-1} = \tilde{f}^{-1}$ and therefore $f = \tilde{f}$.

To show that $g = \tilde{g}$, we apply the same argument with $(F, G, \tilde{F}, \tilde{G})$ in place of $(f, g, \tilde{f}, \tilde{g})$, where

$$F(z) \equiv \frac{g(z)}{g(1)}, \quad \tilde{F}(z) \equiv \frac{\tilde{g}(z)}{\tilde{g}(1)}, \quad G(z) \equiv \frac{f(z)}{g(1)}, \quad \tilde{G}(z) \equiv \frac{\tilde{f}(z)}{\tilde{g}(1)}.$$

The conclusion $F = \tilde{F}$ implies that $g = a\tilde{g}$ for some $a \in (0, \infty)$. Choose $\varepsilon, \delta > 0$ such that $f(\varepsilon) = g(\delta)$, and therefore $f(\varepsilon) + g(0) = f(0) + g(\delta)$ (by (A.2.3)). By (A.2.4), it must also be the case that $\tilde{f}(\varepsilon) = \tilde{g}(\delta)$. Since $f = \tilde{f}$ and $\tilde{g} = ag$, this shows that $a = 1$ and therefore $g = \tilde{g}$, completing the proof.   □

A.3 Concave Additive Representations

A preference $\succ$ on $(\ell, \infty)^N$ is **convex** if for all $y \in (\ell, \infty)^N$, the set $\{x \mid x \succ y\}$ is convex. Preference convexity is a key component of this text's theory as it expresses preference for smoothing across time and states. As pointed out in Section 3.3, a convex preference admitting a utility representation may admit no concave utility representation. This is also true of additive representations: $U(x) \equiv e^{x_1} - e^{-x_2}$ defines an additive utility on $(0, \infty)^2$ that represents a convex preference (uniquely up to positive affine transformations). On the other hand, we have the following positive and rather surprising result.[3]

THEOREM A.3.1 *Suppose the preference $\succ$ on $(\ell, \infty)^N$ is convex and admits an additive utility representation $U : (\ell, \infty)^N \to \mathbb{R}$. Then at least $N - 1$ of the functions U_n defined in (A.2.1) are concave.*

PROOF The theorem follows from Lemma A.3.3.   □

In fact, the argument that follows proves a stronger result in that our standing monotonicity assumption on preferences plays no role and the interval (ℓ, ∞) is replaced by the interval (ℓ, m), for any $m \in (\ell, \infty]$. The remainder of this section shows two lemmas that lead to the proof of Theorem A.3.1. The first lemma is stated a little more generally than

---

[3] The proof of Theorem A.3.1 is based on Yaari (1977), who credited Koopmans with a different proof, as well as Gorman for the twice-differentiable case, in both cases in unpublished papers.

needed in this section. We will use it again in our discussion of expected utility and risk aversion.

LEMMA A.3.2 (Yaari) *Suppose the continuous function* $f: (\ell, m) \to \mathbb{R}$ *is not concave and* $p \in (0,1)$. *Then there exist some* $x^* \in (\ell, m)$ *and small enough scalar* $\varepsilon > 0$ *such that for all* $\delta \in (0, \varepsilon)$,

$$f(x^*) < pf(x^* + (1-p)\,\delta) + (1-p)\,f(x^* - p\delta). \qquad (A.3.1)$$

PROOF Since f is not concave, there exist $x^0, x^1 \in (\ell, m)$ and $\alpha \in (0, 1)$ such that

$$f(x^\alpha) < (1-\alpha)\,f(x^0) + \alpha f(x^1),$$

where $x^\alpha \equiv (1-\alpha)\,x^0 + \alpha x^1$ and $x^0 < x^1$. Let $h: (\ell, m) \to \mathbb{R}$ be the affine function whose graph is the straight line through the points $(x^0, f(x^0))$ and $(x^1, f(x^1))$. Inequality (A.3.1) is satisfied by the given function f if and only if it is satisfied by the function $f - h$. Replacing f by $f - h$, we proceed under the assumption that $f(x^0) = f(x^1) = 0$ and therefore $f(x^\alpha) < 0$. Let $K = \text{argmin}\{f(x) : x \in [x^0, x^1]\}$, a compact set (since f is continuous), and let $x^* = \min K$. Clearly, $x^* \in (x^0, x^1)$ and $f(x^*) < 0$. Choose $\varepsilon > 0$ small enough so that $x^* + (1-p)\varepsilon$ and $x^* - p\varepsilon$ are both valued in $[x^0, x^1]$. The fact that x^* minimizes f on $[x^0, x^1]$ and x^* is the least point in $[x^0, x^1]$ with this property implies, for all $\delta \in (0, \varepsilon)$, the inequalities

$$f(x^*) \le f(x^* + (1-p)\,\delta) \text{ and } f(x^*) < f(x^* - p\delta),$$

which in turn imply (A.3.1). $\qquad\qquad\square$

We close the section with the lemma that contains the essential part of the proof of Theorem A.3.1.

LEMMA A.3.3 *Suppose that the functions* $f, g: (\ell, m) \to \mathbb{R}$ *are continuous and neither of them is concave. Then there exists some* (x^*, y^*) *such that the set*

$$\{(x, y) \in (\ell, m)^2 \mid f(x) + g(y) > f(x^*) + g(y^*)\}$$

is nonempty and not convex.

PROOF Let x^* and ε be selected in terms of f as in Lemma A.3.2 with $p = 1/2$, and let y^* and ε be analogous quantities for g. (Note that we can choose the same ε in both cases.) Clearly, the validity of the lemma's claim does not change if f and g are modified by adding a constant. Replacing f with $f - f(x^*)$ and g with $g - g(y^*)$, we proceed under the assumption that

$$f(x^*) = g(y^*) = 0. \qquad (A.3.2)$$

Letting $\epsilon \equiv \varepsilon/2$, condition (A.3.1) can be restated as

$$f(x^* + a) + f(x^* - a) > 0 \quad \text{for all } a \in (-\epsilon, \epsilon). \qquad (A.3.3)$$

This implies that x^* is not a local maximizer of f. Therefore, the set

$$I_f \equiv \{f(x^* + a) \mid a \in (-\epsilon, \epsilon)\}$$

is an interval (since f is continuous) that includes a subinterval of the form $[0, \delta)$ for some $\delta > 0$. Analogously, we have

$$g(y^* + b) + g(y^* - b) > 0 \quad \text{for all } b \in (-\epsilon, \epsilon), \qquad (A.3.4)$$

and I_g includes a subinterval $[0, \delta)$ for some $\delta > 0$. It follows that $I_f \cap I_g$ is nonempty, which is to say that there exist a and b in $(-\epsilon, \epsilon)$ such that $f(x^* + a) = g(y^* + b)$. The last equality in combination with (A.3.3) and (A.3.4) implies

$$f(x^* - a) + g(y^* + b) > 0 \quad \text{and} \quad f(x^* + a) + g(y^* - b) > 0.$$

We have proved that $S \equiv \{(x, y) \mid f(x) + g(y) > 0\}$ contains the points $(x^* - a, y^* + b)$ and $(x^* + a, y^* - b)$, whose midpoint is (x^*, y^*). By (A.3.2), $(x^*, y^*) \notin S$ and therefore S is not convex. $\qquad \square$

A.4 Scale/Translation-Invariant Representations

In Sections A.1 and A.2, we characterized every preference $\succ$ that admits a utility representation and saw that if $\succ$ admits a utility representation and $N > 2$, it admits an additive utility representation if and only if it is separable. In this section, we take the existence of an additive utility representation as given and show (for any N) that a translation-invariance property of $\succ$ is equivalent to the utility taking an exponential functional form, and a scale-invariance property of $\succ$ is equivalent to the utility taking a power or logarithmic functional form.[4] In the following sections, we will interpret these results as providing ordinal foundations for expected utility with constant absolute risk aversion in the case of translation-invariant preferences, and constant relative risk aversion in the case of scale-invariant preferences.

[4] To my knowledge, the purely ordinal characterizations of Theorems A.4.2 and A.4.3, without any differentiability assumptions, first appear in Skiadas (2013b).

The section's central argument is an application of Theorem A.2.4 on the uniqueness of additive representations and the following characterization of a linear function on the real line.[5]

LEMMA A.4.1 *Suppose that the function* $f : \mathbb{R} \to \mathbb{R}$ *satisfies*

$$f(x + y) = f(x) + f(y) \quad \text{for all } x, y \in \mathbb{R}, \qquad (A.4.1)$$

and there exists some nonempty open interval on which f *is bounded. Then* $f(x) = f(1) x$ *for all* $x \in \mathbb{R}$.

PROOF We must show that $\delta(x) \equiv f(x) - f(1) x = 0$ for all $x \in \mathbb{R}$. Using (A.4.1), we find $\sum_{i=1}^{n} \delta(1/n) = \delta(1) = 0$ and therefore $\delta(1/n) = 0$ for every positive integer n. Similarly, for positive integers m and n, $\delta(m/n) = \sum_{i=1}^{m} \delta(1/n) = 0$. We have shown that $\delta(r) = 0$ if r is a positive rational number. Since $\delta(0 + 0) = \delta(0) + \delta(0)$ and $\delta(0) = \delta(r) + \delta(-r)$, $\delta(r) = 0$ for every rational r. The function δ inherits from f the property that it is bounded on some nonempty open interval (a, b). Given any $x \in \mathbb{R}$, we can find a rational r such that $x + r \in (a, b)$, and since $\delta(x) = \delta(x) + \delta(r) = \delta(x + r)$, δ is bounded on the entire real line. Finally, for all $x \in \mathbb{R}$, the set of all $\delta(nx) = n\delta(x)$, $n = 1, 2, \ldots$, is bounded only if $\delta(x) = 0$. □

The preference $\succ$ on $\mathbb{R}^N$ is said to be **translation invariant** if

$$x \succ y \text{ and } t \in \mathbb{R} \quad \Longrightarrow \quad x + t \succ y + t.$$

THEOREM A.4.2 *Suppose the preference* $\succ$ *on* $\mathbb{R}^N$ *admits an additive utility representation. Then* $\succ$ *is translation invariant if and only if it admits a utility representation of the form*

$$U(x) = \sum_{n=1}^{N} w_n \frac{1 - \exp(-\alpha x_n)}{\alpha}, \quad x \in \mathbb{R}^N, \qquad (A.4.2)$$

for unique $\alpha \in \mathbb{R}$ *and* $w_1, \ldots, w_N \in (0, 1)$ *such that* $\sum_n w_n = 1$. *The convention for* $\alpha = 0$ *is that* $(1 - \exp(-\alpha x)) / \alpha$ *is equal to* x, *which is the limit as* $\alpha \to 0$.

PROOF Suppose U is an additive utility representing the translation-invariant preference $\succ$. Without loss of generality, we assume that $U_n(0) = 0$ for all $n = 1, \ldots, N$ (otherwise, replace U_n by $U_n - U_n(0)$). For any $t \in \mathbb{R}$, the translation invariance of $\succ$ implies that $U(x + t)$ as a

---

[5] For some historical context and related results, see Aczél (2006).

function of $x \in (0, \infty)^N$ defines another additive utility representation of $\succ$. By Theorem A.2.4 and the fact that $U_n(0) = 0$, there exists a function $a : \mathbb{R} \to (0, \infty)$ such that

$$U_n(s + t) = U_n(s) a(t) + U_n(t), \quad s, t \in \mathbb{R}, \quad n = 1, \ldots, N. \quad (A.4.3)$$

Suppose $a(t) = 1$ for all t. By Lemma A.4.1, $U_n(x) = w_n x$, $x \in \mathbb{R}$, for necessarily positive constants w_n, which can be assumed to add up to one after positively scaling the utility, resulting in representation (A.4.2) with $\alpha = 0$.

Suppose instead that $a(y) \neq 1$ for some y. Equation (A.4.3) applied with $(s, t) = (x, y)$ and again with $(s, t) = (y, x)$ implies that

$$U_n(x) = w_n (1 - a(x)), \quad x \in \mathbb{R}, \quad (A.4.4)$$

where $w_n \equiv U_n(y) / (1 - a(y)) \neq 0$ (since U_n is increasing). Equation (A.4.3) with $s = 1$ implies that a is the difference of two increasing functions and is therefore bounded on some open interval. Substituting expression (A.4.4) for U_n into equation (A.4.3) and simplifying, we find that $\log a$ satisfies the assumptions of Lemma A.4.1. It follows that $\log a(x) = -\alpha x$ for some scalar α, and equation (A.4.4) becomes the claimed representation (A.4.2) after positive scaling.

The converse claim is immediate. $\qquad \qquad \square$

The preference $\succ$ on $(0, \infty)^N$ is **scale invariant** if

$$x \succ y \text{ and } s \in (0, \infty) \quad \Longrightarrow \quad sx \succ sy.$$

THEOREM A.4.3 *Suppose the preference $\succ$ on $(0, \infty)^N$ admits an additive utility representation. Then $\succ$ is scale invariant if and only if it admits a utility representation of the form*

$$U(x) = \sum_{n=1}^{N} w_n \frac{x_n^{1-\gamma} - 1}{1 - \gamma}, \quad x \in (0, \infty)^N, \quad (A.4.5)$$

for unique $\gamma \in \mathbb{R}$ and $w_1, \ldots, w_N \in (0, 1)$ such that $\sum_n w_n = 1$. The convention for $\gamma = 1$ is that $\left(x^{1-\gamma} - 1 \right) / (1 - \gamma)$ is equal to $\log x$, which is the limit as $\gamma \to 1$.

PROOF Using the notation $\exp x = (\exp x_1, \ldots, \exp x_N)$, define the preference $\succ^{\exp}$ on $\mathbb{R}^N$ by

$$x \succ^{\exp} y \quad \Longleftrightarrow \quad \exp x \succ \exp y.$$

Note that $\succ^{\exp}$ is translation invariant if and only if $\succ$ is scale invariant. The result follows from Theorem A.4.2 applied to $\succ^{\exp}$, with expression (A.4.5) resulting from (A.4.2) after replacing x_n with $\log x_n$ and $-\alpha$ with $1 - \gamma$. $\quad \square$

A.5 Expected Utility Representations

The rest of this appendix is on expected utility, which is a special type of additive utility. We henceforth refer to $\{1, \ldots, N\}$ as the **state space** and regard every $x \in (\ell, \infty)^N$ as a (state-contingent) **payoff**. An agent expresses preferences over such payoffs, represented by the preference relation $\succ$, not knowing what state will be realized. A **full-support probability** is a vector $P = (P_1, \ldots, P_N) \in (0, 1)^N$ such that $\sum_n P_n = 1$. (Note that here P is a probability *mass* function, while in the main text a probability P refers to the corresponding probability *measure*. Clearly, one determines the other and there should be no confusion.) An **expected utility** is a pair (P, u) of a full-support probability P and an increasing continuous function $u \colon (\ell, \infty) \to \mathbb{R}$. The expected utility (P, u) **represents** the preference $\succ$ if

$$U(x) \equiv \sum_{n=1}^{N} P_n u(x_n), \quad x \in (\ell, \infty)^N, \tag{A.5.1}$$

defines a utility representation U of $\succ$. The utilities (A.4.2) and (A.4.5) are examples of the expected utility form (A.5.1), representing a translation- or scale-invariant preference $\succ$.

The uniqueness of additive representations (up to a positive affine transformation) specializes to expected utilities as follows.

THEOREM A.5.1 *If (P, u) and $(\tilde{P}, \tilde{u})$ are expected utilities representing the same preference, then $\tilde{P} = P$ and there exist constants $a \in (0, \infty)$ and $b \in \mathbb{R}$ such that $\tilde{u} = au + b$.*

PROOF By the uniqueness Theorem A.2.4, there exist $a \in (0, \infty)$ and $b_1, \ldots, b_n \in \mathbb{R}$ such that $\tilde{P}_n \tilde{u} = a P_n u + b_n$ for all n. Adding up over n, we obtain $\tilde{u} = au + b$, where $b \equiv \sum_n b_n$. Therefore, $\tilde{P}_n (au + b) = a P_n u + b_n$ or $(\tilde{P}_n - P_n)au = b_n - \tilde{P}_n b$. The right-hand side of the last equation is a constant but the left-hand side is a strictly monotone function, unless $\tilde{P}_n = P_n$. □

As a corollary of Theorem A.2.3 and the results of Section A.4, we have the following ordinal sufficient conditions for the existence of an expected utility representation.

THEOREM A.5.2 *Assuming $N > 2$, every scale- or translation-invariant separable preference admits an expected utility representation.*

The remainder of this section provides an ordinal characterization of the existence of an expected utility representation, without the scale- or translation-invariance assumption. We introduce a property we call state independence and we refine Theorem A.2.3, where $N > 2$, by showing that $\succ$ is both separable and state independent if and only if it admits an expected utility representation.[6] Clearly, scale or translation invariance implies state independence. The details of this argument are not needed elsewhere in this text and can therefore be omitted.

First, some terminology and notation. The **indifference relation** $\sim$ associated with the preference $\succ$ is defined by

$$x \sim y \quad \Longleftrightarrow \quad (\text{not } x \succ y) \text{ and } (\text{not } y \succ x).$$

We write 1^n for the payoff that is the indicator of $\{n\}$, that is,

$$1^n_k = \begin{cases} 1, & \text{if } k = n; \\ 0, & \text{if } k \neq n. \end{cases}$$

As usual, we identify a scalar $w \in (\ell, \infty)$ and the constant payoff $(w, w, \ldots, w)$. Fixing any distinct states $m, n \in \{1, \ldots, N\}$, suppose the agent is indifferent between the constant payoff w and the same payoff but with the contingent value at state m increased by $z \in (0, \infty)$ and the contingent value at state n decreased by $y \in (0, w - \ell)$. We express this indifference with the notation

$$(z; m_+) =_w (y; n_-) \quad \Longleftrightarrow \quad w \sim w + z1^m - y1^n.$$

We denote the agent's indifference between increasing the constant payoff w by $z \in (0, \infty)$ at state m or by $x \in (0, \infty)$ at state n as follows

$$(z; m_+) =_w (x; n_+) \quad \Longleftrightarrow \quad w + z1^m \sim w + x1^n.$$

We write $(x; n_+) =^m_w (y; n_-)$ if there exists some $z \in (0, \infty)$ such that

$$(z; m_+) =_w (y; n_-) \quad \text{and} \quad (z; m_+) =_w (x; n_+).$$

We write $(x; n_+) \neq^m_w (y; n_-)$ if there exists some $z \in (0, \infty)$ such that

$$(z; m_+) =_w (y; n_-) \quad \text{and} \quad \text{not } (z; m_+) =_w (x; n_+).$$

---

[6] Preferences over probability distributions that admit an expected-utility representation were first characterized by von Neumann and Morgenstern (1944) (see also Herstein and Milnor [1953]). Incorporating a subjective view of probability in the tradition of Ramsey (1926), Savage (1954) developed an axiomatic foundation for expected utility with a nonatomic probability that is uniquely determined by preferences over mappings from states to a set of consequences. Anscombe and Aumann (1963) offered an alternative foundation for expected utility with subjective beliefs that utilizes objective probabilities to calibrate subjective beliefs. The approach presented here is a variant of that in Wakker (1984, 1988, 1989) and is based on Skiadas (1997, 2009).

The preference $\succ$ is **state independent** if [7] for all $w \in (\ell, \infty)$ and distinct states $m, n \in \{2, \ldots, N\}$, there do *not* exist $x \in (0, \infty)$ and $y \in (0, w - \ell)$ such that $(x; 1_+) =_w^m (y; 1_-)$ and $(x; n_+) \neq_w^m (y; n_-)$.

THEOREM A.5.3 *Suppose that $N > 2$ and $\succ$ is a preference that admits a utility representation. Then $\succ$ admits an expected utility representation if and only if it is both separable and state independent.*

PROOF The "only if" part is straightforward. Conversely, suppose $\succ$ is separable and state independent. By Theorem A.2.3, $\succ$ is represented by an additive utility $U = \sum_n U_n$. For every state n, define the open interval $I_n \equiv \{U_n(x) \mid x \in (\ell, \infty)\}$ and the function $f_n : I_1 \mapsto I_n$ such that $U_n(x) = f_n(U_1(x))$ for all $x \in (\ell, \infty)$. Note that f_n is increasing and surjective, and therefore continuous. We will show that each f_n is affine, that is, there exist $a_n \in (0, \infty)$ and $b_n \in \mathbb{R}$ such that

$$f_n(U_1) = a_n U_1 + b_n.$$

Letting $u \equiv U_1$ and $P_n \equiv a_n / \sum_n a_n$ results in an expected utility representation (P, u).

To prove that f_n is affine, we will show that for all $\alpha \in I_1$, there exists $\varepsilon > 0$ such that for all $\delta \in (0, \varepsilon)$,

$$f_n(\alpha + \delta) - f_n(\alpha) = f_n(\alpha) - f_n(\alpha - \delta). \tag{A.5.2}$$

By Yaari's Lemma A.3.2 with $p = 1/2$, this condition implies that both f_n and $-f_n$ are concave, and therefore f_n is affine.

Fix any $\alpha \in I_1$ and let $w \in (\ell, \infty)$ be defined by $U_1(w) = \alpha$. Consider any distinct states $m, n \in \{2, \ldots, N\}$. By the utility continuity, there exists a sufficiently small $\varepsilon > 0$ such that for all $\delta \in (0, \varepsilon)$ there exist positive scalars x, y, z, z' that solve the equations

$$\delta = U_1(w + x) - U_1(w) = U_1(w) - U_1(w - y),$$
$$\delta = U_m(w + z') - U_m(w),$$
$$U_m(w + z) - U_m(w) = U_n(w) - U_n(w - y).$$

The first three equalities imply that $(x; 1_+) =_w^m (y; 1_-)$, and the last one implies that $(z; m_+) =_w (y; n_-)$. The state-independence assumption then requires that $(z; m_+) =_w (x; n_+)$, and therefore

$$U_n(w + x) - U_n(w) = U_m(w + z) - U_m(w).$$

---

[7] Theorem A.5.3 and its proof remain valid if we define state independence to mean that for all distinct $m, n \in \{2, \ldots, N\}$, there exists $\epsilon > 0$ such that for all $x, y \in (0, \epsilon)$, it is not the case that $(x; 1_+) =_w^m (y; 1_-)$ and $(x; n_+) \neq_w^m (y; n_-)$.

This proves that

$$U_n(w + x) - U_n(w) = U_n(w) - U_n(w - y),$$

which in turn implies (A.5.2). □

A.6 Expected Utility and Risk Aversion

We close with a discussion of risk aversion in the context of expected utility, with the central theme being the connection between utility concavity and risk aversion. As is customary in the related literature, we use the term "more risk averse" to mean "no less risk averse," and "risk averse" to mean "risk averse or risk neutral."

Consider a preference $\succ$ on $(\ell, \infty)^N$ with expected utility representation (P, u). The corresponding **certainty equivalent (CE)** v is the unique utility representation of $\succ$ satisfying $v(w) = w$ for all $w \in (\ell, \infty)$. Let $\mathbb{E}$ denote the expectation operator under P:

$$\mathbb{E}z \equiv \sum_{n=1}^{N} z_n P_n, \quad z \in \mathbb{R}^N.$$

The CE $v \colon (\ell, \infty)^N \to (\ell, \infty)$ is necessarily given as $v = u^{-1}\mathbb{E}u$, meaning $v(x) = u^{-1}(\mathbb{E}u(x))$ for all $x \in (\ell, \infty)^N$. An agent maximizing v is indifferent between x and a constant payoff $v(x)$. For given probability P, a lower CE value $v(x)$ represents higher risk aversion, since, given the same odds, the agent is willing to settle for a lower sure payoff in place of x. We therefore define an expected utility CE $\tilde{v} = \tilde{u}^{-1}\mathbb{E}\tilde{u}$ to be **more risk averse** than v if

$$\tilde{v}(x) \leq v(x) \text{ for all } x \in (\ell, \infty)^N,$$

which we state simply as $\tilde{v} \leq v$. The condition is characterized as follows, where $f \circ u$ denotes function composition: $(f \circ u)(x) = f(u(x))$.

THEOREM A.6.1 *Suppose (P, u) and $(P, \tilde{u})$ are expected utilities, with corresponding CEs $v \equiv u^{-1}\mathbb{E}u$ and $\tilde{v} \equiv \tilde{u}^{-1}\mathbb{E}\tilde{u}$. Then $\tilde{v} \leq v$ if and only if $\tilde{u} = f \circ u$ for a concave function $f \colon u(\ell, \infty) \to \mathbb{R}$.*

PROOF Define the function $f \colon u(\ell, \infty) \to \mathbb{R}$ so that $\tilde{u} = f \circ u$. If f is concave, then $\tilde{v} \leq v$ by Jensen's inequality:

$$\tilde{u}(\tilde{v}(x)) = \mathbb{E}f(u(x)) \leq f(\mathbb{E}u(x)) = \tilde{u}(v(x)), \quad x \in (\ell, \infty)^N.$$

Suppose f is *not* concave and let $p \equiv P_1 > 0$. By Yaari's Lemma A.3.2, there exist $w \in (\ell, \infty)$ and $\varepsilon > 0$ such that for all $\delta \in (0, \varepsilon)$,

$$f(u(w)) < pf(u(w) + (1 - p)\delta) + (1 - p)f(u(w) - p\delta). \quad \text{(A.6.1)}$$

Pick $\delta > 0$ small enough so that $x \in (\ell, \infty)^N$ is well defined by

$$u(x_1) = u(w) + (1 - p)\delta, \quad u(x_n) = u(w) - p\delta, \quad n \geq 2,$$

which implies $w = v(x)$. By inequality (A.6.1), $\tilde{u}(w) < \mathbb{E}\tilde{u}(x)$ and therefore $v(x) < \tilde{v}(x)$, proving that it is not the case that $\tilde{v} \leq v$. $\quad\square$

EXAMPLE A.6.2 (Multiple-prior representation of risk aversion) Consider the expected utility CEs $v \equiv u^{-1}\mathbb{E}u$ and $\tilde{v} \equiv \tilde{u}^{-1}\mathbb{E}\tilde{u}$ satisfying, for a given constant $\alpha \in (0, \infty)$,

$$\tilde{u} = f \circ u \quad \text{where} \quad f(x) \equiv \frac{1 - \exp(-\alpha x)}{\alpha}, \quad x \in \mathbb{R}.$$

For example, this condition is satisfied for $u(x) = x$ and $\tilde{u} = f$, as in Theorem A.4.2, and also for

$$u(x) = \log(x) \quad \text{and} \quad \tilde{u}(x) = \frac{x^{1-\gamma} - 1}{1 - \gamma}, \text{ where } \gamma \equiv 1 + \alpha,$$

as in Theorem A.4.3.

Given a probability Q, we write

$$\frac{dQ}{dP} \equiv \left(\frac{Q_1}{P_1}, \ldots, \frac{Q_N}{P_N}\right),$$

for the **density** of Q with respect to P, which is useful in relating expectation under Q, denoted by $\mathbb{E}^Q$, to expectation under P:

$$\mathbb{E}^Q z \equiv \sum_{n=1}^{N} z_n Q_n = \sum_{n=1}^{N} z_n \frac{Q_n}{P_n} P_n = \mathbb{E}\left[z \frac{dQ}{dP}\right], \quad z \in \mathbb{R}^N. \quad \text{(A.6.2)}$$

The probability of an expected utility representation is often referred to as a *prior*, in the Bayesian sense of belief prior to the arrival of information. The increase in risk aversion from $v \equiv u^{-1}\mathbb{E}u$ to $\tilde{v} \equiv \tilde{u}^{-1}\mathbb{E}\tilde{u}$ is sometimes expressed in the literature (often under the rubric of *robustness* or *ambiguity aversion*) as the multiple-prior specification

$$u(\tilde{v}(x)) = \min_{Q \in \mathcal{Q}} \left\{\mathbb{E}^Q u(x) + \frac{1}{\alpha}\mathbb{E}^Q \log\left(\frac{dQ}{dP}\right)\right\}, \quad \text{(A.6.3)}$$

where $\mathcal{Q}$ denotes the set of all full-support probabilities. The quantity $\mathbb{E}^{\mathcal{Q}} \log(dQ/dP)$, known as the *relative entropy* of Q with respect to P, is minimized for $Q = P$.

To show identity (A.6.3), let us first reduce it to its essentials. Using the assumption $\tilde{u} = f \circ u$ and the fact that the CE $f^{-1}\mathbb{E}f$ is invariant to a positive affine transformation of f, we have

$$u(\tilde{v}(x)) = f^{-1}\mathbb{E}f(u(x)) = -\frac{1}{\alpha} \log\big(\mathbb{E}\exp(-\alpha u(x))\big).$$

After the change of variables $z \equiv \alpha u(x)$, it follows that identity (A.6.3) is equivalent to

$$- \log\big(\mathbb{E}\exp(-z)\big) = \min_{Q \in \mathcal{Q}} \mathbb{E}^{\mathcal{Q}}\left[z + \log\left(\frac{dQ}{dP}\right)\right]. \tag{A.6.4}$$

Let $g(z) \equiv -\exp(-z)$, $z \in \mathbb{R}$, and define the convex dual[8]

$$g^*(y) \equiv \max_{z \in \mathbb{R}} \{g(z) - yz\} = y \log y - y, \quad y \in (0, \infty).$$

The key behind (A.6.4) is the fact that

$$g(z) = \min_{y \in (0,\infty)} \{zy + g^*(y)\}. \tag{A.6.5}$$

Let $\phi(z)$ denote the CE on the left-hand side of (A.6.4) and let $\tilde{\phi}(z)$ denote the CE on the right-hand side of (A.6.4). Utilizing identities (A.6.2) and (A.6.5), the definitions of g and g^*, and the fact that every $y \in (0, \infty)^N$ can be decomposed as $y = sdQ/dP$, $s \in (0, \infty)$, $Q \in \mathcal{Q}$, we have

$$g(\phi(z)) = \mathbb{E}g(z) = \min_{y \in (0,\infty)^N} \mathbb{E}\left[zy + g^*(y)\right]$$

$$= \min_{s \in (0,\infty),\, Q \in \mathcal{Q}} \mathbb{E}\left[zs\frac{dQ}{dP} + g^*\left(s\frac{dQ}{dP}\right)\right]$$

$$= \min_{s \in (0,\infty)} \{\tilde{\phi}(z)s + g^*(s)\} = g(\tilde{\phi}(z)).$$

We have shown that $\phi = \tilde{\phi}$, which is identity (A.6.4), completing the proof of identity (A.6.3). $\diamondsuit$

---

[8] This type of convex duality also appears in Section 3.5. For each $y > 0$, $g^*(y)$ is the vertical axis intercept of the line of slope y that is tangent to the graph of g. That $g(z) \leq g^*(y) + yz$ for all z corresponds to the fact that this line lies above the graph of g. The equality $g(z) = g^*(y) + yz$ is obtained for the value $y = g'(z)$ that is the slope of the tangent line at $(z, g(z))$.

Suppose the preference $\succ$ admits an expected utility representation (P, u). By Theorem A.5.1, the probability P is uniquely determined by $\succ$ and defines the **risk-neutral** CE $\mathbb{E}$. As a corollary of Theorem A.6.1, we have the following characterization of the condition that $u^{-1}\mathbb{E}u$ is **more risk averse than risk neutral**.

COROLLARY A.6.3 *For every expected utility (P, u), u is concave if and only if $u^{-1}\mathbb{E}u(x) \leq \mathbb{E}x$ for all $x \in (0, \infty)^N$.*

By Theorem A.3.1, the concavity of u is also equivalent to the convexity of the preference $\succ$, which can be thought of as **preference for diversification**. We define $\succ$ to be risk averse if whenever it rejects adding a payoff to a sure payoff, it also rejects adding every scaled-up version of the same payoff.[9] More precisely, $\succ$ is **risk averse** if for all $w \in (\ell, \infty)$, provided $\theta x \in (\ell - w, \infty)^N$,

$$w \succ w + x \text{ and } \theta \in (1, \infty) \implies w \succ w + \theta x. \tag{A.6.6}$$

This condition defines aversion to a specific risk x. Implicit in expected utility is a sense in which risk aversion is risk-source independent,[10] which leads to the equivalence of risk aversion to the concavity of u, and hence the equivalence of all three notions just introduced: risk aversion, preferences for diversification, and more risk averse than risk neutral.

THEOREM A.6.4 *A preference with expected utility representation (P, u) is risk averse if and only if u is concave.*

PROOF Suppose u is concave, $w \in (\ell, \infty)$ and $\theta \in (1, \infty)$. Since $w + x$ is a convex combination of w and $w + \theta x$, if $\mathbb{E}u(w + \theta x) \geq u(w)$ then $\mathbb{E}u(w + x) \geq u(w)$, which is the contrapositive of condition (A.6.6) and therefore proves that $\succ$ is risk averse.

Conversely, suppose $\succ$ is risk averse. We show that u is concave assuming $N = 2$, which entails no loss of generality. We do so by confirming that $u^{-1}\mathbb{E}u \leq \mathbb{E}$ and applying Corollary A.6.3. Suppose instead that for some $x \in (\ell, \infty)^2$, $u^{-1}\mathbb{E}u(x) > \mathbb{E}x$. Define $\mu \equiv \mathbb{E}x$ and $\hat{x} \equiv x - \mu$, and therefore

$$\mathbb{E}u(\mu + \hat{x}) > u(\mu) \quad \text{and} \quad \mathbb{E}\hat{x} = 0. \tag{A.6.7}$$

---

[9] To my knowledge, this definition of risk aversion and Theorem A.6.4 first appeared in Skiadas (2009). The proof given here is somewhat shorter thanks to Lemma A.3.2 of Yaari (1977).

[10] For a specification of scale-invariant preferences with source-dependent risk aversion, see Skiadas (2013b, 2015).

We claim that this condition implies that u is convex on $(\mu - \epsilon, \mu + \epsilon)$ for some $\epsilon > 0$. Suppose instead u is not convex on $(\mu - \epsilon, \mu + \epsilon)$ for all $\epsilon > 0$. Since there are only two states, Yaari's Lemma A.3.2 applied to $-u$, for all $n = 1, 2, \ldots,$ implies that there exist $\mu_n \in (\mu - 1/n, \mu + 1/n)$ and $\delta_n \in (0, 1)$ such that $u(\mu_n) > \mathbb{E}u(\mu_n + \delta_n \hat{x})$, hence $u(\mu_n) > \mathbb{E}u(\mu_n + \hat{x})$ by risk aversion. Since u is continuous and $\lim_{n \to \infty} \mu_n = \mu$, $u(\mu) \geq \mathbb{E}u(\mu + \hat{x})$ in contradiction to (A.6.7). So let $\epsilon > 0$ be such that u is convex on $(\mu - \epsilon, \mu + \epsilon)$. Since u is also (strictly) increasing, it follows that the derivative u' exists and is positive at all but at most countably many points of $(\mu - \epsilon, \mu + \epsilon)$.[11] Since u is continuous, in condition (A.6.7), we can slightly perturb μ if necessary to some value w such that $u'(w)$ exists and is positive, while it is still the case that $\mathbb{E}u(w + \hat{x}) > u(w)$. We can then slightly decrease $\hat{x}$ to $\tilde{x}$ so that $\mathbb{E}u(w + \tilde{x}) \geq u(w)$, $\mathbb{E}\tilde{x} < 0$ and $\tilde{x}_n \neq 0$ for every state n. Risk aversion implies that for all $\delta \in (0, 1)$, $\mathbb{E}u(w + \delta\tilde{x}) \geq u(w)$ and therefore

$$\mathbb{E}\left[\frac{u(w + \delta\tilde{x}) - u(w)}{\delta\tilde{x}}\tilde{x}\right] \geq 0.$$

Letting $\delta \downarrow 0$ gives $u'(w)\mathbb{E}\tilde{x} \geq 0$, a contradiction, since $u'(w) > 0$ and $\mathbb{E}\tilde{x} < 0$ by construction. $\qquad\square$

Consider now an expected utility (P, u), where u belongs to the set C^2 of all twice continuously differentiable real-valued functions on (ℓ, ∞). The scale- or translation-invariant cases of Section A.4 are examples. More generally, every expected utility is "close" to a smooth version in the following sense.

REMARK A.6.5 Suppose $x \in (\ell, \infty)^N$ is perceived as $x + \varepsilon$, where ε is an arbitrarily small noise term that is stochastically independent of x and continuously distributed over a compact interval with a density f_ε that is

---

[11] Suppose u is strictly increasing and convex on an interval (a, b) and $w \in (a, b)$. For all small enough $\delta > 0$, define the ordered slopes

$$\Delta_+(\delta) \equiv \frac{u(w + \delta) - u(w)}{\delta} \geq \frac{u(w) - u(w - \delta)}{\delta} \equiv \Delta_-(\delta).$$

As $\delta \downarrow 0$, $\Delta_+(\delta)$ monotonically decreases to a limit defining the right derivative $u'_+(w)$, and $\Delta_-(\delta)$ monotonically increases to a limit defining the left derivative $u'_-(w)$. Clearly, $u'_+(w) > 0$ and $u'_+(w) \geq u'_-(w)$. The intervals $(u'_-(w), u'_+(w))$ are nonoverlapping as w ranges over (a, b) and therefore at most countably many of them are nonempty. This proves that for all except at most countably many values of $w \in (a, b)$, $u'(w) = u'_-(w) = u'_+(w) > 0$.

twice continuously differentiable. The expected utility of x can be thought of as the reduced form of the expected utility of $x + \varepsilon$, and therefore

$$u(x) \equiv \int_{\mathbb{R}} u_\varepsilon(x + y) \, f_\varepsilon(y) \, dy = \int_{\mathbb{R}} u_\varepsilon(z) \, f_\varepsilon(z \dot{-} x) \, dz.$$

For continuous but otherwise arbitrary u_ε, the assumed properties of f_ε imply that $u \in C^2$. $\Diamond$

The **coefficient of absolute risk aversion** associated with $u \in C^2$ is the function

$$A^u \equiv -\frac{u''}{u'}. \tag{A.6.8}$$

Our principal motivation for introducing this function is the approximation of the expected utility of small Brownian risks, discussed in Section 3.8. Just like the CE $u^{-1}\mathbb{E}u$, A^u is invariant to positive affine transformations of u and determines u up to a positive affine transformation. (To show the last claim, write (A.6.8) as $A^u(x) \, dx = -d \log u'(x)$ and integrate twice.) Given P, the function A^u is therefore uniquely associated with the corresponding CE $u^{-1}\mathbb{E}u$ and hence the preference represented by (P, u). The translation-invariant representation of Theorem A.4.2 corresponds to constant absolute risk aversion: $A^u(x) = \alpha$ for all $x \in \mathbb{R}$. The function $x \mapsto x A^u(x)$ is known as the **coefficient of relative risk aversion**. The scale-invariant representation of Theorem A.4.3 corresponds to constant relative risk aversion: $x A^u(x) = \gamma$ for all $x \in (0, \infty)$.

By Theorem A.6.4, for $u \in C^2$, the expected utility (P, u) is risk averse if and only if $u'' \le 0$ if and only if $A^u \ge 0$. (The inequalities are to be interpreted as holding on the entire domain, for example, $A^u \ge 0$ means $A^u(x) \ge 0$ for all $x \in (\ell, \infty)$.) By Theorem A.6.1, for $u, \tilde{u} \in C^2$, the expected utility $(P, \tilde{u})$ is more risk averse than the expected utility (P, u) (in the sense that $\tilde{u}^{-1}\mathbb{E}\tilde{u} \le u^{-1}\mathbb{E}u$) if and only if $A^{\tilde{u}} \ge A^u$. To prove this claim, note that if $\tilde{u} = f \circ u$ then $A^{\tilde{u}} = A^u + \left(A^f \circ u\right) u'$ and f is concave if and only if $A^f \ge 0$.

Appendix B

Elements of Convex Analysis

This appendix reviews some basic linear algebra and convex analysis concepts, with optimization principles as the main focus. The emphasis is on the finite-dimensional case, but from a perspective that is amenable to infinite-dimensional extensions.[1] As in the main text, $\equiv$ means equal by definition.

B.1 Vector Spaces

The analysis of this appendix is all carried out in the context of an inner product space X, which is a vector space together with an inner product, which in turn defines a norm. We introduce these concepts in order in this section and the next one.

Vector spaces in this text are defined relative to the field $\mathbb{R}$ of the real numbers or **scalars**. A **vector** or **linear** space is a triple of a set X, whose elements are called **vectors** or **points**, an **addition** operation assigning to each $(x, y) \in X \times X$ a vector $x + y$, and a **scaling** operation assigning to each $(\alpha, x) \in \mathbb{R} \times X$ a vector αx, provided that for all $x, y, z \in X$ and $\alpha, \beta \in \mathbb{R}$, the following conditions are satisfied.

- $x + y = y + x$; $(x + y) + z = x + (y + z)$; and there exists a vector 0 with the property: for all $x \in X$, $x + 0 = x$ and there exists $-x \in X$ such that $x + (-x) = 0$.
- $(\alpha\beta) x = \alpha(\beta x)$ and $1x = x$.
- $\alpha(x + y) = \alpha x + \alpha y$ and $(\alpha + \beta) x = \alpha x + \beta x$.

---

[1] Treatments of finite-dimensional convex optimization theory include Rockafellar (1970), Bertsekas (2003) and Boyd and Vandenberghe (2004), the latter providing an introduction to optimization algorithms. Infinite-dimensional extensions can be found in Luenberger (1969), Zeidler (1985), Dunford and Schwartz (1988), Ekeland and Témam (1999) and Bonnans and Shapiro (2000).

The **zero vector** 0 and **additive inverses** $-x$ are necessarily unique. The **vector difference** between x and y is $x - y \equiv x + (-y)$.

It is customary to refer to a vector space X, while it is implied that addition and scaling operations are also specified along with X. Two vector spaces X and $\tilde{X}$ are **isomorphic** if there is a bijection (one-to-one and onto function) $\phi \colon X \to \tilde{X}$ that preserves the addition and scaling operations: For all $x, y \in X$ and $\alpha \in \mathbb{R}$,

$$\phi(x + y) = \phi(x) + \phi(y) \quad \text{and} \quad \phi(\alpha x) = \alpha \phi(x).$$

In these two equations, addition and scaling on the left-hand side refer to operations in X and those on the right-hand side refer to operations in $\tilde{X}$. Such a function ϕ is a **vector space isomorphism**. The composition of two (vector space) isomorphisms and the inverse of an isomorphism are also isomorphisms.

A canonical vector space example is d-**dimensional Euclidean space**, for positive integer d, defined as the cartesian product $\mathbb{R}^d$ with coordinate-wise addition and scaling operations: $(x + y)_i = x_i + y_i$ and $(\alpha x)_i = \alpha x_i$ for all $i \in \{1, \dots, d\}$. The set L of all random variables on a given state space Ω is another example of a vector space, where a random variable is a real-valued function on Ω and addition and scaling are defined statewise: $(x + y)(\omega) = x(\omega) + y(\omega)$ and $(\alpha x)(\omega) = \alpha x(\omega)$ for all $\omega \in \Omega$. If $\Omega = \{\omega_1, \dots, \omega_d\}$, then L is isomorphic to $\mathbb{R}^d$. A vector space isomorphism can "forget" aspects not relevant to the vector space structure. For example, the vector space $\mathbb{R}^{m \times n}$ of all $m \times n$ matrices with real entries, with the usual addition and scaling operations, is isomorphic to mn-dimensional Euclidean space, but the shape of a matrix is not preserved. The point of an isomorphism is that we can identify any two isomorphic structures in contexts where only the isomorphically preserved structure is relevant. For example, throughout this text, we identify $\mathbb{R}^{1 \times d}$ and $\mathbb{R}^{d \times 1}$ with the Euclidean space $\mathbb{R}^d$ if the shape of the vector is irrelevant.

We henceforth fix a reference vector space X. Unless otherwise specified, a **vector** is an element of X. A **linear subspace** of X is any subset Y of X that is closed with respect to addition and scaling: For all $x, y \in Y$ and $\alpha \in \mathbb{R}$, $x + y \in Y$ and $\alpha x \in Y$. Equivalently, a linear subspace of X is a subset of X that is a vector space in its own right, with the suitably restricted addition and scaling operations inherited from X. A **linear combination** of the vectors $x_1, \dots, x_n$ is an expression of the form $\alpha_1 x_1 + \cdots + \alpha_n x_n$, for scalar α_i, as well as the vector this expression evaluates to (recursively in the obvious way). The term *linear combination* will always refer to the linear combination of finitely many vectors. If all the α_i are zero, we call the linear combination **trivial**. The **linear span** of a set of vectors S, denoted span(S), is the intersection of all linear subspaces that include S, which equals the set of all linear combinations of elements of S. We say that span(S) is the linear subspace **generated** or **spanned** by S or the elements of S. If $S = \{x_1, \ldots, x_n\}$, we also write span($x_1, \ldots, x_n$) for span(S). We call a set of vectors S, as well as its elements, **linearly independent** if every linear combination of vectors in S that equals the zero vector is trivial, a condition that is equivalent to the unique representation (up to reordering of terms) of each element of span(S) as a linear combination of elements of S, and is also equivalent to the nonexistence of a vector x in S that can be expressed as a linear combination of vectors in $S\backslash\{x\}$.

A **basis** of the vector space X is a linearly independent set of vectors that spans X. The space X is **finite dimensional** if it has a finite basis. Given an arbitrary ordering $B_1, \ldots, B_d$ of a finite basis, we define the column matrix $B \equiv (B_1, \ldots, B_d)'$, which we also call a **basis**, and write $\sigma^x \in \mathbb{R}^{1 \times d}$ for the row matrix of the **coordinates** of $x \in X$ relative to B:

$$x = \sigma^x B \equiv \sigma^x_1 B_1 + \cdots + \sigma^x_d B_d \equiv \sum_{i=1}^{d} \sigma^x_i B_i.$$

The mapping $x \mapsto \sigma^x$ defines a vector space isomorphism from X to d-dimensional Euclidean space. For $n > m$, n-dimensional Euclidean space is not isomorphic to m-dimensional Euclidean space, since the linear independence on n vectors in $\mathbb{R}^n$ is not preserved when linearly mapped to $\mathbb{R}^m$. This shows that every finite-dimensional space X is isomorphic to a unique Euclidean space, whose dimension d defines the **dimension** of X and equals the number of elements of every basis of X. The vector space $\{0\}$ has **dimension zero**, and a vector space that is not finite dimensional is **infinite dimensional**.

A **functional** is a function of the form $f : X \to \mathbb{R}$. The functional f is **linear** if $f(x + y) = f(x) + f(y)$ and $f(\alpha x) = \alpha f(x)$ for all $x, y \in X$ and $\alpha \in \mathbb{R}$. Given a finite basis $B = (B_1, \ldots, B_d)'$ and a linear functional f, we use the notation

$$f(B) \equiv (f(B_1), \ldots, f(B_d))' \in \mathbb{R}^{d \times 1}.$$

The vector $f(B)$ determines the entire function f, since $x = \sigma^x B$ implies $f(x) = \sigma^x f(B) = \sigma^x \cdot f(B)$. The dot of the last expression denotes the **Euclidean inner product**:

$$x \cdot y \equiv \sum_{i=1}^{d} x_i y_i, \quad x, y \in \mathbb{R}^d.$$

The mapping $(x, y) \mapsto x \cdot y$ is symmetric $(x \cdot y = y \cdot x)$, linear in x (hence linear in y, by symmetry), and positive definite $(x \cdot x \geq 0$, with equality only if $x = 0$). These three properties lead us into a more-general notion of an inner product.

B.2 Inner Products

Abstracting away the essential properties of the Euclidean inner product, we define a (real) **inner product** $\langle \cdot \,|\, \cdot \rangle$ on the vector space X as a mapping that assigns to each $(x, y) \in X \times X$ a scalar $\langle x \,|\, y \rangle$ such that for all $x, y, z \in X$ and $\alpha \in \mathbb{R}$,

- $\langle x \,|\, y \rangle = \langle y \,|\, x \rangle$;
- $\langle x + y \,|\, z \rangle = \langle x \,|\, z \rangle + \langle y \,|\, z \rangle$ and $\langle \alpha x \,|\, y \rangle = \alpha \langle x \,|\, y \rangle$;
- $\langle x \,|\, x \rangle \geq 0$, with $\langle x \,|\, x \rangle = 0$ only if $x = 0$.

An **inner product space** is a vector space together with an inner product on this space. Similarly to vector spaces, it is customary to refer to an inner product space X, with the understanding that there is a specified inner product that is associated with the vector space X.

EXAMPLE B.2.1 A symmetric positive definite matrix $Q \in \mathbb{R}^{d \times d}$ defines an inner product on $\mathbb{R}^d$: $\langle x \,|\, y \rangle \equiv \sum_{i,j} x_i Q_{ij} y_j = xQy'$, where x and y are treated as row vectors. The Euclidean inner product is obtained if Q is the identity matrix. $\diamond$

Two inner product spaces X and $\tilde{X}$ (each with an implied inner product) are said to be **isomorphic** if there is a vector space isomorphism $\phi \colon X \to \tilde{X}$ that preserves inner products:

$$\langle \phi(x) \,|\, \phi(y) \rangle = \langle x \,|\, y \rangle, \quad x, y \in X,$$

where the right-hand side is an inner product in X and the left-hand side is an inner product in $\tilde{X}$. The function ϕ is an **inner product space isomorphism**. As with vector space isomorphisms, the set of isomorphisms between two inner product spaces is closed with respect to inverses and function composition.

EXAMPLE B.2.2 Let $\mathbb{E}$ denote an expectation operator relative to a full-support probability on a finite state space Ω. The set L of all random variables on Ω with the inner product $\langle x \,|\, y \rangle \equiv \mathbb{E}[xy]$ is an inner product space, which is isomorphic to a Euclidean inner product space. Covariance agrees with this inner product on the subspace of zero-mean random variables, but is not an inner product on L. $\diamond$

EXAMPLE B.2.3 Suppose X is finite dimensional. A basis B of X defines the inner product $\langle x \mid y \rangle \equiv \sigma^x \cdot \sigma^y$, where $x = \sigma^x B$ and $y = \sigma^y B$. Later, we will establish that every inner product on X has this representation for some basis, and therefore X is isomorphic to a Euclidean inner product space. $\diamond$

The **norm induced** by the inner product $\langle \cdot \mid \cdot \rangle$ is the function $\| \cdot \| : X \to \mathbb{R}$ defined by

$$\|x\| \equiv \sqrt{\langle x \mid x \rangle}, \quad x \in X.$$

The value $\|x\|$ defines the **norm of** x, which we informally think of as the *length* of x.

An inner product space isomorphism ϕ is an **isometry**, meaning it preserves norms: $\|\phi(x)\| = \|x\|$ for all $x \in X$ (with the norm symbols referring to norms in their respective spaces). Conversely, every isometry also preserves inner products, since

$$\|x + y\|^2 = \|x\|^2 + 2\langle x \mid y \rangle + \|y\|^2. \tag{B.2.1}$$

The vectors x and y are **orthogonal** if $\langle x \mid y \rangle = 0$, a condition that should be visualized as x and y forming a right angle. Identity (B.2.1) implies that the vectors x and y are orthogonal if and only if they satisfy the **Pythagorean identity**

$$\|x + y\|^2 = \|x\|^2 + \|y\|^2.$$

Toward a geometric interpretation of an inner product, consider a nonzero vector x and let $\hat{x} \equiv \|x\|^{-1}x$, which is the unique positively scaled version of x whose norm is one. For any other vector y, $\langle \hat{x} \mid y \rangle$ is the value of the scalar α that minimizes $\|y - \alpha\hat{x}\|$. In other words,

$$\bar{y} \equiv \langle \hat{x} \mid y \rangle \hat{x} = \frac{\langle x \mid y \rangle}{\langle x \mid x \rangle} x$$

is the point of the line span(x) closest to y, a condition that defines $\bar{y}$ as the **projection** of y on span(x). To confirm the expression for $\bar{y}$, use identity (B.2.1) and the fact that $\|\hat{x}\| = 1$ to find

$$\|y - \alpha\hat{x}\|^2 = \|y\|^2 - \|\bar{y}\|^2 + \left(\alpha - \langle \hat{x} \mid y \rangle \right)^2.$$

The quadratic is clearly minimized for $\alpha = \langle \hat{x} \mid y \rangle$. The corresponding minimum is equal to $\|y - \bar{y}\|^2 = \|y\|^2 - \|\bar{y}\|^2$, which is a Pythagorean identity equivalent to the orthogonality of $y - \bar{y}$ to $\bar{y}$ and therefore x. This shows that the orthogonality of $y - \bar{y}$ to x uniquely characterizes a point $\bar{y} \in \text{span}(x)$ as the projection of y on $\text{span}(x)$.

Suppose now y is also nonzero and let $\hat{y} \equiv \|y\|^{-1}y$. The vectors $\langle \hat{x} \mid \hat{y} \rangle \hat{x}$ and $\hat{y}$ form an orthogonal triangle whose hypotenuse $\hat{y}$ has unit length and therefore $\langle \hat{x} \mid \hat{y} \rangle \hat{x}$ has length less than one, or one if $\hat{y} = \pm\hat{x}$. This suggests the **Cauchy–Schwarz inequality**

$$|\langle x \mid y \rangle| \leq \|x\|\|y\|, \tag{B.2.2}$$

which combined with identity (B.2.1) implies the **triangle inequality**

$$\|x + y\| \leq \|x\| + \|y\|. \tag{B.2.3}$$

PROPOSITION B.2.4 *In an inner product space, all vectors x and y satisfy the Cauchy–Schwarz inequality* (B.2.2), *with equality holding if and only if one of the two vectors is on the linear span of the other.*

PROOF If either vector is zero, it is on the linear span of the other, and the Cauchy–Schwarz inequality holds trivially as an equality. Suppose x and y are nonzero. Since $\hat{y} - \langle \hat{x} \mid \hat{y} \rangle \hat{x}$ is orthogonal to $\hat{x}$, the Pythagorean identity implies

$$0 \leq \|\hat{y} - \langle \hat{x} \mid \hat{y} \rangle \hat{x}\|^2 = \|\hat{y}\|^2 - \|\langle \hat{x} \mid \hat{y} \rangle \hat{x}\|^2 = 1 - \langle \hat{x} \mid \hat{y} \rangle^2,$$

which implies $\langle \hat{x} \mid \hat{y} \rangle^2 \leq 1$, and hence the Cauchy–Schwarz inequality. Equality holds if and only if $\hat{y} - \langle \hat{x} \mid \hat{y} \rangle \hat{x} = 0$ if and only if $y \in \text{span}(x)$ if and only if $x \in \text{span}(y)$. $\qquad\square$

EXAMPLE B.2.5 In the context of Example B.2.2, consider the vector space $\{x - \mathbb{E}x \mid x \in L\}$ with the covariance inner product. In this case, the Cauchy–Schwarz inequality states that the correlation coefficient of two random variables lies in the interval $[-1, 1]$. $\diamond$

For the remainder of this appendix, X is an inner product space with implied inner product $\langle \cdot \mid \cdot \rangle$. For every $x \in X$,

$$f(y) = \langle x \mid y \rangle, \quad y \in X,$$

defines a linear functional f. In this case, x is the (necessarily unique) **Riesz representation** of f. The effect of applying a linear functional with Riesz representation x to a vector y can be visualized as being the result of first projecting y onto the line defined by x, taking the length of that projection, signing it based on whether y forms an acute or oblique angle with x, and finally scaling the result by the length of x.

If X is finite dimensional, every linear functional f admits a Riesz representation, which can be expressed concretely in terms of a basis

$B = (B_1, \ldots, B_d)'$. To simplify notation, we extend the inner product notation to matrices of vectors, with the usual matrix scaling, addition and multiplication rules. In particular,

$$\langle x \mid B' \rangle \equiv \big(\langle x \mid B_1 \rangle, \ldots, \langle x \mid B_d \rangle \big),$$

and $\langle B \mid B' \rangle$ is the **Gram matrix** of B, defined as the $d \times d$ matrix whose (i, j) entry is $\langle B_i \mid B_j \rangle$:

$$\langle B \mid B' \rangle \equiv \begin{pmatrix} \langle B_1 \mid B_1 \rangle & \langle B_1 \mid B_2 \rangle & \cdots & \langle B_1 \mid B_d \rangle \\ \langle B_2 \mid B_1 \rangle & \langle B_2 \mid B_2 \rangle & \cdots & \langle B_2 \mid B_d \rangle \\ \vdots & \vdots & \ddots & \vdots \\ \langle B_d \mid B_1 \rangle & \langle B_d \mid B_2 \rangle & \cdots & \langle B_d \mid B_d \rangle \end{pmatrix}.$$

PROPOSITION B.2.6 *Suppose $B = (B_1, \ldots, B_d)'$ is a basis of X and for all $x \in X$, $\sigma^x \in \mathbb{R}^{1 \times d}$ is defined by $x = \sigma^x B$. Then the Gram matrix $\langle B \mid B' \rangle$ is symmetric and positive definite, and*

$$\langle x \mid y \rangle = \sigma^x \langle B \mid B' \rangle \sigma^{y\prime}, \quad x, y \in X. \tag{B.2.4}$$

Every linear functional f has a Riesz representation: For all $y \in X$,

$$f(y) = \langle x \mid y \rangle, \quad \text{where } x = f(B)' \langle B \mid B' \rangle^{-1} B.$$

Consequently, $\sigma^x = \langle x \mid B' \rangle \langle B \mid B' \rangle^{-1}$ for all $x \in X$.

PROOF The linearity of the inner product in each of its arguments implies (B.2.4). By the symmetry of the inner product, $\langle B \mid B' \rangle$ is a symmetric matrix. By the positive definiteness of the inner product, for all $\sigma \in \mathbb{R}^{1 \times d}$, $\sigma \langle B \mid B' \rangle \sigma' = \langle \sigma B \mid \sigma B \rangle \geq 0$, with equality holding if and only if $\sigma B = 0$ if and only if $\sigma = 0$ (since B is a basis). This proves that the matrix $\langle B \mid B' \rangle$ is positive definite (hence invertible). For any linear functional f, $f(y) = f(\sigma^y B) = f(B)' \sigma^{y\prime}$. Comparing to (B.2.4), it follows that x is the Riesz representation of f if and only if $f(B)' = \sigma^x \langle B \mid B' \rangle$ if and only if $x = f(B)' \langle B \mid B' \rangle^{-1} B$, in which case $\sigma^x = \langle x \mid B' \rangle \langle B \mid B' \rangle^{-1}$, since $f(B)' = \langle x \mid B' \rangle$. □

A corollary is that every d-dimensional inner product space is isomorphic to an inner product space of the form of Example B.2.1. The isomorphism is the mapping $x \mapsto \sigma^x$, and the matrix Q of Example B.2.1

is the Gram matrix $\langle B \mid B' \rangle$. The basis B is said to be **orthonormal** if the corresponding Gram matrix is an identity matrix:

$$\langle B_i \mid B_j \rangle = \begin{cases} 1 & \text{if } i = j, \\ 0 & \text{if } i \neq j. \end{cases} \tag{B.2.5}$$

Note that the orthonormality condition (B.2.5) implies the linear independence of the vectors B_i (since $x = \sum_i \alpha_i B_i$ implies $\alpha_i = \langle x \mid B_i \rangle$).

PROPOSITION B.2.7 *Every finite-dimensional inner product space has an orthonormal basis and is therefore isomorphic to a Euclidean inner product space.*

PROOF We construct an orthonormal basis recursively.[2] Start with any nonzero vector x and let $B_1 \equiv \|x\|^{-1} x$. For the recursive step, suppose we have already constructed orthonormal vectors $B_1, \ldots, B_n$ and let $L \equiv \text{span}(B_1, \ldots, B_n)$. If $n = d$, the dimension of X, then $L = X$ and the construction is complete. If $n < d$, choose any $x \in X \backslash L$, and define $x_L \equiv \sum_{i=1}^n \langle x \mid B_i \rangle B_i$ (which, as we will see in Section B.5, is the projection of x on L). By the orthonormality condition (B.2.5), $x - x_L$ is orthogonal to each of $B_1, \ldots, B_n$. We can therefore define $B_{n+1} \equiv \|x - x_L\|^{-1} (x - x_L)$ to construct $n + 1$ orthonormal vectors, completing the recursive step. $\square$

B.3 Some Basic Topological Concepts

We continue in the context of an inner product space X with induced norm $\| \cdot \|$, which we use to define convergence and related basic topological concepts. The function $\|\cdot\| : X \to \mathbb{R}_+$ is positively homogeneous: $\|\alpha x\| = |\alpha| \|x\|$ for all $\alpha \in \mathbb{R}$; positive definite: $\|x\| \geq 0$, with $\|x\| = 0$ only if $x = 0$; and satisfies the triangle inequality (B.2.3), which is equivalent to

$$\left| \|x\| - \|y\| \right| \leq \|x - y\| . \tag{B.3.1}$$

These three properties more broadly define a **norm** on a vector space (which may or may not be induced by an inner product).

A **sequence in** a set S is a function of the form $x \colon \{1, 2, \ldots\} \to S$. We write x_n for the sequence as well as its value at n, which we think of as the nth element in the sequence. A sequence x_n in X **converges** to some **limit** $x \in X$ if for all $\varepsilon > 0$, there exists an integer N such that $n > N$ implies $\|x_n - x\| < \varepsilon$. In this case, the sequence x_n is said to be **convergent**. A

---

[2] This recursive construction, known as *Gram–Schmidt orthogonalization*, is mainly of theoretical value. Its numerical implementation can be unstable and better alternatives exist. See, for example, the discussion of this point in Meyer (2004).

function $f \colon D \to \mathbb{R}$, where $D \subseteq X$, is **continuous at** $x \in D$ if for all sequences x_n in D, the convergence of x_n to $x \in D$ implies the convergence of $f(x_n)$ to $f(x)$. A function is **continuous** if it is continuous at every point of its domain. Inequality (B.3.1) shows that a norm is continuous. Thanks to identity (B.2.1), norm continuity implies the continuity of an inner product that induces the norm.

PROPOSITION B.3.1 *Suppose x_n and y_n are sequences in X converging to x and y, respectively. Then $\langle x_n \mid y_n \rangle$ converges to $\langle x \mid y \rangle$.*

If X is finite dimensional, a linear functional f is continuous, since it has a Riesz representation x and therefore

$$|f(y) - f(z)| \le \|x\| \, \|y - z\| \tag{B.3.2}$$

by the Cauchy–Schwarz inequality. A linear functional on an infinite-dimensional space may not be continuous.

EXAMPLE B.3.2 Suppose X is the inner product space of all sequences x in $\mathbb{R}$ such that $x_n = 0$ for all but finitely many values of n, with the inner product $\langle x \mid y \rangle = \sum_n x_n y_n$. The linear functional $f(x) \equiv \sum_n n x_n$ is not continuous. $\diamond$

A sequence x_n in X is **Cauchy** if for all $\varepsilon > 0$, there exists an integer N such that $m, n > N$ implies $\|x_m - x_n\| < \varepsilon$. The triangle inequality implies that *every convergent sequence is Cauchy.* Example B.3.3 shows that there are Cauchy sequences that do not converge. Intuitively, a Cauchy sequence should converge to something, but if that something is not in X, then the sequence is not convergent. As we will see shortly, this is not an issue in finite-dimensional spaces.

EXAMPLE B.3.3 Suppose X is the vector space of all continuous functionals on the unit interval with the inner product $\langle x \mid y \rangle = \int_0^1 x(t) \, y(t) \, dt$. The sequence $x_n(t) = 1/(1 + nt)$, $t \in [0, 1]$, is Cauchy but does not converge in X. $\diamond$

In a finite-dimensional space, the convergence or Cauchy property of a sequence is equivalent to the respective property of the sequence's coordinates relative to any given basis, no matter what inner product is used.

PROPOSITION B.3.4 *Suppose $B_1, \ldots, B_d$ is a basis of X. For scalar σ_n^i and σ^i, the sequence $\sum_i \sigma_n^i B_i$ is Cauchy (resp. converges to $\sum_i \sigma^i B_i$) if and only if for every coordinate i, the sequence σ_n^i is Cauchy (resp. converges to σ^i).*

PROOF Suppose σ_n^i is Cauchy for all i. By the triangle inequality (B.2.3), $\left\| \sum_i \sigma_m^i B_i - \sum_i \sigma_n^i B_i \right\| \leq \sum_i \left| \sigma_m^i - \sigma_n^i \right| \|B_i\|$ and therefore $\sum_i \sigma_n^i B_i$ is Cauchy. Conversely, the mapping from a vector to its jth coordinate is a linear functional f. Inequality (B.3.2), where x is the Riesz representation of f, implies that $\sigma_n^j \equiv f\left(\sum_i \sigma_n^i B_i \right)$ is Cauchy if $\sum_i \sigma_n^i B_i$ is Cauchy. The claim in parentheses follows by the same argument, but with *convergent* in place of *Cauchy* and σ in place of σ_m. □

A subset S of X is **closed** if every *convergent* sequence in S converges to a point in S. The set S is **complete** if every *Cauchy* sequence in S converges to a point in S. Since every convergent sequence is Cauchy, all complete sets are closed. A **Hilbert space** is an inner product space X such that the set X is complete (relative to the norm induced by the inner product). Note that a subset of a complete set is complete if and only if it is closed, and therefore the complete subsets of a Hilbert space are its closed subsets. Example B.3.3 is an instance of an inner product space that is *not* a Hilbert space, and the corresponding set X is an example of a closed set that is not complete. On the other hand, *every finite-dimensional inner product space is a Hilbert space*. By Proposition B.3.4, the completeness of a finite-dimensional space is implied by the completeness of $\mathbb{R}$, which is a fundamental property of the real line. (One of the constructions of the real numbers from the rational numbers is as equivalence classes of Cauchy sequences of rationals, thus constructing a *completion* of the rationals.)

Since all inner product space isomorphisms are isometries, convergence, the Cauchy property, continuity, the property of being closed, completeness, as well as compactness (discussed below) are all properties that are preserved by an inner product space isomorphism. By Proposition B.2.7, every finite-dimensional inner product space is isomorphic to a Euclidean space, and therefore all these properties in a finite-dimensional space are equivalent to corresponding properties in a Euclidean space. There is a great variety of useful (nonisomorphic) infinite-dimensional Hilbert spaces. One example is the space of all finite-variance random variables on an infinite probability space with the inner product $\langle x \mid y \rangle = \mathbb{E}[xy]$.

A **subsequence** of a sequence $x: \{1, 2, \ldots\} \rightarrow S$ is a function of the form $x \circ n$, for some strictly increasing function $n: \{1, 2, \ldots\} \rightarrow \{1, 2, \ldots\}$ (where $x \circ n$ denotes function composition: $(x \circ n)(k) = x_{n(k)}$, $k = 1, 2, \ldots$). A subset S of X is **compact** if every sequence in S has a subsequence that converges to a point in S. Note that a compact set is complete and therefore closed. A compact set S is also **bounded**, meaning $\sup \{\|x\| \mid x \in S\} < \infty$. If S were unbounded, there would exist a sequence

x_n in S such that $\|x_n\| > n$ for all n, which precludes the existence of a convergent subsequence. On the real line, a closed bounded interval is compact since it can be sequentially subdivided into halves that contain infinitely many points of a given sequence, leading to the selection of a subsequence that is Cauchy and therefore convergent (by the completeness of the real line). By virtue of Proposition B.3.4, this type of argument can be extended to $\mathbb{R}^d$ by applying it to each coordinate sequentially. Since every finite-dimensional space is isomorphic to $\mathbb{R}^d$, the following is true.

PROPOSITION B.3.5 *In a finite-dimensional space, a set is compact if and only if it is closed and bounded.*

EXAMPLE B.3.6 The space ℓ_2 of all sequences $x \colon \{1, 2, \dots\} \to \mathbb{R}$ such that $\sum_{n=1}^{\infty} x(n)^2 < \infty$ with $\langle x \mid y \rangle \equiv \sum_{n=1}^{\infty} x(n)\, y(n)$ can be shown to be a Hilbert space. The sequence x_n in ℓ_2, where $x_n(n) = 1$ and $x_n(m) = 0$ for all $m \neq n$, is in the closed and bounded set $B \equiv \{x \mid \|x\| \leq 1\}$, but has no convergent subsequence. In fact, B is not compact in all infinite-dimensional spaces. $\diamond$

Our interest in compactness is due to the fact that over a compact set, a continuous function achieves its supremum and infimum.

PROPOSITION B.3.7 *Suppose that S is a compact subset of X and the function $f \colon S \to \mathbb{R}$ is continuous. Then there exist $x^*, x_* \in S$ such that $f(x^*) \geq f(x) \geq f(x_*)$ for all $x \in S$.*

PROOF Let x_n be a sequence such that $f(x_n)$ converges to $\sup f$. By the compactness of S, there exists a subsequence of x_n converging to some $x^* \in S$. Since f is continuous, $f(x^*) = \sup f$ and therefore $f(x^*) \geq f(x)$ for all $x \in S$. The same argument applied to $-f$ completes the proof. $\qquad\square$

Proposition B.3.7 implies that in a finite-dimensional space all norms are topologically equivalent, in the sense of the following proposition, which implies that on a finite-dimensional vector space, all norms define the same convergent sequences, Cauchy sequences, continuous functions, closed sets, complete sets and compact sets. The situation is quite different with infinite-dimensional spaces, where different norms can define dramatically different notions of convergence.

PROPOSITION B.3.8 *Suppose X is finite dimensional, and $\|\cdot\|'$ is a norm on X, that is, any positively homogeneous, positive definite function $\|\cdot\|' \colon X \to \mathbb{R}_+$ satisfying the triangle inequality (not necessarily induced* *by an inner product). Then there exist constants k and K such that k* $\|x\| \leq$ $\|x\|' \leq K \|x\|$ *for all* $x \in X$.

PROOF The norm $\|\cdot\|'$ is continuous on $B \equiv \{x \mid \|x\| \leq 1\}$, which is closed and bounded, and hence compact. Let k and K be the minimum and maximum of $\|\cdot\|'$ over B. Since $\|x\|^{-1} x \in B$ for every nonzero $x \in X$, the claim follows. □

We conclude the section with an overview of some related topological concepts and notation. Consider any subset S of X. The **closure** of S, denoted by $\bar{S}$, is the set of every point x for which there exists some sequence in S that converges to x. Note that the set S is closed if and only if $S = \bar{S}$. The **interior** of S, denoted by S^0, is the set of all $x \in S$ such that for every sequence x_n converging to x, there exists an N such that $n > N$ implies $x_n \in S$. The **boundary** of S is the set $\bar{S} \backslash S^0$. The set S is **open** if $S = S^0$, or equivalently, if its complement $X \backslash S$ is closed.

The following properties can be verified as an exercise. The empty set and X are both open and closed. The union of finitely many closed sets is closed, and the intersection of finitely many open sets is open. Arbitrary intersections of closed sets are closed, and arbitrary unions of open sets are open. The set of all open subsets of X is known as the *topology* of X. It turns out that convergence and related properties discussed in this section can all be specified entirely in terms of the space's topology, rather than sequences. *General topology* allows an arbitrary set of open sets respecting the preceding properties, in which case sequences are not sufficient in characterizing topological properties (see, for example, Dudley (2002)). We have no need for this extension in this text.

B.4 Convexity

The purpose of this section is to introduce the central notions of convexity and concavity, which underlie the analysis of the rest of this appendix, as well as some basic topological implications of convexity assumptions. As always, we take as given the reference inner product space X.

A set $C \subseteq X$ is **convex** if the line segment connecting any two points in C lies entirely within C: For all $x, y \in C$,

$$\alpha \in (0, 1) \implies \alpha x + (1 - \alpha) y \in C.$$

Two important special types of convex set that arise in this text are convex cones and linear manifolds. C is a **cone** if for all $x \in C, \alpha \in [0, \infty)$ implies $\alpha x \in C$. Note that a cone C is convex if and only if $x, y \in C$ implies $x + y \in C$; and a convex cone C is a linear subspace if and only if $x \in C$

implies $-x \in C$. A set $M \subseteq X$ is a **linear manifold** if the line through any two points in M lies in M: For all $x, y \in M$,

$$\alpha \in \mathbb{R} \implies \alpha x + (1 - \alpha) y \in M.$$

Clearly, every linear manifold is convex. The **translation** of $S \subseteq X$ by the vector x is defined and denoted by $x + S \equiv S + x \equiv \{x + y \mid y \in S\}$. If M is a linear manifold and $m \in M$, then $M - m$ is a linear subspace. Conversely, the translation of any linear subspace is a linear manifold. Therefore, $M \subseteq X$ is a linear manifold if and only if it takes the form $M = m + L$ for some vector m and linear subspace L. In this case, the **dimension** of M is the dimension of L. By Proposition B.3.4, a finite-dimensional linear manifold is complete and therefore closed. In contrast, an infinite-dimensional linear subspace may not be closed. For instance, in the context of Example B.3.6, the linear subspace of all sequences x such that $x(n) = 0$ for all but finitely many values of n is not closed.

A function $f \colon C \to \mathbb{R}$, where $C \subseteq X$, is **concave** if C is convex and for all $x, y \in C$,

$$\alpha \in (0, 1) \implies f(\alpha x + (1 - \alpha) y) \geq \alpha f(x) + (1 - \alpha) f(y).$$

The function f is **convex** if $-f$ is concave. Note that a functional $f \colon X \to \mathbb{R}$ is linear if and only if it is both concave and convex. We saw in the last section that a linear functional is necessarily continuous if X is finite dimensional. This fact generalizes as follows.

THEOREM B.4.1 *Suppose X is finite dimensional, C is a convex subset of X, and the function $f \colon C \to \mathbb{R}$ is concave. Then f is continuous at every point of the interior of C.*

PROOF Since every finite-dimensional inner product space is isomorphic to a Euclidean space, we assume that $X = \mathbb{R}^d$ with the Euclidean inner product. Given any $x \in C^0$, choose $\alpha, \beta \in \mathbb{R}^d$ such that $\alpha_i < x_i < \beta_i$ for all i and $[\alpha, \beta] \equiv \prod_{i=1}^{d} [\alpha_i, \beta_i] \subseteq C$. A point in $[\alpha, \beta]$ is **extreme** if each of its coordinates equals α_i or β_i. (For example, for $d = 2$, $[\alpha, \beta]$ is a rectangle and its extreme points are the four corners.) Concavity of f implies that for all $y \in [\alpha, \beta]$, there exists an extreme point $\bar{y} \in [\alpha, \beta]$ such that $f(y) \geq f(\bar{y})$. (To see that, think of a plot of $f(y)$ on $[\alpha_i, \beta_i]$ as a function of its ith coordinate and note that the minimum is achieved at one of the endpoints.) Since $[\alpha, \beta]$ has finitely many extreme points, f is minimized by one of them. Let b denote the minimum value, and let $r > 0$ be small enough so that $B(x; r) \equiv \{y \mid \|y - x\| \leq r\} \subseteq [\alpha, \beta]$. Fixing any $y \in B(x; r)$, let $u \equiv (y - x) / \|y - x\|$ and $\phi(\alpha) \equiv f(x + \alpha u)$ for all $\alpha \in [-r, r]$. We then have the inequalities

$$K \equiv \frac{\phi(0) - b}{r} \geq \frac{\phi(0) - \phi(-r)}{r}$$

$$\geq \frac{\phi(\|y - x\|) - \phi(0)}{\|y - x\|} \geq \frac{\phi(r) - \phi(0)}{r}$$

$$\geq \frac{b - \phi(0)}{r} = -K.$$

The fact that ϕ is bounded below by b justifies the first and last inequalities. The three middle expressions represent slopes that decrease from left to right because $\phi \colon [-r, r] \to \mathbb{R}$ is concave. This proves that $|f(y) - f(x)| \leq K \|y - x\|$ for all $y \in B(x; r)$. □

Convex complete bounded sets can play the role compact sets did in Section B.3, even in the infinite-dimensional case, where compactness is less helpful.[3] Given a sequence x_n, we write $\mathrm{conv}(x_n, x_{n+1}, \dots)$ for the set of all finite linear combinations of the form $\sum_i \alpha_i x_{n_i}$, where $\alpha_i > 0$, $\sum_i \alpha_i = 1$ and $n_i \geq n$ for all i. Equivalently, $\mathrm{conv}(x_n, x_{n+1}, \dots)$ is the intersection of all convex sets that contain the points $x_n, x_{n+1}, \dots$

LEMMA B.4.2 *Suppose x_n is a sequence in a subset of X that is convex, complete and bounded. Then there exists a convergent sequence y_n such that $y_n \in \mathrm{conv}(x_n, x_{n+1}, \dots)$ for all n.*

PROOF Since x_n lies in a bounded set,

$$L_n \equiv \inf \{ \|y\| \mid y \in \mathrm{conv}(x_n, x_{n+1}, \dots) \}$$

is finite, increasing in n, and converges to the finite limit $L \equiv \sup_n L_n$ as $n \to \infty$. Choose $y_n \in \mathrm{conv}(x_n, x_{n+1}, \dots)$ such that $\|y_n\| < L_n + 1/n$ for all n. We will show that the sequence y_n is Cauchy and hence convergent, since x_n is assumed to lie in a convex complete set. Given any $\epsilon > 0$, choose N such that $1/N < \epsilon$ and $L_n > L - \epsilon$ for all $n > N$. For all $m > n > N$, we have $(1/2)(y_m + y_n) \in \mathrm{conv}(x_n, x_{n+1}, \dots)$ and therefore $\|(1/2)(y_m + y_n)\| \geq L_n > L - \epsilon$ and

$$\|y_m - y_n\|^2 = 2\|y_m\|^2 + 2\|y_n\|^2 - \|y_m + y_n\|^2$$
$$< 2(L_m + 1/m)^2 + 2(L_n + 1/n)^2 - 4(L - \epsilon)^2$$
$$< 4(L + \epsilon)^2 - 4(L - \epsilon)^2 = 16L\epsilon.$$

The Cauchy property of y_n follows. □

---

[3] The argument that follows originates in Komlós (1967). Here we follow the exposition of Beiglböck et al. (2012), who show a useful application in stochastic analysis.

The following result is a corollary of Proposition B.3.7 in the finite-dimensional case, but is remarkable in that it applies equally well in the infinite-dimensional case.

PROPOSITION B.4.3 *Suppose the nonempty set $C \subseteq X$ is convex, complete and bounded, and the function $f : C \to \mathbb{R}$ is convex and continuous. Then there exists $y \in C$ such that $f(y) \le f(x)$ for all $x \in C$.*

PROOF Choose the sequence x_n in C so that $f(x_n)$ converges to $M \equiv \inf\{f(x) \mid x \in C\}$. Let y_n be as in Lemma B.4.2, converging to $y \in C$ (since C is closed). Given any $\epsilon > 0$, choose N large enough so that $n > N$ implies $f(x_n) < M + \epsilon$. Fixing $n > N$, suppose that $y_n = \sum_i \alpha_i x_{n_i}$ for finitely many $\alpha_i > 0$ such that $\sum_i \alpha_i = 1$ and $n_i \ge n$. Convexity of C and f implies $y_n \in C$ and

$$M \le f(y_n) \le \sum_i \alpha_i f\left(x_{n_i}\right) < \sum_i \alpha_i (M + \epsilon) = M + \epsilon.$$

This shows that the sequence $f(y_n)$ converges to M and also converges to $f(y)$, since f is continuous. Therefore, $M = f(y)$.   □

B.5 Projections on Convex Sets

In Section B.2 we characterized the projection of a vector on a line by an orthogonality condition. As the geometric intuition suggests, this argument generalizes to projections on convex sets. The vector x_C is a **projection** of the vector x on the set $C \subseteq X$ if

$$x_C \in C \quad \text{and} \quad \|x - x_C\| \le \|x - c\| \text{ for all } c \in C.$$

In other words, x_C is a point of C that is closest to x. A projection x_C on a convex set C may not exist (for example, take X to be the real line, $C = (0, 1)$ and $x = 2$), but, as we will now show, if x_C exists it is unique, and if C is complete (Cauchy sequences in C converge in C) then x_C does exist. A central concern is the dual characterization of a projection. We say that a vector y **supports** the set C at $x_C \in C$ if

$$\langle y \mid c - x_C \rangle \ge 0 \quad \text{for all } c \in C. \tag{B.5.1}$$

In other words, y makes an acute angle with every vector connecting x_C to a point in C. Drawing a picture suggests that if x_C is the projection of x on the convex set C, then $y \equiv x_C - x$ supports C at x_C, which is precisely the dual characterization of a projection.

The central result on projections on convex sets follows. Note that the existence statement applies in particular if X is a Hilbert space and C is closed, since then C is necessarily complete.

THEOREM B.5.1 (Projection theorem) *Suppose C is a convex subset of the inner product space X. Then the following are true for all vectors $x, y \in X$.*

(1) The vector x_C is a projection of x on C if and only if $x_C \in C$ and $x_C - x$ supports C at x_C.

(2) If x_C is a projection of x on C and y_C is a projection of y on C, then $\|x_C - y_C\| \leq \|x - y\|$.

(3) If a projection of x on C exists, it is unique.

(4) If C is nonempty and complete, then the projection of x on C exists.

PROOF (1) Given $c, x_C \in C$, define $x^\alpha \equiv x_C + \alpha(c - x_C) \in C$ for all $\alpha \in [0, 1]$ (and therefore $x^0 = x_C$ and $x^1 = c$). The quadratic

$$\|x - x^\alpha\|^2 = \|x - x_C\|^2 + 2\alpha \langle x_C - x \mid c - x_C \rangle + \alpha^2 \|c - x_C\|^2$$

is minimized at $\alpha = 0$ if and only if $\langle x_C - x \mid c - x_C \rangle \geq 0$.

(2) Let $\delta \equiv y - x$ and $\delta_C \equiv y_C - x_C$. The support condition of part (1) requires $\langle x_C - x \mid \delta_C \rangle \geq 0$ and $\langle y - y_C \mid \delta_C \rangle \geq 0$. Adding up the two inequalities, we find $\langle \delta - \delta_C \mid \delta_C \rangle \geq 0$ and therefore

$$\|\delta\|^2 = \|\delta - \delta_C\|^2 + 2\langle \delta - \delta_C \mid \delta_C \rangle + \|\delta_C\|^2 \geq \|\delta_C\|^2.$$

(3) Let $x = y$ in part (2).

(4) For an arbitrary $c \in C$, let $r \equiv \|c - x\|$, which ensures that the bounded set $C_r \equiv C \cap \{z \mid \|z - x\| \leq r\}$ is nonempty. Since C_r is also convex and complete (being a closed subset of C), the projection x_C of x on C_r exists by Proposition B.4.3. Clearly, x_C is also the projection of x on C. □

For the rest of this section, our focus is on projections on linear manifolds, where the support condition becomes an orthogonality condition. A vector x is **orthogonal** to the linear manifold M if x is orthogonal to $y - z$ for all $y, z \in M$. The **orthogonal to M subspace**, denoted by $M^\perp$, is the linear subspace of all vectors that are orthogonal to M. Note that $M^\perp = (x + M)^\perp$ for all $x \in X$. A linear manifold M can be expressed as $M = m + L$ for a linear subspace L, in which case $M^\perp = L^\perp = \{x \mid \langle x \mid y \rangle = 0 \text{ for all } y \in L\}$. A vector x supports a linear manifold M at some point of M if and only if $x \in M^\perp$. As a consequence, Theorem B.5.1 applied to linear manifolds gives

COROLLARY B.5.2 (Orthogonal projection theorem) *Suppose M is a linear manifold in X. A point x_M is the projection of the vector x on M if and only if $x_M \in M$ and $x - x_M \in M^\perp$. If M is complete (for example, finite dimensional), then every vector x has a unique decomposition $x = x_M + y$ where $x_M \in M$ and $y \in M^\perp$.*

The following is an example of what can go wrong if M is not complete (and therefore infinite dimensional).

EXAMPLE B.5.3 Consider the vector space $C[0, 1]$ of all continuous functions of the form $x \colon [0, 1] \to \mathbb{R}$ with the inner product $\langle x \mid y \rangle = \int_0^1 x(t)\, y(t)\, dt$. If it existed, the projection of the zero vector on the linear manifold $M \equiv \{x \in C[0, 1] \mid x(0) = 1\}$ would minimize $\int_0^1 x(t)^2\, dt$ subject to the constraint $x(0) = 1$. Such a minimum does not exist within $C[0, 1]$. $\diamondsuit$

A useful extension of Corollary B.5.2 is the following geometrically intuitive statement, which is, however, not valid without the completeness assumption.[4]

PROPOSITION B.5.4 *Suppose $M = x + L$ for a vector x and a complete linear subspace L. Then $M^{\perp\perp} = L$.*

PROOF Since $M^\perp = L^\perp$, it suffices to show that $L = L^{\perp\perp}$. That $L \subseteq L^{\perp\perp}$ (whether L is complete or not) is immediate from the definitions. Conversely, suppose L is complete and $x \in L^{\perp\perp}$. Let $x = x_L + y$, where $x_L \in L$ and $y \in L^\perp$, and therefore $\langle x \mid y \rangle = \langle x_L \mid y \rangle + \langle y \mid y \rangle$ and $\langle x_L \mid y \rangle = 0$. Since $x \in L^{\perp\perp}$, we also have $\langle x \mid y \rangle = 0$. The last three equalities imply that $\langle y \mid y \rangle = 0$ and therefore $x = x_L \in L$. $\qquad\square$

Orthogonality and projections are closely related to the notion of a Riesz representation. Suppose f is a nonzero linear functional and n is a unit-norm vector that is orthogonal to the functional's **null space** $N \equiv \{x \mid f(x) = 0\}$. Then $f(n)\, n$ is the Riesz representation of f. Indeed, for all $x \in X$, the vector $f(x)\, n - f(n)\, x$ is in N and is therefore orthogonal to n, a condition that rearranges to $f(x) = \langle f(n)\, n \mid x \rangle$. This observation leads to the following existence argument (which does not require X to be finite dimensional).

---

[4] Although not needed in this text, it is not hard to show that for an arbitrary linear subspace L of a Hilbert space, $L^{\perp\perp}$ is the closure of L.

THEOREM B.5.5 *In a Hilbert space, a linear functional has a Riesz representation if and only if it is continuous.*

PROOF A linear functional with a Riesz representation must be continuous thanks to the Cauchy–Schwarz inequality. Conversely, suppose f is a continuous linear functional with null space N. If $X = N$, the claim is trivial. Otherwise, pick any $x \in X \backslash N$. Since N is closed and X is assumed complete, N is complete. By Corollary B.5.2, $x = x_N + y$, where $x_N \in N$ and $y \in N^\perp$. Since $n \equiv y / \|y\|$ is orthogonal to N and has unit norm, $f(n)\, n$ is the Riesz representation of f. $\qquad\square$

The projection of a Riesz representation on a linear subspace corresponds to the functional's restriction on the linear subspace.

PROPOSITION B.5.6 *Suppose* $f(y) = \langle x \mid y \rangle$ *for all* $y \in X$, *and* f_L *is the restriction of* f *on the linear subspace* L. *Then* x_L *is the Riesz representation of* f_L *in* L *if and only if it is the projection of* x *on* L.

PROOF The joint requirement $x_L \in L$ and $f(y) = \langle x_L \mid y \rangle$ for all $y \in L$ is equivalent to $x_L \in L$ and $x - x_L$ is orthogonal to L. Corollary B.5.2 completes the proof. $\qquad\square$

Projections on a finite-dimensional linear subspace can be simply expressed in terms of a basis.

PROPOSITION B.5.7 *Suppose* $B = (B_1, \ldots, B_d)'$ *is a basis for the linear subspace* L. *Then* $\langle x \mid B' \rangle \langle B \mid B' \rangle^{-1} B$ *is the projection of* x *on* L.

PROOF By Proposition B.5.6, the projection of x on L is the Riesz representation of the restriction of the linear functional $f(y) \equiv \langle x \mid y \rangle$ on L, which is given by Proposition B.2.6 as $f(B)' \langle B \mid B' \rangle^{-1} B$. (Alternatively, one can directly check the orthogonality condition.) $\qquad\square$

A useful application of Proposition B.5.7 is to problems that take the form of minimizing a norm subject to linear constraints.

PROPOSITION B.5.8 *Suppose* $B = (B_1, \ldots, B_d)'$ *is a matrix of linearly independent vectors and* $b \in \mathbb{R}^{d \times 1}$. *The vector* $b' \langle B \mid B' \rangle^{-1} B$ *is the value of* x *that minimizes* $\|x\|$ *subject to* $\langle B \mid x \rangle = b$ *and* $x \in X$.

PROOF Let $L \equiv \operatorname{span}(B)$ and $M \equiv \{x \in X \mid \langle B \mid x \rangle = b\}$. Recall that, by Proposition B.3.4, L is complete and therefore, by Proposition B.5.4, $L^{\perp\perp} = L$. Pick any $x \in M$. Since $M = x + L^\perp$, it follows that $M^\perp = L$. The point 0_M is the projection of the zero vector on M if and only if $0_M \in M \cap M^\perp$, if and only if $x - 0_M \in L^\perp$ and $0_M \in L$, if and only if 0_M is the projection of x on L, if and only if $0_M = \langle x \mid B' \rangle \langle B \mid B' \rangle^{-1} B$ (by Proposition B.5.7). Since $\langle B \mid x \rangle = b$, the result follows. $\qquad\square$

B.6 Supporting Hyperplanes and (Super)gradients

Continuing in the context of the inner product space X, we extend the projection argument of Section B.5 by essentially allowing the projected point to approach the boundary of a convex set, thus obtaining a supporting vector without reference to a projection. More precisely, consider a set $C \subseteq X$ and a vector $\bar{c}$ on the boundary of C, but not necessarily in C. We say that the vector y **supports** C at $\bar{c}$ if

$$\langle y \mid \bar{c} \rangle = \inf \{ \langle y \mid c \rangle \mid c \in C \}. \tag{B.6.1}$$

The term is consistent with the usage of Section B.5 since (B.6.1) reduces to (B.5.1) if $\bar{c} = x_C \in C$. Condition (B.6.1) can be visualized as the inclusion of $\bar{C}$ in the half-space $\{ x \mid \langle y \mid x \rangle \geq \langle y \mid \bar{c} \rangle \}$ with $\bar{C}$ touching the hyperplane $\{ x \mid \langle y \mid x \rangle = \langle y \mid \bar{c} \rangle \}$ at $\bar{c}$. (A **hyperplane** is a linear manifold whose orthogonal subspace is one dimensional.)

THEOREM B.6.1 (Supporting hyperplane theorem) *Suppose X is finite dimensional,[5] and the vector $\bar{c}$ is on the boundary of a convex set $C \subseteq X$. Then there exists a nonzero vector y that supports C at $\bar{c}$.*

PROOF Let x_n be a sequence in $X \backslash \bar{C}$ that converges to $\bar{c}$. By Theorem B.5.1, the projection $\bar{x}_n$ of x_n on $\bar{C}$ exists, the sequence $\bar{x}_n$ converges to $\bar{c}$ (since $\| \bar{x}_n - \bar{c} \| \leq \| x_n - \bar{c} \|$), and $y_n \equiv \bar{x}_n - x_n$ satisfies $\langle y_n \mid \bar{x}_n \rangle \leq \langle y_n \mid c \rangle$ for all $c \in C$. Dividing the last inequality by the norm of y_n, we can assume the inequality with $\| y_n \| = 1$. The set $\{ y \mid \| y \| = 1 \}$ is closed and bounded and hence compact (assuming X is finite dimensional). There is, therefore, a subsequence of y_n that converges to some unit-norm vector y. By Proposition B.3.1, we can take the limit along this subsequence to conclude that $\langle y \mid \bar{c} \rangle \leq \langle y \mid c \rangle$ for all $c \in C$. Since $\bar{c}$ is on the boundary of C, there exists a sequence c_n in C converging to $\bar{c}$, and therefore $\langle y \mid c_n \rangle$ converges to $\langle y \mid \bar{c} \rangle$. These two conditions together confirm the support condition (B.6.1). $\qquad\square$

---

[5] The assumption of finite dimensions cannot be omitted: Suppose X is the space of all sequences x such that $\sum_{n=1}^{\infty} x_n^2 < \infty$, with $\langle x \mid y \rangle \equiv \sum_{n=1}^{\infty} x_n y_n$. The set $C \equiv \{ x \in X \mid x_n \geq 0 \text{ for all } n \}$ has an empty interior, and if $\bar{c} \in C$ is such that $\bar{c}_n > 0$ for all n, there does not exist a nonzero vector that supports C at $\bar{c}$.

A useful application of the projection Theorem B.5.1 and the supporting hyperplane Theorem B.6.1 is the following separation result.

THEOREM B.6.2 (Separating hyperplane theorem) *Suppose X is finite dimensional, the sets $A, B \subseteq X$ are convex and $A \cap B = \emptyset$. Then there exists a nonzero vector y such that*

$$\inf_{a \in A} \langle y \,|\, a \rangle \geq \sup_{b \in B} \langle y \,|\, b \rangle. \tag{B.6.2}$$

PROOF The convex set

$$C \equiv A - B = \{ a - b \,|\, a \in A, b \in B \}$$

does not contain the zero vector 0. If $0 \notin \bar{C}$, the projection y of 0 on C satisfies the support condition $\langle y \,|\, (a - b) - y \rangle \geq 0$ for all $a \in A$ and $b \in B$, which implies (B.6.2) as a strict inequality. If $0 \in \bar{C}$, let y be a nonzero vector supporting C at 0. The support condition

$$0 = \inf \{ \langle y \,|\, a - b \rangle \,|\, a \in A, b \in B \}$$

implies (B.6.2) as an equality. $\qquad\square$

Applying the support condition to a set defined as the region below the graph of a function leads to the concept of a supergradient. Consider any function $f \colon C \to \mathbb{R}$, where $C \subseteq X$. The vector y is a **supergradient** of f at $x \in C$ if it satisfies the **gradient inequality**:

$$f(x + h) \leq f(x) + \langle y \,|\, h \rangle \text{ for all } h \text{ such that } x + h \in C.$$

The **superdifferential** of f at x, denoted by $\partial f(x)$, is the set of all supergradients of f at x. The supergradient property can be visualized as a support condition in the space $X \times \mathbb{R}$ with the inner product

$$\langle (x_1, \alpha_1) \,|\, (x_2, \alpha_2) \rangle = \langle x_1 \,|\, x_2 \rangle + \alpha_1 \alpha_2, \quad x_i \in X, \, \alpha_i \in \mathbb{R}. \tag{B.6.3}$$

The (strict) **subgraph** of f is the set

$$\mathrm{sub}(f) \equiv \{ (x, \alpha) \in C \times \mathbb{R} \,|\, \alpha < f(x) \}. \tag{B.6.4}$$

The definitions imply that

$$y \in \partial f(x) \iff (y, -1) \text{ supports } \mathrm{sub}(f) \text{ at } (x, f(x)).$$

THEOREM B.6.3 *Suppose X is finite dimensional, the set $C \subseteq X$ is convex, the function $f \colon C \to \mathbb{R}$ is concave and x is in the interior of C. Then the superdifferential $\partial f(x)$ is nonempty, convex and compact.*

PROOF By the supporting hyperplane Theorem B.6.1, there exists nonzero
$(y, -\beta) \in X \times \mathbb{R}$ that supports sub(f) at $(x, f(x))$:

$$\langle y \mid x \rangle - \beta f(x) = \inf \{ \langle y \mid c \rangle - \beta \alpha \mid \alpha < f(c), \; c \in C, \; \alpha \in \mathbb{R} \}.$$

Since the left-hand side is finite, $\beta \geq 0$. If $\beta = 0$, then y supports C at x,
contradicting the assumption $x \in C^0$. Therefore, $\beta > 0$ and $\beta^{-1} y \in \partial f(x)$,
which proves $\partial f(x) \neq \emptyset$. It follows easily from the definitions that $\partial f(x)$
is also convex and closed. Finally, we show that $\partial f(x)$ is bounded and
therefore compact. By Theorem B.4.1, we can choose $\varepsilon > 0$ and $K \in \mathbb{R}$
such that $\|h\| = \varepsilon$ implies $x + h \in C$ and $f(x + h) > K$. For all nonzero
$y \in \partial f(x)$, let $h \equiv - \|y\|^{-1} \varepsilon y$ and note that

$$\varepsilon \|y\| = -\langle y \mid h \rangle \leq f(x) - f(x + h) < f(x) - K.$$

This proves that $\partial f(x)$ is bounded.   □

The **directional derivative** of f at x in the direction h is the limit

$$f'(x; h) \equiv \lim_{\alpha \downarrow 0} \frac{f(x + \alpha h) - f(x)}{\alpha}.$$

The **gradient** $\nabla f(x)$ of f at x is the Riesz representation of the functional
$f'(x; \cdot)$, which is characterized by the condition

$$f'(x; h) = \langle \nabla f(x) \mid h \rangle \text{ for all } h \in X.$$

Of course, neither the gradient nor a directional derivative need to exist,
but under the assumptions of Theorem B.6.3, we have the following
characterization.

THEOREM B.6.4 *Suppose X is finite dimensional, $C \subseteq X$ is convex,
$f : C \to \mathbb{R}$ is concave and x is in the interior of C. Then the directional
derivative $f'(x; h)$ exists and is finite for all $h \in X$, and*

$$\partial f(x) = \big\{ y \in X \mid \langle y \mid h \rangle \geq f'(x; h) \text{ for all } h \in X \big\}. \tag{B.6.5}$$

*The gradient $\nabla f(x)$ exists if and only if $\partial f(x)$ is a singleton, in which case
$\partial f(x) = \{ \nabla f(x) \}$.*

PROOF Given any $h \in X$, consider the slopes

$$\Delta(\alpha) \equiv \frac{f(x + \alpha h) - f(x)}{\alpha}, \quad x + \alpha h \in C, \; \alpha \in \mathbb{R}.$$

The concavity of f implies that $\Delta(\alpha)$ is nonincreasing in α. Choosing
$\varepsilon > 0$ such that $x + \alpha h \in C$ for all $\alpha \in [-\varepsilon, \varepsilon]$, it follows that

$$f'(x; h) \equiv \lim_{\alpha \downarrow 0} \Delta(\alpha) = \sup \{ \Delta(\alpha) \mid \alpha \in (0, \varepsilon) \} \leq \Delta(-\varepsilon). \tag{B.6.6}$$

This proves that $f'(x; h)$ exists and is finite.

Suppose $y \in \partial f(x)$. Then $\langle y \mid h \rangle \geq \Delta(\alpha)$ for all $\alpha \in (0, \varepsilon)$, and therefore $\langle y \mid h \rangle \geq f'(x; h)$ by (B.6.6). Conversely, suppose $y \notin \partial f(x)$ and therefore $\Delta(1) \equiv f(x + h) - f(x) > \langle y \mid h \rangle$ for some h. Since Δ is decreasing, (B.6.6) implies $f'(x; h) \geq \Delta(1) > \langle y \mid h \rangle$. This proves (B.6.5).

To show the final claim, suppose $\nabla f(x)$ exists and $y \in \partial f(x)$. By (B.6.5), $\langle \nabla f(x) - y \mid h \rangle \leq 0$ for all $h \in X$. Letting $h \equiv \nabla f(x) - y$, it follows that $y = \nabla f(x)$. Conversely, suppose $\partial f(x) = \{y\}$. We will show that $y = \nabla f(x)$ by confirming that $f'(x; h) = \langle y \mid h \rangle$ for all h. Consider any h. The fact that $f'(x; h) \geq \Delta(\alpha)$ if $\alpha > 0$ and $f'(x; h) \leq \Delta(\alpha)$ if $\alpha < 0$ implies

$$f(x + \alpha h) \leq f(x) + \alpha f'(x; h), \quad \alpha \in [-\varepsilon, \varepsilon].$$

It follows that, in the space $X \times \mathbb{R}$, the subgraph of f, as defined in (B.6.4), does not intersect the line segment

$$\left\{ \left(x + \alpha h, f(x) + \alpha f'(x; h) \right) \mid \alpha \in [-\varepsilon, \varepsilon] \right\}.$$

By the separating hyperplane Theorem B.6.2, with the inner product (B.6.3), there exists nonzero $(p, \beta) \in X \times \mathbb{R}$ such that

$$\inf \left\{ \langle p \mid x + \alpha h \rangle + \beta \left(f(x) + \alpha f'(x; h) \right) \mid \alpha \in [-\varepsilon, \varepsilon] \right\}$$
$$\geq \sup \left\{ \langle p \mid \tilde{x} \rangle + \beta \alpha \mid (\tilde{x}, \alpha) \in \mathrm{sub}(f) \right\}.$$

Clearly, this separation condition can only hold if $\beta > 0$. The right-hand side is at least as large as $\langle p \mid x \rangle + \beta f(x)$, which is also obtained as the expression on the left-hand side with $\alpha = 0$. The coefficient of α on the left-hand side must therefore vanish: $\langle p \mid h \rangle + \beta f'(x; h) = 0$. It follows that $f'(x; h) = \langle y \mid h \rangle$, where $y = -\beta^{-1} p$ is a consequence of the fact that the separation condition reduces to the gradient inequality. $\qquad\square$

B.7 Optimality Conditions

We conclude with some basic results relating to constrained optimization problems of the form

$$F(\delta) \equiv \sup \{ f(x) \mid g(x) \leq \delta, \ x \in C \}, \quad \delta \in \mathbb{R}^n, \qquad (B.7.1)$$

for given $C \subseteq X$ and functions $f : C \to \mathbb{R}$ and $g : C \to \mathbb{R}^n$, where n is a positive integer. With the convention $\sup \emptyset = -\infty$, equation (B.7.1) defines the function $F : \mathbb{R}^n \to [-\infty, +\infty]$. Although the parameter δ seems redundant, it will play an important role.

Assuming $F(0)$ is finite, the **superdifferential** of F at zero is

$$\partial F(0) \equiv \left\{ \lambda \in \mathbb{R}^n \mid F(\delta) \leq F(0) + \lambda \cdot \delta \text{ for all } \delta \in \mathbb{R}^n \right\} \subseteq \mathbb{R}^n_+.$$

The inclusion in $\mathbb{R}^n_+$ is due to the monotonicity of F: For all $\lambda \in \partial F(0)$, if $\delta \geq 0$ then $\lambda \cdot \delta \geq F(\delta) - F(0) \geq 0$, and therefore $\lambda \geq 0$.

The **Lagrangian** function associated with f and g is

$$\mathcal{L}(x, \lambda) \equiv f(x) - \lambda \cdot g(x), \quad x \in C, \quad \lambda \in \mathbb{R}^n.$$

LEMMA B.7.1 *Assuming $F(0) < \infty$ and $\lambda \in \mathbb{R}^n_+$, $\lambda \in \partial F(0)$ if and only if $F(0) = \sup \{\mathcal{L}(x, \lambda) \mid x \in C\}$.*

PROOF In $\mathbb{R}^n \times \mathbb{R}$ with the Euclidean inner product, $\lambda \in \partial F(0)$ if and only if $(\lambda, -1)$ supports at $(0, F(0))$ the set

$$\text{sub}(F) \equiv \left\{ (\delta, \alpha) \in \mathbb{R}^n \times \mathbb{R} \mid \alpha < F(\delta) \right\}. \tag{B.7.2}$$

Similarly, $F(0) = \sup \{\mathcal{L}(x, \lambda) \mid x \in C\}$ if and only if $(\lambda, -1)$ supports at $(0, F(0))$ the set

$$S \equiv \left\{ (\delta, \alpha) \in \mathbb{R}^n \times \mathbb{R} \mid \text{for some } x \in C, \ g(x) \leq \delta \text{ and } \alpha < f(x) \right\}.$$

Finally, note that $\text{sub}(F) = S$. □

Assuming that the supremum defining $F(0)$ is a maximum and $\partial F(0)$ is nonempty, the following main result provides a way of converting the constrained optimization problem defining $F(0)$ to an unconstrained problem. The intuitive idea is that the parameter λ, known as a **Lagrange multiplier**, provides an appropriate pricing of the constraint $g \leq 0$. If the constraint $g_i(x) \leq 0$ is slack, then the corresponding price λ_i must be zero. This leads to the **complementary slackness** conditions $g_i(x) < 0 \implies \lambda_i = 0$, which, given the inequalities $g(x) \leq 0$ and $\lambda \geq 0$, are equivalent to $\lambda \cdot g(x) = 0$.

THEOREM B.7.2 *Assuming $x \in C$, $g(x) \leq 0$ and $\lambda \in \mathbb{R}^n$, the following two conditions are equivalent.*

(1) $f(x) = F(0)$ and $\lambda \in \partial F(0)$.
(2) $\mathcal{L}(x, \lambda) = \max_{y \in C} \mathcal{L}(y, \lambda), \quad \lambda \cdot g(x) = 0, \quad \lambda \geq 0.$

PROOF Suppose condition (1) holds. We have already seen that $\lambda \in \partial F(0)$ implies $\lambda \geq 0$. Since $g(x) \leq 0$, $\mathcal{L}(x, \lambda) \geq f(x) = F(0)$. By Lemma B.7.1, $F(0) \geq \mathcal{L}(x, \lambda)$. Therefore, $\mathcal{L}(x, \lambda) = f(x) = F(0)$. The fact that $\mathcal{L}(x, \lambda) = f(x)$ implies $\lambda \cdot g(x) = 0$. The fact that $\mathcal{L}(x, \lambda) = F(0)$ and Lemma B.7.1 imply $\mathcal{L}(x, \lambda) = \max_{y \in C} \mathcal{L}(y, \lambda)$.

Conversely, suppose condition (2) holds. The assumption $\lambda \cdot g(x) = 0$ implies $\mathcal{L}(x, \lambda) = f(x)$. For all $y \in C$ such that $g(y) \leq 0$,

$$f(y) \leq \mathcal{L}(y, \lambda) \leq \mathcal{L}(x, \lambda) = f(x).$$

Therefore, $f(x) = F(0)$, and $\lambda \in \partial F(0)$ follows by Lemma B.7.1. $\quad\square$

The following lemma gives easy to check sufficient conditions for $\partial F(0)$ to be nonempty.

LEMMA B.7.3 *Suppose that $F(0) < \infty$, C and g are convex, f is concave, and there exists some $x \in C$ such that $g_i(x) < 0$ for all i. Then $\partial F(0)$ is nonempty.*

PROOF The concavity of f and convexity of g imply that the set $\mathrm{sub}(F)$, defined in (B.7.2), is convex. (The proof of this claim is a matter of applying the definitions and is left to the reader.) By the supporting hyperplane Theorem B.6.1, $\mathrm{sub}(F)$ is supported at $(0, F(0))$ by some nonzero $(\lambda, -\beta)$, and therefore $\alpha < F(\delta)$ implies $-\beta F(0) \leq \lambda \cdot \delta - \alpha\beta$ for all $\delta \in \mathbb{R}^n$. If $\beta < 0$, setting $\delta = 0$ leads to a contradiction. The existence of an x such that $g_i(x) < 0$ for all i ensures that $F(\delta) > -\infty$ for all sufficiently small δ. As a consequence, if $\beta = 0$, the support condition implies $0 \leq \lambda \cdot \delta$, where $\delta \equiv -\varepsilon\lambda$ for some $\varepsilon > 0$, a contradiction. Therefore $\beta > 0$ and $\beta^{-1}\lambda \in \partial F(0)$. $\quad\square$

The optimality condition (2) of Theorem B.7.2 is *global* since it utilizes the objective and constraint functions over their entire domain. Another type of optimality condition is *local*, providing implications of the fact that small feasible perturbations from an optimum cannot improve the objective. For example, assuming $x \in C^0$ (the interior of C) and the gradient $\nabla f(x)$ exists,

$$f(x) = \max\{f(y) \mid y \in C\} \implies \nabla f(x) = 0. \tag{B.7.3}$$

This is because $\phi(\alpha) \equiv f(x + \alpha\nabla f(x))$, for all α near zero, is maximized at zero and therefore $\phi'(0) = \langle \nabla f(x) \mid \nabla f(x) \rangle = 0$. For the converse of implication (B.7.3) to be true, we need some global regularity condition so that we can deduce a global condition from a local one. For example, if f is assumed to be concave and $\nabla f(x) = 0$, the gradient inequality implies that $f(y) \leq f(x)$ for all $y \in C$.

The preceding argument applies in particular to the function $\mathcal{L}(\cdot, \lambda)$ of optimality condition (2) of Theorem B.7.2. Assuming $x \in C^0$ and the existence of $\nabla f(x)$ and $\nabla g(x)$,

$$\mathcal{L}(x, \lambda) = \max_{y \in C} \mathcal{L}(y, \lambda) \implies \nabla f(x) = \lambda \cdot \nabla g(x),$$

and the converse is true if C and g are convex and f is concave. This leads us to the **Kuhn–Tucker conditions**:

$$\nabla f(x) = \lambda \cdot \nabla g(x), \quad \lambda \cdot g(x) = 0, \quad \lambda \geq 0. \tag{B.7.4}$$

The sufficiency of the Kuhn–Tucker conditions for optimality under convexity assumptions is covered by Theorem B.7.2 and the gradient inequality. A necessity argument via Theorem B.7.2 and Lemma B.7.3 requires global convexity assumptions. These are redundant in the following local argument, under a regularity condition requiring that at the given optimum, there exists a marginal deviation that makes all constraints slack.

THEOREM B.7.4 *Suppose $x \in C^0$, $g(x) \leq 0$, $f(x) = F(0)$, and the gradients $\nabla f(x)$ and $\nabla g(x)$ exist. Under the regularity condition that there exists some vector h such that $g_i(x) + \langle \nabla g_i(x) \mid h \rangle < 0$ for all i, the Kuhn–Tucker conditions (B.7.4) are satisfied for some $\lambda \in \mathbb{R}^n$.*

PROOF Given the assumed feasibility and optimality of x,

$$g_i(x) + \langle \nabla g_i(x) \mid h \rangle < 0 \text{ for all } i \implies \langle \nabla f(x) \mid h \rangle \leq 0. \tag{B.7.5}$$

To show this claim, consider any direction h for which the condition on the left side of implication (B.7.5) holds. Let $\varepsilon_i > 0$ be such that $g_i(x + \alpha h) \leq 0$ for all $\alpha \in [0, \varepsilon_i]$. If $g_i(x) < 0$, the existence of such an ε_i follows from the continuity of g_i (which is implied by the existence of a gradient). If $g_i(x) = 0$, such an an ε_i exists because $g_i(x + \alpha h)/\alpha$ converges to $\langle \nabla g_i(x) \mid h \rangle < 0$ as $\alpha \downarrow 0$. Letting $\varepsilon \equiv \min\{\varepsilon_1, \ldots, \varepsilon_n\}$, it follows that $g(x + \alpha h) \leq 0$ for all $\alpha \in [0, \varepsilon]$. Since $f(x) = F(0)$, $f(x + \alpha h) \leq f(x)$ for all $\alpha \in [0, \varepsilon]$, and therefore

$$\langle \nabla f(x) \mid h \rangle = \lim_{\alpha \downarrow 0} \frac{f(x + \alpha h) - f(x)}{\alpha} \leq 0,$$

completing the proof of condition (B.7.5).

Define $A \equiv (-\infty, 0)^n \times (0, \infty)$ and the set B of all $(b, \beta) \in \mathbb{R}^n \times \mathbb{R}$ for which there exists $h \in X$ such that

$$g(x) + \langle \nabla g(x) \mid h \rangle \leq b \quad \text{and} \quad \langle \nabla f(x) \mid h \rangle \geq \beta.$$

Condition (B.7.5) implies $A \cap B = \emptyset$. By the separating hyperplane Theorem B.6.2, there exists some nonzero $(-\lambda, \mu) \in \mathbb{R}^n \times \mathbb{R}$ and some scalar s such that

$$(a, \alpha) \in A \quad \Longrightarrow \quad -\lambda \cdot a + \mu \alpha \geq s, \qquad (B.7.6)$$

$$(b, \beta) \in B \quad \Longrightarrow \quad -\lambda \cdot b + \mu \beta \leq s. \qquad (B.7.7)$$

Condition (B.7.6) implies $\lambda, \mu \geq 0$ and $s \leq 0$. Given that, (B.7.7) implies

$$-\lambda \cdot (g(x) + \langle \nabla g(x) \mid h \rangle) + \mu \langle \nabla f(x) \mid h \rangle \leq 0, \text{ all } h \in X. \qquad (B.7.8)$$

The case $\mu = 0$ is ruled out by the constraint regularity condition. After rescaling of λ, we can therefore assume that (B.7.8) holds with $\mu = 1$. Since $\lambda \geq 0$ and $g(x) \leq 0$, condition (B.7.8) with $h = 0$ implies that $\lambda \cdot g(x) = 0$. Finally, condition (B.7.8) with $\mu = 1$, $\lambda \cdot g(x) = 0$ and $h = \nabla f(x) - \lambda \cdot \nabla g(x)$ implies $\nabla f(x) = \lambda \cdot \nabla g(x)$. $\qquad \square$

The ideas behind Theorem B.7.4 also apply to equality constraints. For example, consider the problem of finding an $x \in C$ that maximizes $f(x)$ subject to $\langle B \mid x \rangle = b$, where $B \equiv (B_1, \ldots, B_n)'$ and $b \in \mathbb{R}^{n \times 1}$. Suppose $x \in C^0$, $\nabla f(x)$ exists, $\langle B \mid x \rangle = b$ and

$$f(x) = \max \{ f(y) \mid \langle B \mid y \rangle = b, \ y \in C \}.$$

Consider any $h \in \operatorname{span}(B)^\perp$ and choose $\varepsilon > 0$ so that $x + \alpha h \in C$ for $\alpha \in [-\varepsilon, \varepsilon]$. Since $\langle B \mid x + \alpha h \rangle = b$, the function $\phi(\alpha) \equiv f(x + \alpha h)$, $\alpha \in [-\varepsilon, \varepsilon]$, is maximized at zero and $\phi'(0) = \langle \nabla f(x) \mid h \rangle = 0$. We have shown that $\nabla f(x) \in \operatorname{span}(B)^{\perp\perp} = \operatorname{span}(B)$, and therefore $\nabla f(x) \in \operatorname{span}(B)$ is a necessary local optimality condition. By virtue of the gradient inequality, it is also a sufficient optimality condition (given the feasibility of x) under the additional global assumption that C is convex and f is concave. The argument can be extended to equality constraints of the form $g(x) = 0$, with $\nabla g(x)$ playing the role of B in the preceding argument, but we have no need for such an extension in this text.

References

Aczél, J. *Lectures on Functional Equations and Their Applications.* Dover Publications, Mineola, NY, 2006. 195

Anscombe, F. J. and Aumann, R. J. A definition of subjective probability. *Annals of Mathematical Statistics,* 34:199–205, 1963. 198

Applebaum, D. *Lévy Processes and Stochastic Calculus.* Cambridge University Press, Cambridge, UK, 2004. 77

Arrow, K. J. Le rôle des valeurs boursières pour la répartition la meilleure des risques. *Econométrie, Colloques Internationaux du Centre National de la Recherche Scientifique,* 40:41–47, 1953. 15, 112

Arrow, K. J. The role of securities in the optimal allocation of risk bearing. *Review of Economic Studies,* 31:91–96, 1964. 15, 112

Arrow, K. J. *Aspects of the Theory of Risk Bearing.* Yrjo Jahnssonin Saatio, Helsinki, 1965. 156

Arrow, K. J. *Essays in the Theory of Risk Bearing.* North Holland, London, 1971. 60, 156

Back, K. E. *Asset Pricing and Portfolio Choice Theory.* Oxford University Press, New York, NY, second ed., 2017. xi

Beiglböck, M., Schachermayer, W., and Veliyev, B. A short proof of the Doob–Meyer theorem. *Stochastic Processes and Their Applications,* 122:1204–1209, 2012. 219

Bertsekas, D. P. *Convex Analysis and Optimization.* Athena Scientific, Belmont, MA, 2003. 206

Black, F. and Scholes, M. The pricing of options and corporate liabilities. *Journal of Political Economy,* 3:637–654, 1973. 93

Bonnans, J. F. and Shapiro, A. *Perturbation Analysis of Optimization Problems.* Springer-Verlag, New York, NY, 2000. 206

Boyd, S. and Vandenberghe, L. *Convex Optimization.* Cambridge University Press, Cambridge, UK, 2004. 206

Cinlar, E. *Probability and Stochastics.* Springer-Verlag, New York, NY, 2010. 77

Cox, J. and Ross, S. The valuation of options for alternative stochastic processes. *Journal of Financial Economics,* 3:145–166, 1976. 60

Cox, J., Ross, S., and Rubinstein, M. Option pricing: A simplified approach. *Journal of Financial Economics,* 7:229–263, 1979. 38

Dalang, R., Morton, A., and Willinger, W. Equivalent martingale measures and no-arbitrage in stochastic securities market models. *Stochastics and Stochastic Reports*, 29:185–201, 1990. 16

Debreu, G. *Theory of Value*. Cowles Foundation Monograph, Yale University Press, New Haven, CT, 1959. 15, 112

Debreu, G. *Mathematical Economics: Twenty Papers of Gerard Debreu*. Cambridge University Press, New York, NY, 1983. 188, 190

Delbaen, F. and Schachermayer, W. *The Mathematics of Arbitrage*. Springer-Verlag, New York, NY, 2006. 16

DeMarzo, P. and Skiadas, C. Aggregation, determinacy, and informational efficiency for a class of economies with asymmetric information. *Journal of Economic Theory*, 80:123–152, 1998. 181

DeMarzo, P. and Skiadas, C. On the uniqueness of fully informative rational expectations equilibria. *Economic Theory*, 13:1–24, 1999. 181

Dreyfus, S. Richard Bellman on the birth of dynamic programming. *Operations Research*, 50(1), 2002. 24

Drèze, J. Market allocation under uncertainty. *European Economic Review*, 15:133–165, 1971. 60

Dudley, R. M. *Real Analysis and Probability*. Cambridge University Press, New York, NY, 2002. 162, 217

Duffie, D. *Dynamic Asset Pricing Theory*. Princeton University Press, Princeton, NJ, third ed., 2001. xi

Duffie, D. and Epstein, L. G. Stochastic differential utility. *Econometrica*, 60:353–394, 1992a. 157, 158, 163

Duffie, D. and Epstein, L. G. Asset pricing with stochastic differential utility. *Review of Financial Studies*, 5:411–436, 1992b. 163

Duffie, D. and Skiadas, C. Continuous-time security pricing: A utility gradient approach. *Journal of Mathematical Economics*, 23:107–131, 1994. 163

Dunford, N. and Schwartz, J. T. *Linear Operators, Part I, General Theory*. Wiley, Hoboken, NJ, 1988. 206

Ekeland, I. and Témam, R. *Convex Analysis and Variational Problems*. SIAM, Philadelphia, PA, 1999. 206

Epstein, L. and Zin, S. Substitution, risk aversion, and the temporal behavior of consumption and asset returns: A theoretical framework. *Econometrica*, 57:937–969, 1989. 136

Hansen, L. P. and Jagannathan, R. Implications of security market data for models of dynamic economies. *Journal of Political Economy*, 99:225–262, 1991. 59

Harrison, M. J. and Kreps, D. M. Martingale and arbitrage in multiperiod securities markets. *Journal of Economic Theory*, 20:381–408, 1979. 60

Herstein, I. N. and Milnor, J. An axiomatic approach to measurable utility. *Econometrica*, 21:291–297, 1953. 198

Jacod, J. and Protter, P. *Discretization of Processes*. Springer-Verlag Berlin, Heidelberg, 2012. 77

Jacod, J. and Shiryaev, A. N. *Limit Theorems for Stochastic Processes*. Springer-Verlag, Berlin, Heidelberg, second ed., 2003. 30, 77, 79, 80, 81

Kabanov, Y. M. and Kramkov, D. O. No-arbitrage and equivalent martingale measure: An elementary proof of the Harrison–Pliska theorem. *Theory of Probability and its Applications*, 39:523–527, 1994. 16

Komlós, J. A generalization of a problem of Steinhaus. *Acta Mathematica Hungarica*, 18:217–229, 1967. 219

Krantz, D. H., Luce, R. D., Suppes, P., and Tversky, A. *Foundations of Measurement*, volume I. Academic Press, Inc., San Diego, CA, 1971. 190

Kreps, D. M. Arbitrage and equilibrium in economies with infinitely many commodities. *Journal of Mathematical Economics*, 8:15–35, 1981. 16

Kreps, D. M. and Porteus, E. Temporal resolution of uncertainty and dynamic choice theory. *Econometrica*, 46:185–200, 1978. 157, 183

Lintner, J. The valuation of risk assets and the selection of of risky investments in stock portfolios and capital budgets. *Review of Economics and Statistics*, 47:13–37, 1965. 114

Lucas, R. E. B. Asset prices in an exchange economy. *Econometrica*, 46:1429–1446, 1978. 145

Luenberger, D. G. *Optimization by Vector Space Methods*. Wiley, New York, NY, 1969. 206

Markowitz, H. Portfolio selection. *Journal of Finance*, 7:77–91, 1952. 54

Merton, R. C. Lifetime portfolio selection under uncertainty: The continuous time case. *Review of Economics and Statistics*, 51:247–257, 1969. 167

Merton, R. C. Optimum consumption and portfolio rules in a continuous-time model. *Journal of Economic Theory*, 3:373–413, 1971. Erratum 6 (1973):213–214. 167

Merton, R. C. The theory of rational option pricing. *Bell Journal of Economics and Management Science*, 4:141–183, 1973. 95

Meyer, C. D. *Matrix Analysis and Applied Linear Algebra*. SIAM, Philadelphia, PA, 2004. 213

Mörters, P. and Peres, Y. *Brownian Motion*. Cambridge University Press, New York, NY, 2010. 79

Mossin, J. Equilibrium in a capital asset market. *Econometrica*, 34:768–783, 1966. 114

Pardoux, E. and Peng, S. Adapted solution of a backward stochastic differential equation. *Systems and Control Letters*, 14:55–61, 1990. 157

Pratt, J. W. Risk aversion in the small and in the large. *Econometrica*, 32:122–136, 1964. 156

Ramsey, F. P. Truth and probability. In H. E. Kyburg, Jr. and H. E. Smokler, editors, *Studies in Subjective Probability (1980)*. Robert E. Krieger Publishing Company, New York, NY, 1926. 198

Rendleman, Jr. R. J. and Bartter, B. J. Two-state option pricing. *The Journal of Finance*, 34:1093–1110, 1979. 38

Reny, P. J. A simple proof of the nonconcavifiability of functions with linear not-all-parallel contour sets. *Journal of Mathematical Economics*, 49:506–508, 2013. 119

Revuz, D. and Yor, M. *Continuous Martingales and Brownian Motion*. Springer, New York, NY, third ed., 1999. 79

Rockafellar, R. T. *Convex Analysis*. Princeton University Press, Princeton, NJ, 1970. 137, 206

Ross, S. A. A simple approach to the valuation of risky streams. *Journal of Business*, 51:453–475, 1978. 16

Savage, L. J. *The Foundations of Statistics*. Dover Publications (1972), New York, NY, 1954. 198

Schachermayer, W. A Hilbert-space proof of the fundamental theorem of asset pricing. *Insurance Mathematics and Economics*, 11:249–257, 1992. 16

Schroder, M. and Skiadas, C. Optimal consumption and portfolio selection with stochastic differential utility. *Journal of Economic Theory*, 89:68–126, 1999. 157

Schroder, M. and Skiadas, C. An isomorphism between asset pricing models with and without linear habit formation. *Review of Financial Studies*, 15:1189–1221, 2002. 177

Schroder, M. and Skiadas, C. Optimal lifetime consumption-portfolio strategies under trading constraints and generalized recursive preferences. *Stochastic Processes and Their Applications*, 108:155–202, 2003. 160

Schroder, M. and Skiadas C. Lifetime consumption-portfolio choice under trading constraints and nontradeable income. *Stochastic Processes and their Applications*, 115:1–30, 2005. 160

Sharpe, W. F. Capital asset prices: A theory of market equilibrium under conditions of risk. *Journal of Finance*, 19:425–442, 1964. 114

Sharpe, W. F. *Investments*. Prentice Hall, Englewood Cliffs, NJ, 1978. 38

Skiadas, C. Subjective probability under additive aggregation of conditional preferences. *Journal of Economic Theory*, 76:242–271, 1997. 198

Skiadas, C. Recursive utility and preferences for information. *Economic Theory*, 12:293–312, 1998. 157, 184

Skiadas, C. Robust control and recursive utility. *Finance and Stochastics*, 7:475–489, 2003. 137

Skiadas, C. *Asset Pricing Theory*. Princeton University Press, Princeton, NJ, 2009. xi, 52, 148, 198, 203

Skiadas, C. Scale-invariant asset pricing and consumption/portfolio choice with general attitudes toward risk and uncertainty. *Mathematics and Financial Economics*, 7:431–456, 2013a. 151

Skiadas, C. Scale-invariant uncertainty-averse preferences and source-dependent constant relative risk aversion. *Theoretical Economics*, 8:59–93, 2013b. 194, 203

Skiadas, C. Smooth ambiguity aversion toward small risks and continuous-time recursive utility. *Journal of Political Economy*, 121:775–792, 2013c. 156

Skiadas, C. Dynamic choice with constant source-dependent relative risk aversion. *Economic Theory*, 3:393–422, 2015. 137, 203

Stoer, J. and Witzgall, C. *Convexity and Optimization in Finite Dimensions*. Springer-Verlag, New York, Heidelberg, Berlin, 1970. 16

Strotz, R. H. Myopia and inconsistency in dynamic utility maximization. *Review of Economic Studies*, 23:165–180, 1957. 111

von Neumann, J. and Morgenstern, O. *Theory of Games and Economic Behavior*. Princeton University Press, Princeton, NJ, 1944. 198

Wakker, P. P. Cardinal coordinate independence for expected utility. *Journal of Mathematical Psychology*, 28:110–117, 1984. 198

Wakker, P. P. The algebraic versus the topological approach to additive representations. *Journal of Mathematical Psychology*, 32:421–435, 1988. 190, 198

Wakker, P. P. *Additive Representations of Preferences*. Kluwer, Dordrecht, The Netherlands, 1989. 190, 198

Walras, L. *Eléments d'Economie Pure*. Corbaze, Lausanne. English translation: *Elements of Pure Economics*, R. D. Irwin, Homewood, IL (1954), 1874. 112

Weil, P. The equity premium puzzle and the risk-free rate puzzle. *Journal of Monetary Economics*, 24:401–421, 1989. 136

Weil, P. Non-expected utility in macroeconomics. *The Quarterly Journal of Economics*, 105:29–42, 1990. 136

Whitt, W. Proofs of the martingale FCLT. *Probability Surveys*, 4:268–302, 2007. 79

Xing, H. Consumption investment optimization with Epstein–Zin utility in incomplete markets. *Finance and Stochastics*, 21:227–262, 2017. 157

Yaari, M. E. A note on separability and quasiconcavity. *Econometrica*, 45:1183–1186, 1977. 192, 203

Yan, J. A. Caracterisation d'une class d'ensembles convexes de l^1 ou h^1. *Lecture Notes in Mathematics*, 784:220–222, 1980. 16

Zeidler, E. *Nonlinear Functional Analysis and its Applications III: Variational Methods and Optimization*. Springer-Verlag, New York, NY, 1985. 206

Index

adapted
  market, 9
  process, 5, 6
  strategy, 27
additive dynamic utility
  inadequacy, 126, 129
additive utility, 189
  concave, 192
  exponential, 195
  power or logarithmic, 196
  representation theorem, 190
  scale-invariant (SI), 196
  translation-invariant (TI), 195
  uniqueness, 190
aggregator, 132
  concave, 140
  conditional, 130
  continuous time, 157
  proportional, 140
  scale invariant (SI), 136
algebra, 2
  generated, 2, 5
allocation policy, 150
  continuous time, 160
  Markovian, 153
  optimal, 151, 161
arbitrage, 7
Arrow–Debreu equilibrium, 113
Arrow–Pratt approximation, 156

backward stochastic differential equation
  (BSDE), 92, 95, 157, 161
basis, 208
  dynamically orthonormal, 68
  orthonormal, 213
Bayes' rule, 44
Bellman equation, 24, 185
beta error, 101
beta pricing, 51, 52, 58, 96, 97
  proxy, 52

binomial model, 38, 40, 101
  high-frequency limit, 103
boundary, 217
bounded, 215
Brownian motion, 77, 79
  geometric, 86
  Girsanov's theorem, 89, 91
  Lévy's characterization, 80, 81
  reflection principle, 82
  time-changed, 82
BSDE, 92, 95, 157, 161
budget equation, 29, 31, 32
  continuous time, 87

capital asset pricing model (CAPM), 114, 179, 183
cash flow, 7
  Arrow, 15
  desirable, 106
  dominant, 21
  generated, 12, 29
  marketed, 8, 9
  minimum variance frontier, 115
  positive, 13
  present value, 8
  traded, 7
Cauchy sequence, 214
Cauchy–Schwarz inequality, 211
central planner, 124
  additive utilities, 126
certainty equivalent (CE), 133, 200
  concave, 151
  conditional, 130
  derivative, 134
  expected utility, 131, 134
  homogeneous of degree one, 136
  scale-invariant (SI), 136, 151
change of numeraire, 31
change of unit of account, 32, 87, 137, 159
closed set, 215, 217

Printed in the United States
by Baker & Taylor Publisher Services